ULTRAFILTRATION and MICROFILTRATION HANDBOOK

ULTRAFILTRATION
and
MICROFILTRATION
HANDBOOK

Munir Cheryan, Ph.D.

University of Illinois
Urbana, Illinois, USA

TECHNOMIC
PUBLISHING CO., INC.
LANCASTER · BASEL

Ultrafiltration and Microfiltration Handbook

a **TECHNOMIC**®publication

Published in the Western Hemisphere by
Technomic Publishing Company, Inc.
851 New Holland Avenue, Box 3535
Lancaster, Pennsylvania 17604 U.S.A.

Distributed in the Rest of the World by
Technomic Publishing AG
Missionsstrasse 44
CH-4055 Basel, Switzerland

Printed in the United States of America
10 9 8 7 6 5 4 3 2 1

Main entry under title:
   Ultrafiltration and Microfiltration Handbook

A Technomic Publishing Company book
Bibliography: p.
Includes index p. 517

Library of Congress Catalog Card No. 97-62251
ISBN No. 1-56676-598-6

TABLE OF CONTENTS

PREFACE

Going back over the events in the membrane world since the first edition of *Ultrafiltration Handbook*, one cannot help being pleasantly surprised at the remarkable progress in many aspects of this ubiquitous technology. The development of the Sourirajan-Loeb synthetic membrane in 1960 provided a valuable separation tool to the process industries, but it faced considerable resistance in its early days. The situation is different today: membranes are more robust, modules and equipment are better designed (if the feedstream can be pumped, the chances are one or more of the modules available today will be able to handle it), and we have a better understanding of the fouling phenomenon and how to minimize its effects. Most important, costs have come down significantly, partly because of maturing of the technology and partly because of competition from an increasing number of membrane suppliers and original equipment manufacturers (OEM). Simultaneously, several company mergers and marketing alliances occurred that should provide a firmer footing from a business viewpoint. Developments in nanofiltration, gas separations, pervaporation, and bipolar membrane electrodialysis have widened the applicability of membranes, thus attracting even more attention. This revision of the *Ultrafiltration Handbook* is an attempt to catch up with these developments. The main themes remain the same and familiar to readers of the previous edition, but each chapter has been updated and revised while keeping the "handbook" flavor intact.

One major change in this edition starts with its title: *microfiltration* has been added to recognize it as an important member of the family of membrane technologies. Purists may argue that microfiltration (MF) is essentially the same as ultrafiltration (UF), with the difference being only in pore size. On the other hand, end users and membrane manufacturers tend to view them as distinct enough to justify separate treatment. I have tried to merge the two views since both are correct, but for different reasons. The scientific principles and much of the equipment may be the same, but these two sister technologies differ in operating strategies, mathematical modeling, and applications. A unified approach has been taken in earlier chapters, and distinctions are drawn in later chapters, especially when describing specific applications.

I have followed the same format as the first edition. Chapter 1 is a brief history of membranes, definitions, and basic thermodynamic principles. Chapter 2 reviews membrane chemistry and materials. The objective is not to teach membrane manufacture or design, but what membranes are designed to do. Unlike the early days when most membrane development was done by a few companies, there are numerous public and private institutions, universities, and independent research organizations involved today. This has lifted the veil of secrecy and improved manufacturing techniques to the extent that membranes are now considered to be a commodity. The trend today is towards specialization: many companies offer only membranes and/or modules of a certain type while relying on OEMs to provide system design and engineering. This is why Chapter 3 assumes even greater importance. Quality control and properties of membranes, inasmuch as they affect their potential use, are now the shared responsibility of the end user. Chapter 4 reviews mathematical models that will be useful in understanding the effect of process parameters on system performance. Here also, the emphasis is on factors the end user will need to consider when designing a membrane process.

Listing all the changes that have occurred over the past decade in equipment and module design (Chapter 5) has been a daunting task. Some of the companies that were major players a decade ago have ceased to exist or have been merged out of existence. This is part of the risk in a technology that is rapidly changing, not only to users of the equipment (where will they get replacement parts and support from?), but to authors of books targeted at the end user. Rather than attempt to describe each manufacturer's equipment in detail, the approach in this book has been to describe general operating principles of each type of equipment, with commercial examples being used to illustrate selected design features. Chapter 6 deals with an area of crucial importance: membrane fouling. A more general approach has been taken instead of the case study approach of the first edition. This is partly because of a better understanding of this vexing problem and also in order to be useful in as many applications as possible. Cleaning has been discussed in greater detail in this edition. Chapter 7 focuses on process design aspects, with expanded coverage of system design and cost calculations.

Like the previous edition, Chapter 8 forms the bulk of this book. At that time, I noted that the bias towards citing biologically oriented examples was probably because of the special interests of the author, rather than a reflection of actual usage of ultrafiltration. Although the range of applications of MF and UF has widened, it now appears that these bio-industries did indeed constitute the major market for UF and MF and will continue to be important for the foreseeable future. In contrast, chemical and petroleum industry applications are few. It is likely that water treatment and environmental applications will see the greatest growth in the next decade.

In order to serve readers with a variety of backgrounds and to keep this book as practical as possible, I have not delved too deeply into the theoretical aspects of the technology. Appendix C contains a list of books that provide greater depth in these areas. I have also minimized the use of jargon in order to be readily comprehensible to the novice, but sometimes it is unavoidable. A list of abbreviations at the beginning of the book and the glossary of terms at the end should be useful in this regard. Appendix A provides names and addresses of some membrane manufacturers (with the caveat that inclusion in this list should not be interpreted as an endorsement nor should omission be taken to mean otherwise). Appendix B contains conversion factors (to help translate English engineering units to the metric and vice versa).

Numerous individuals working for membrane manufacturers, engineering companies, and end users have continued to educate me in this exciting technology. Interacting with them has expanded my knowledge and appreciation of what it takes for this technology to succeed in the marketplace as much as scholarly papers from academic institutions helped elucidate the scientific principles. This subject has long ceased to be a "laboratory curiosity" or an "emerging technology." This, in turn, has generated vast numbers of papers and books over the past decade. I may have summarized, simplified, or omitted contributions of several distinguished workers in this area and perhaps not cited them individually. It should not be construed as ignoring or minimizing their contributions or those of the legions of scientists, engineers, and marketing people who may not publish papers but have done much to move this technology forward.

I am once again indebted to my graduate students and research associates for their enthusiasm and doing much of the experimental work while we were learning the art of membrane technology. Technomic Publishing Company did a magnificent job of converting essentially classroom notes into a widely used reference book with the first edition. They were extraordinarily patient waiting for this long-overdue revision. Needless to say, the most important element has been my family. This book is dedicated to them in appreciation for their support and for sharing the joys and tribulations that accompanied my professional life and the writing of this book.

MUNIR CHERYAN
Urbana, Illinois

LIST OF ABBREVIATIONS

ACFF	affinity cross-flow filtration
AFM	atomic force microscopy
ATD	antitelescoping device
ATP	adenosine $5'$-triphosphate
BOD	biochemical oxygen demand
BSA	bovine serum albumin
Btu	British thermal units
CA	cellulose acetate
CD	continuous diafilitration
cfu	colony forming units
CGM	corn gluten meal
CIP	clean-in-place
CMC	carboxylmethyl cellulose (Section 8.E)
CMC	critical micelle concentration (Section 8.D.7)
CMP	caseinomacropeptide
Co-A	coenzyme A
COD	chemical oxygen demand
CPF	co-current permeate flow
CR	cross-rotating
CSTR	continuous, stirred-tank reactor
CTA	cellulose triacetate
DAF	dissolved air flotation
d.b.	dry basis
DBP	disinfection by-product
DD	discontinuous diafiltration
DE	dextrose equivalent or diatomaceous earth
DESC	dead-end stirred cell
DF	diafiltration
DMF	dimethylformamide
DS	degree of substitution
E-coat	electrocoat
ED	electrodialysis

EDTA	ethylenediaminetetraacetic acid
FESEM	field emission scanning electron microscopy
FFA	free fatty acid
FIP	formed-in-place
FOG	fats, oils, and greases
FRP	fiberglass reinforced plastic
GFD	gallons per square foot per day
gpd	gallons per day
gpm	gallons per minute
HFF	hollow fiber fermenter
HFER	hollow fiber enzyme reactor
HIMA	Health Industry Manufacturers Association
IgG	immunoglobulin G
IPA	isopropyl alcohol
IPC	isophthaloyl chloride
JTU	Jackson Turbidity Units
LMH	liters per square meter per hour
LRV	log reduction value
LWC	low-weight cardboard
MEUF	micellar-enhanced ultrafiltration
MF	microfiltration
MJ	Megajoule
MPD	m-phenylene diamine
MRB	membrane recycle bioreactor
MW	molecular weight
MWCO	molecular weight cut-off
NAD	nicotinamide adenine dinucleotide
NADP	nicotinamide adenine dinucleotide phosphate
NF	nanofiltration
NMWCO	nominal molecular weight cut-off
NMWL	nominal molecular weight limit
NTU	nephelometric turbidity unit
OEM	original equipment manufacturer
ONPG	o-nitrophenyl-β-D-galactopyranoside
PA	polyamide
PAC	powdered activated carbon
PAN	polyacrylonitrile
PBW	periodic backwash
PEG	polyethylene glycol
PEI	polyethylenimine
PES	polyethersulfone
PI	polyimide
PLC	programmable logic controller

PP	polypropylene
PS	polysulfone
PTFE	polytetrafluoroethylene
PV	pervaporation
PVA	polyvinyl alcohol
PVC	polyvinyl chloride
PVDF	polyvinylidene fluoride
PVP	polyvinylpyrrolidone
QAC	quartenary ammonium compound
RBC	red blood cells
RC	regenerated cellulose
RO	reverse osmosis
RPM	revolutions per minute
RVPF	rotary vacuum precoat filter
SBR	styrene butadiene rubber
SCR	solute concentration ratio
SDS	sodium dodecylbenzene sulfonate
SEC	size exclusion chromatography
SEM	scanning electron microscope
SS	stainless steel
SS	suspended solids
SSL	spent sulfite liquor
TDI	toluene 2,4 diisocyanate
TDS	total dissolved solids
TEM	transmission electron microscope
TFC	thin-film composite
THM	trihalomethane
TMC	trimesoyl chloride
TMP	transmembrane pressure
TOC	total organic carbon
TPH	total petroleum hydrocarbon
TS	total solids
UF	ultrafiltration
UPW	ultra-pure water
UTP	uniform transmembrane pressure
VCR	volume concentration ratio
V-SEP	vibratory shear enhanced processing
WCR	weight concentration ratio
WPC	whey protein concentrate

CHAPTER 1

Introduction

1.A.
DEFINITION AND CLASSIFICATION OF MEMBRANE SEPARATION PROCESSES

Filtration is defined as the separation of two or more components from a fluid stream based primarily on size differences. In conventional usage, it usually refers to the separation of solid immiscible particles from liquid or gaseous streams. Membrane filtration extends this application further to include the separation of dissolved solutes in liquid streams and for separation of gas mixtures.

The primary role of a membrane is to act as a selective barrier. It should permit passage of certain components and retain certain other components of a mixture. By implication, either the permeating stream or the retained phase should be enriched in one or more components. In its broadest sense a membrane could be defined as "a region of discontinuity interposed between two phases" (Hwang and Kammermeyer 1975), or as a "phase that acts as a barrier to prevent mass movement but allows restricted and/or regulated passage of one or more species through it" (Lakshminarayanaiah 1984). By these definitions, a membrane can be gaseous, liquid, or solid or combinations of these. Membranes can be further classified by (a) nature of the membrane—natural versus synthetic; (b) structure of the membrane—porous versus nonporous, its morphological characteristics, or as liquid membranes; (c) application of the membrane—gaseous phase separations, gas–liquid, liquid–liquid, etc.; (d) mechanism of membrane action—adsorptive versus diffusive, ion-exchange, osmotic, or nonselective (inert) membranes.

Membranes can also physically or chemically modify the permeating species (as with ion-exchange or biofunctional membranes), conduct electric current, prevent permeation (e.g., in packaging or coating applications), or regulate the rate of permeation (as in controlled release technology). Thus, membranes may be either passive or reactive, depending on the membrane's ability to alter the chemical nature of the permeating species (Lloyd 1985). Ionogenic groups and pores in the membrane confer properties such as *permselectivity* and *semipermeability*.

1

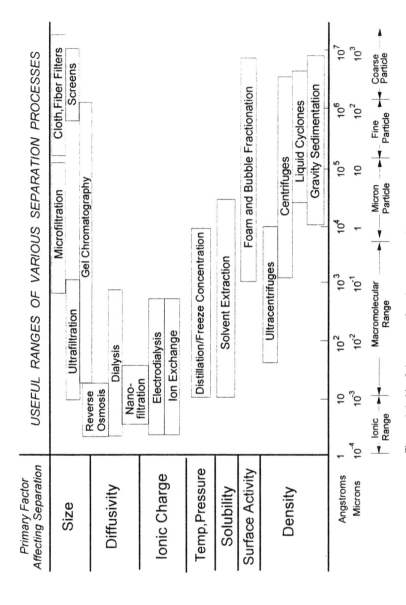

Figure 1.1. Useful ranges of various separation processes.

2

Table 1.1. Characteristics of membrane processes.

Process	Driving Force	Retentate	Permeate
Osmosis	Chemical potential	Solutes, water	Water
Dialysis	Concentration difference	Large molecules, water	Small molecules, water
Microfiltration	Pressure	Suspended particles, water	Dissolved solutes, water
Ultrafiltration	Pressure	Large molecules, water	Small molecules, water
Nanofiltration	Pressure	Small molecules, divalent salts, dissociated acids, water	Monovalent ions, undissociated acids, water
Reverse osmosis	Pressure	All solutes, water	Water
Electrodialysis	Voltage/current	Nonionic solutes, water	Ionized solutes, water
Pervaporation	Pressure	Nonvolatile molecules, water	Volatile small molecules, water

Figure 1.1 shows a classification of various separation processes based on particle or molecular size and the primary factor affecting the separation process. The major membrane separation processes—reverse osmosis (RO), nanofiltration (NF), ultrafiltration (UF), microfiltration (MF), dialysis, electrodialysis (ED), and pervaporation (PV)—cover a wide range of particle/molecular sizes and applications. Among membrane separation processes, the distinction between the various processes is somewhat arbitrary and has evolved with usage and convention. Table 1.1 shows the characteristics of various membrane processes. Osmosis (to be discussed in detail in Section 1.C.) is the transport of solvent through a semipermeable membrane from the dilute solution side to the concentrated solution side of the membrane. It is driven by chemical potential differences between the water on either side of the membrane. With an ideal semipermeable membrane, only water should permeate through the membrane. The common laboratory technique of dialysis, on the other hand, is primarily a technique for purifying macromolecules, such as desalting of proteins, and the primary driving force is the difference in concentration of the permeable species between the solution in the dialysis bag and outside the bag. Electrodialysis relies primarily on voltage or electromotive force and ion-selective membranes to effect a separation between charged ionic species.

What distinguishes the more common pressure-driven membrane processes—microfiltration, ultrafiltration, nanofiltration, and reverse osmosis—is the application of hydraulic pressure to speed up the transport process. However, the nature of the membrane itself controls which components permeate and which

Membrane Separations

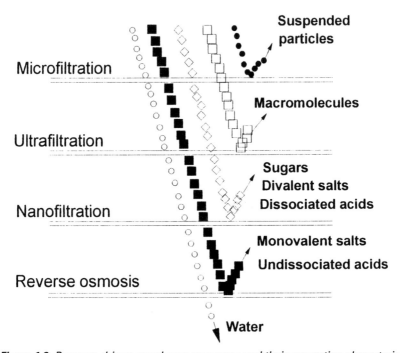

Figure 1.2. *Pressure-driven membrane processes and their separation characteristics.*

are retained, as shown in Figure 1.2. In its ideal definition, reverse osmosis retains *all* components other than the solvent (e.g., water) itself, while ultrafiltration retains only macromolecules or particles larger than about 10–200 Å (about 0.001–0.02 μm). Microfiltration, on the other hand, is designed to retain particles in the "micron" range, that is, suspended particles in the range of 0.10 μm to about 5 μm (particles larger than 5–10 μm are better separated using conventional cake filtration methods). Thus, in its broadest sense, reverse osmosis is essentially considered to be a dewatering technique, while ultrafiltration can be looked at as a method for simultaneously purifying, concentrating, and fractionating macromolecules or fine colloidal suspensions. Microfiltration is used mainly as a clarification technique, separating suspended particles from dissolved substances, provided the particles meet the size requirements for microfiltration membranes.

Nanofiltration is a relatively new process that uses charged membranes with pores that are larger than RO membranes, but too small to allow permeation of many organic compounds such as sugars. They also have a useful property in

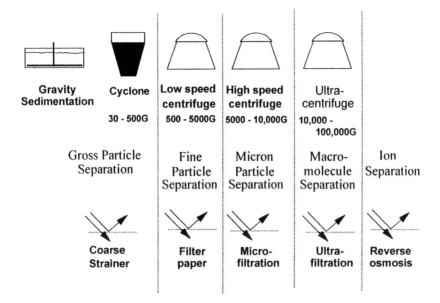

Figure 1.3. *Comparison of centrifugation and filtration processes.*

that they can separate dissociated forms of a compound from the undissociated form; e.g., organic acids such as lactic, citric, and acetic pass through easily at low pH but are rejected at higher pH when in their salt forms (Raman et al. 1994).

In terms of versatility, centrifugation is perhaps the only method to match membrane technology (Figure 1.3). However, an absolute requirement for centrifugal processes is the existence of a suitable density difference between the two phases that are to be separated, in addition to the two phases being immiscible. Membrane separation processes have no such requirement; indeed, the real value of membranes is that they permit separation of dissolved molecules down to the ionic range, provided the appropriate membrane is used.

Figure 1.4 shows some typical examples of components that fall under these four processes. Membranes are usually classified according to the size of the separated components, and thus particle sizes in MF applications are specified in microns (i.e., μm). However, with UF membranes, it is customary to refer to the "molecular weight cut-off" (MWCO) instead of particle size per se. In the early days of membrane technology, UF membranes were characterized by studying the relative permeabilities of proteins and polyethylene glycols, which were characterized in terms of their molecular weights. Even though it is known that molecular weight alone does not determine the size of a protein and, indeed, many manufacturers use dextrans rather than proteins to characterize UF membranes (as discussed in Chapter 3), this terminology is still used, sometimes prefixed with the word *nominal*, as in NMWCO. Thus, UF covers "particles"

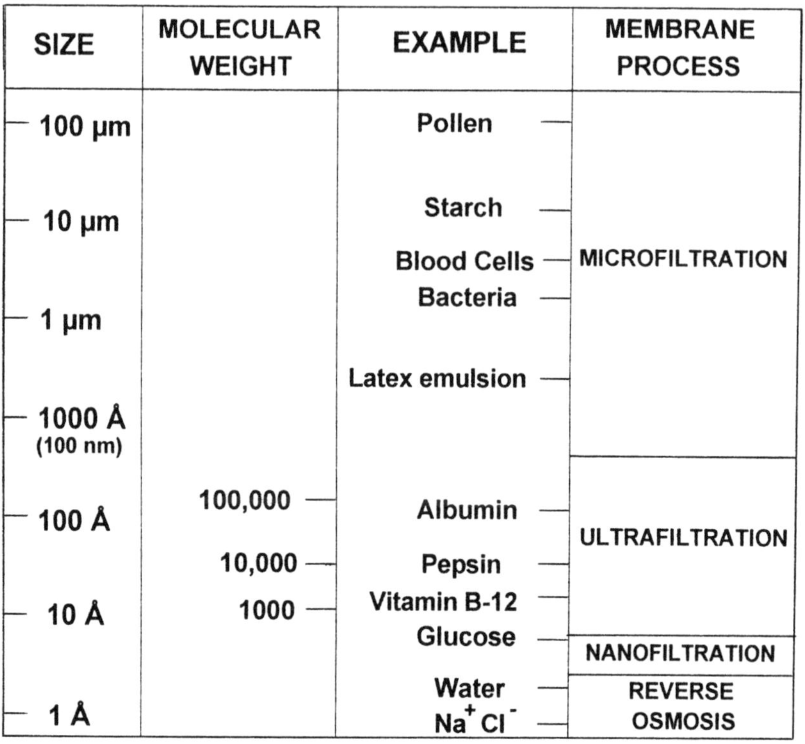

SIZE	MOLECULAR WEIGHT	EXAMPLE	MEMBRANE PROCESS
— 100 μm		Pollen —	
— 10 μm		Starch —	
		Blood Cells —	MICROFILTRATION
— 1 μm		Bacteria —	
		Latex emulsion —	
— 1000 Å (100 nm)			
— 100 Å	100,000 —	Albumin —	ULTRAFILTRATION
	10,000 —	Pepsin —	
— 10 Å	1000 —	Vitamin B-12 —	
		Glucose —	NANOFILTRATION
		Water —	REVERSE
— 1 Å		Na^+ Cl^- —	OSMOSIS

Figure 1.4. Typical examples of solutes separated by membrane processes.

and molecules that range from about 1000 in molecular weight to about 500,000 daltons.

When first developed in the 1960s, UF and RO—and later joined by their sister pressure-driven membrane processes, MF and NF—constituted the first continuous molecular separation processes that do not involve a phase change or interphase mass transfer. This is perhaps what is most exciting when considering applications in food, pharmaceutical, and biological processing. In its simplest form, as shown in Figure 1.5, membrane technology consists merely of pumping the feed solution under pressure over the surface of a membrane of the appropriate chemical nature and physical configuration. In MF and UF, the pressure gradient across the membrane would force solvent and smaller species through the pores of the membrane, while the larger molecules/particles would be retained. Thus, one feedstream is split into two product streams. The retained stream (referred to as "retentate" or "concentrate") will thus be enriched in the retained macromolecules, while the fraction going through the membrane (referred to as the "permeate") will be depleted of the macromolecules. The retentate will also contain some of the permeable solutes. In fact, the permeable

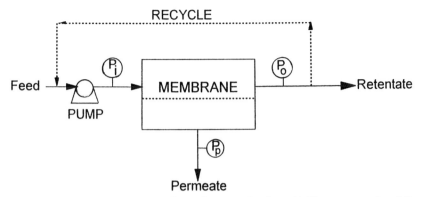

Figure 1.5. *Operating principle of membrane technology. Unlike a conventional fil-tration process, which operates in a "dead-end" mode, membranes are configured to be operated in the "cross-flow" mode, where the feed is pumped over the surface of the membrane, resulting in two product streams. Further details are provided in Chapter 4.*

solutes may be at the very same or higher concentration than in the perme-ate stream, depending on how the membrane separates or "rejects" that solute. However, since the retentate now forms a much smaller volume than the feed, there has, in effect, been a "purification" of the retained species. This will be explained in detail in Chapter 7.

Because ultrafiltration deals with the separation of fairly large molecules, such as natural polymers like proteins, starch and gums, and colloidally dis-persed compounds such as clays, paints, pigments, latex particles, etc., the osmotic pressures involved in ultrafiltration processes are fairly low (see Sec-tion 1.C. for osmotic pressure calculations). In contrast, pressures involved in reverse osmosis would be fairly high, of the order of 500–1500 psi (35–100 bar), in order to overcome the high osmotic pressures of the small solutes. NF, used as it is for desalting and deacidification, has lower osmotic pressure to work against and thus will need lower operating pressures of 150–450 psi (10–30 bar). UF and MF, on the other hand, would thus need fairly low pressures for operation, which would lower equipment and operating (pumping) costs by a considerable margin.

A further advantage of membrane technology, as compared to conventional dewatering processes, is the absence of a change in phase or state of the sol-vent during the dewatering process. Evaporation and freeze concentration are common dewatering techniques used for liquid products. Evaporation requires the input of about 1000 Btu/lb of water evaporated (540 kcal/kg) while freezing requires about 144 Btu/lb water frozen, merely to effect the change in state of water from liquid to vapor and liquid to solid, respectively. Since membrane separations do not require a change in state of the solvent to effect a dewatering,

Table 1.2. Comparison of energy requirements and costs between
evaporation and membrane processes.

Process	Evaporation	Membrane
Whole milk (2.2×)	136 kcal/kg (MVR)	17 kcal/kg (RO)
Cheese whey (3×)	$380,000/year	$130,000/year
18,000 lb/hour	(double effect)	(RO)
Corn steep liquor (6 to 50%	$1.2 million/year	$390,000/year
TS) 300 gpm	(MVR)	(RO to 14% TS, then MVR)
Gelatin (2 to 18% TS)	$516,200/year	$186,750/year
20 tons/hour	(4-effect)	(UF)

Source: Data from Cheryan and Nichols (1992); Koch Membrane Systems (1987)

this should result in considerable savings in energy. Some examples of energy and cost savings are shown in Table 1.2.

It should be borne in mind, however, when comparing membrane processes to evaporation, that saving energy does not necessarily imply a savings in cost. Considerable economies in energy usage are possible in evaporation by the use of multiple effects and mechanical vapor recompression. Furthermore, the unit energy cost for steam can be much lower than electricity. For example, steam costs in the United States today are typically $5/1000 lb ($11/ton) and electricity costs $0.05 per kWh. The unit energy cost of steam is $0.515 per 1000 Btu ($0.488 per megajoule) and electricity is $1.46 per 1000 Btu ($1.382 per MJ). Thus, membranes must use 65% less energy in order to compete with a steam-based dewatering process. That is why it is more important to compare actual costs, as shown in Table 1.2 for the cheese whey, corn steep liquor, and gelatin examples, rather than just energy consumption.

A less obvious advantage is that no complicated heat transfer or heat-generating equipment is needed, and the membrane operation, which requires only electrical energy to drive the pump motor, can be situated far from the prime power-generating plant. No additional steam capacity need be installed to handle the UF/RO unit. A further advantage over evaporators is that no condensers (and the huge condenser cooling water supply needed for its operation) are needed, thus avoiding related problems like thermal pollution and overloading of sewage treatment systems. In fact, if a reverse osmosis system is used in tandem with an MF or UF plant to treat the UF permeate, the plant would get high-quality water as one of the by-products of the operation.

Another advantage of membrane processes is that they can be operated at ambient temperatures, even though there may be frequent occasions when it is necessary to operate at considerably lower temperatures (e.g., to prevent microbial growth problems or denaturation of heat-sensitive components) or higher temperatures (e.g., to minimize microbial growth problems; to lower viscosity of the retentate, which lowers pumping cost; to improve mass transfer and flux). Thus, thermal or oxidative degradation problems common to evaporation processes can be avoided. Finally, since small molecules should normally freely pass through UF and MF membranes, their concentration on either side of the membrane should be the same during processing and about equal to that in the original feed solution. Thus, there should be minimal changes in the micro-environment during UF and MF, i.e., no changes in pH or ionic strength, a particular advantage when isolating and purifying proteins.

There are some limitations to membrane processes. None can take the solutes to dryness. In fact, membrane processes are quite limited in their upper solids limits. In RO, it is frequently the osmotic pressure of the concentrated solutes that limits the process. In the case of UF and MF, it is rarely the osmotic pressure of the retained macromolecules, but rather the low mass transfer rates obtained with concentrated macromolecules and the high viscosity that makes pumping of the retentate difficult. As an example, current technology permits skim milk to be concentrated economically by multiple effect evaporation to about 50% total solids, while the best obtained to date by RO is about 30% total solids and by UF about 42% total solids. Other problems that plagued early membrane applications—fouling of membranes, poor cleanability, and restricted operating conditions—have been overcome through the development of superior membrane materials and improved module design. This has vastly enlarged the applicability of MF and UF in the food, pharmaceutical, biological, and chemical processing industries.

1.B.
HISTORICAL DEVELOPMENTS

The phenomenon of osmosis, which is the transport of water or solvent through a semipermeable membrane (defined as a membrane that is permeable to solvent and impermeable to solutes), has been known about since 1748, when Abbe Nollet observed that water diffuses from a *dilute* solution to a more concentrated one when separated by a semipermeable membrane (Boddeker 1995; Lonsdale 1982). Dutorchet is credited with introducing the term *osmosis* to characterize spontaneous liquid flow across permeable partitions. Later, in 1845, Matteucci and Cima observed that these membranes tended to be anisotropic in nature; that is, their behavior was different, depending on which side of the membrane faced the feed solution. Schmidt also observed the same phenomenon in 1856.

In 1855, Fick developed the first synthetic membrane, made apparently of nitrocellulose. Two years later, Traube also prepared artificial membranes, while Pfeffer reported in 1877 the successful manufacture of membranes made by precipitating copper ferrocyanide in the pores of porcelain. The first quantitative measurements of diffusion phenomena and osmotic pressure were made using these early membranes. Interest also focused on membranes made of "collodion," a term commonly used for cellulosic polymers. Basically, the procedure for making these membranes was as follows: nitrocellulose was dissolved in a suitable solvent, such as alcohol-ether or acetic acid, and the solution poured on a flat surface. The solvents were allowed to evaporate. Perhaps the first reference to the use of a membrane for separations is by Graham in 1854, who used it as a dialyser to separate a solution into its components.

Bechhold, around the year 1907, then developed methods for controlling the pore size of these collodion membranes, apparently by controlling the rate of evaporation of the solvents and by water washing of the film. He was the first to suggest using air pressure for improving permeation rates and also developed methods for measuring pore diameters using air pressure and surface tension measurements. He is generally credited with coining the term *ultrafiltration*.

The period of 1870–1920 saw the rapid development of theories of thermodynamics of solutions, most notably those of van't Hoff and his theory of dilute solutions and Gibbs, whose work led to the primary relationship between osmotic pressure and other thermodynamic properties. Membrane filters became commercially available in 1927 from the Sartorius Company in Germany, manufactured using the Zsigmondy process. In 1931, Elford developed methods for sterilizing membrane filters using ultraviolet radiation.

Up until 1945, membrane filters were used primarily for removal of microorganisms and particles from liquid and gaseous streams, for diffusion studies, and for sizing of macromolecules. German scientists also developed methods for culturing bacterial cells on membranes. In 1951, Goetz imprinted grid lines on filters to facilitate counting bacterial colonies. These gridded membranes are now used extensively for water analysis. In 1957, the U.S. Public Health Service officially adopted the membrane filtration procedure for drinking water analysis.

Simultaneously with these developments in microfiltration membranes, there was considerable interest in developing membranes for reverse osmosis applications, especially for desalination of seawater and purification of brackish water. Obviously, the semipermeable membranes used by the early investigators (linings of pig's bladders, onion skins, etc.) would be impractical for these purposes due to the high pressures necessary for practical desalination, which would be of the order of 500–1000 psi (35–70 bar). The MF membranes commercially available at that time were also obviously unsuitable because of their large pore sizes.

In the early 1950s, Samuel Yuster, of the University of California, Los Angeles, had predicted that, based on the Gibbs adsorption isotherm, it should be possible to produce fresh water from brine. Srinivasa Sourirajan, who also worked at the same university, reported some success with this concept (Yuster et al. 1958) using commercially available homogenous membranes (homogenous from an ultrastructural point of view). He used a hand-operated pump, and it is reported that it took a few days to produce a few milliliters of fresh water (Loeb 1981). Around the same time, Reid and Breton (1959) independently obtained promising results using a homogenous cellulose acetate membrane.

At this juncture, the general conclusions by most researchers in the reverse osmosis area were that, to obtain commercially feasible flux or dewatering rates, the most practical route would be to reduce the thickness of the membrane. From 1958 to 1960, Sourirajan, now joined by Sidney Loeb, attempted to modify commercial cellulose acetate membranes by heating them under water, in the apparent hope that this would expand the pores and that the pores would remain open when the membrane cooled, thus increasing flux. But exactly the opposite happened: heating contracted the pores. When the heat treatment was performed with commercially available ultrafiltration membranes, it caused the pores to shrink, which increased the rejection of salts but also resulted in much higher flux than before. This heating or annealing process created a phenomenon known as "anisotropy" or "asymmetry" in the ultrastructure of the membrane; i.e., the behavior of the membrane was different, depending on which side of the membrane faced the feed solution, an observation that had been made over 100 years ago with natural membranes.

Figures 1.6–1.9 show the ultrastructure of a typical asymmetric membrane. The anisotropic nature of the Loeb-Sourirajan membrane is characterized by a thin "skin" on one surface of the membrane, usually 0.1–0.2 μm thick, while the main body of the membrane is sponge-like in nature with extremely porous voids. Since the major resistance to mass transport through the membrane is the thickness of the membrane, and the effective thickness of the anisotropic membrane is now of the order of 0.1–0.2 μm, instead of the 100–200 μm of the old homogeneous membranes, this resulted in fairly high flux. The rejection of salt remained high, however, due to the decreased effective pore size, also a result of the annealing process. This single development of the asymmetric membrane by Loeb and Sourirajan is what converted a hitherto laboratory curiosity into a practical and viable unit operation that is unmatched in its versatility for the widest possible range of applications.

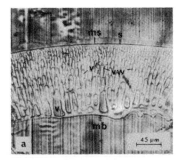

Figure 1.6. *Light microscope view of the cross section of an asymmetric membrane: ms = membrane surface, v = voids, vw = void wall (adapted from Cheryan and Merin 1980).*

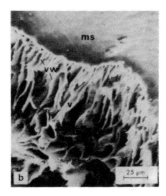

Figure 1.7. Scanning electron micrograph of an asymmetric ultrafiltration membrane: ms = membrane surface, v = voids, vw = void wall (adapted from Cheryan and Merin 1980).

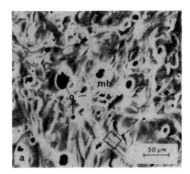

Figure 1.8. Scanning electron micrograph of the bottom side of an asymmetric membrane. The "pores" or openings on this side of the membrane are much larger to minimize resistance of solvent transport (adapted from Merin 1979).

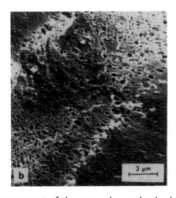

Figure 1.9. Enlargement of the area shown in the box in Figure 1.8.

Since that time, several major commercial developments in membrane science have taken place. The most notable include inorganic membranes, presently dominated by ceramic membranes, and which saw large-scale commercial applications in the early 1980s. Nanofiltration membranes found its own successful niche during this decade, as did large-scale gas separations. As we come to the end of the 20th century, new developments in pervaporation (so termed and observed by Kober as early as 1917) and bipolar membranes for electrodialysis will further widen the applicability of membrane technology in a wide variety of industries.

1.C.
PHYSICAL CHEMISTRY OF MEMBRANE SEPARATIONS

1.C.1.
CHEMICAL POTENTIAL AND OSMOSIS

All thermodynamic relationships used to correlate physicochemical properties of a system with thermodynamic parameters stem from the Gibbs free energy equation, which in its simplest form can be expressed as:

$$G = H - TS \tag{1.1}$$

$$H = E + PV \tag{1.2}$$

where

$G =$ Gibbs free energy
$H =$ enthalpy
$T =$ absolute temperature
$S =$ entropy
$E =$ internal energy
$P =$ pressure
$V =$ volume

In differential form, these equations can be written as

$$dG = dH - T\,dS - S\,dT \tag{1.3}$$

$$dH = dE + P\,dV + V\,dP \tag{1.4}$$

or

$$dG = dE + P\,dV + V\,dP - T\,dS - S\,dT \tag{1.5}$$

From the first law of thermodynamics, we can write

$$dE = dq + dw \tag{1.6}$$

where q is the heat produced and w is the work done. Assuming the change in the system is reversible, by the second law,

$$dq - T \, dS = 0 \tag{1.7}$$

Assuming further that only "PV" work is allowed (i.e., no electric or magnetic fields present) and no change in composition of the system

$$dw + P \, dV = 0 \tag{1.8}$$

Combining Equations (1.5) through (1.8), we get the final result as

$$dG = V \, dP - S \, dT \tag{1.9}$$

For "open" systems, i.e., one in which matter and energy may leave and enter, the earlier equations must be modified by adding terms relating changes in the mass of a system. Equation (1.9) will then become

$$dG = -S \, dT + V \, dP + \mu_1 \, dn_1 + \mu_2 \, dn_2 + \mu_3 \, dn_3 + \cdots \tag{1.10}$$

where

μ = chemical potential, by definition, of component 1, 2, 3 . . .

n = number of moles of component 1, 2, 3 . . .

$$dG = -S \, dT + V \, dP + \mu_i \, dn_i \tag{1.11}$$

Thus, by definition,

$$\mu_i = (\partial G / \partial n_i)_{T,P,n_j} \tag{1.12}$$

where i denotes the ith component of interest and j denotes all other components. Unlike other intensive thermodynamic properties such as pressure and temperature (i.e., those not dependent on the size of the system), chemical potential cannot be physically "felt" the way heat and force can, which results in some difficulty for the novice when trying to grasp its significance. In simple

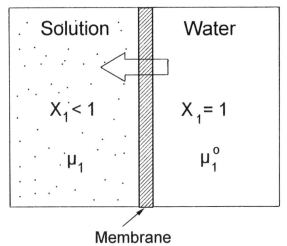

Figure 1.10. *The osmosis phenomenon. The two compartments are separated by an ideal semipermeable membrane. Arrow shows direction of water transport under the chemical potential gradient.*

terms, chemical potential μ can be looked at as being to chemical energy what temperature is to heat energy, pressure is to mechanical energy (e.g., flow of fluids), and voltage, or emf, is to electrical energy. Thus, chemical potential is essentially a driving force expressed as a change in the free energy of the system as a result of the change in the composition of the system.

The application of these concepts is shown in Figure 1.10, which shows two compartments separated by a semipermeable membrane. The right one contains a very dilute solution, or the pure solvent, and the other contains a solute dissolved in the solvent. The standard chemical potential is defined as the free energy change per mole of substance formed, consumed, or transferred from one phase to another in its standard state. The standard state is usually defined as being 1 atmosphere pressure at a particular temperature (e.g., 20°C) and in a certain reference form, usually the pure state of the component. For aqueous systems, this is pure water. Thus, in Figure 1.10, the pure solvent compartment, containing a mole fraction of water (X_1) of 1 would have a chemical potential designated by μ_1^0, while the solution compartment, with a mole fraction of water less than 1, would have a lower chemical potential of μ_1.

Physically speaking, the highest energy form of water is when it is in its pure state. Adding any material or impurity increases the entropy (since we create disorder in the system when a solute is added). The Gibbs energy Equation (1.1) states that this will result in a decrease in free energy. In other words, the chemical potential of water in a solution is always lower than when it is in a pure state. This means that the water in the right compartment has a greater chemical potential than the water in the left compartment. Since the two compartments are separated by a semipermeable membrane, which in the ideal case is permeable only to the water and not to the solute, the natural tendency would be for the water to flow in the downward direction of the driving force. Thus, the water would be transported from the right compartment to the left. This is the phenomenon of *osmosis*, the movement of solvent from the dilute solution to the more concentrated solution. In the ideal case, with no pressure effects on either side of the membrane, the water would diffuse until the chemical potentials on both sides of the membrane were equal. In theory, this should never happen because of the presence of the solute in the solution compartment, resulting in X_1 always being less than 1. In practice, the increase in height of the liquid column in the left (solution) compartment would create a hydraulic pressure against the membrane, and the water would stop diffusing through the membrane when the pressure developed would just balance the chemical potential difference. This is shown schematically in Figure 1.11.

1.C.2.
VAPOR PRESSURE

The two compartments depicted in Figure 1.10 will also have differences in vapor pressure as a result of differences in solvent concentration. The vapor pressure of a solution is always less than that of the pure solvent. This is best expressed according to Raoult's law as follows:

$$P = X_1 P^0 \tag{1.13}$$

where P is the vapor pressure of the solution and P^0 the vapor pressure of the pure solvent at that temperature. As mentioned earlier, to prevent passage of solvent from the pure solvent side to the solution side, we need to apply a pressure on the solution side equal to the osmotic pressure difference. However, this osmotic pressure is *not* the difference in the vapor pressures ($P^0 - P$). It is important to remember that, in osmosis (and reverse osmosis), we must overcome not only hydraulic and mechanical resistances, but also try to achieve chemical equilibrium. In other words, the criterion for osmotic equilibrium is that the *chemical potential of the solvent* should be the same on both sides of the membrane, rather than the "pressures" being the same.

1.C.3.
OSMOTIC PRESSURE AND CHEMICAL POTENTIAL

In order to develop a relationship between osmotic pressure, chemical potential, and parameters that can be easily measured experimentally, we need to make two assumptions: (1) the solvent vapor behaves ideally and Raoult's law applies, and (2) the liquid is incompressible.

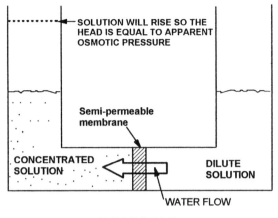

OSMOSIS

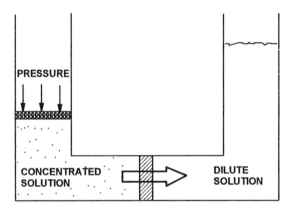

REVERSE OSMOSIS

Figure 1.11. *The phenomena of osmosis and reverse osmosis.*

At constant temperature and composition, we can rewrite Equation (1.11) as

$$(\partial G/\partial P)_{T,n_i} = V \qquad (1.14)$$

$$\left(\frac{\partial^2 G}{\partial n_i \partial P}\right)_{T,n_i} = \left(\frac{\partial V}{\partial n_i}\right)_{T,P,n_j} \qquad (1.15)$$

By definition,

$$(\partial V/\partial n_i)_{T,P,n_j} = \bar{V}_i \qquad (1.16)$$

where $\bar{V}_i$ is the partial molar volume of component i, which is the increase in volume per mole of component i when an infinitesimal amount of i is added. From Equations (1.12), (1.15), and (1.16),

$$\bar{V}_i = (\partial \mu_i / \partial P)_{T, n_i, n_j} \tag{1.17}$$

or

$$d\mu_i = \bar{V}_i \, dP \tag{1.18}$$

Equation (1.18) shows that the chemical potential of a system can be changed by changing the external applied pressure. Furthermore, since a solution and its vapor are in equilibrium, we can also substitute the ideal gas law into Equation (1.18) and obtain

$$d\mu_i = RT \frac{dP_i}{P_i} \tag{1.19}$$

Equation (1.19) states that a change in vapor pressure, due to a change in the concentration of solute or solvent, for example, will result in a change in chemical potential. Both Equations (1.18) and (1.19) relate changes in chemical potential μ for infinitesimally small changes in pressure for an ideal solution.

The following boundary conditions can be used when integrating Equation (1.19): for ideal solutions, $\mu = \mu_i^0$ when $P = P^0$, and $\mu = \mu_1$ when $P = P$. Thus, after integrating Equation (1.19):

$$\mu_i - \mu_i^0 = RT \ln P/P^0 \tag{1.20}$$

Or, for aqueous solutions, denoting $i = 1$ for water,

$$\mu_1^0 - \mu_1 = -RT \ln P/P^0 \tag{1.21}$$

Substituting Equation (1.13) into Equation (1.21),

$$\mu_1^0 - \mu_1 = -RT \ln X_1 \tag{1.22}$$

In physical terms, the above equation states that, since the mole fraction of water in a solution is always less than 1, the term $(\ln X_1)$ is negative, which means the right-hand side of Equation (1.22) is always a positive quantity. Thus, $\mu_1^0 > \mu_1$ and the natural phenomenon will be for water to flow from the pure water side to the solution side. To overcome this natural tendency, the chemical potential difference has to be overcome by applying external pressure to the solution side, in order to raise its chemical potential. Thus, the governing equation will now be a combination of Equations (1.22) and (1.18):

$$\mu_1 - \mu_1^0 = RT \ln X_1 + \int_{P_0}^{P^*} \bar{V} \, dP \tag{1.23}$$

where P^* is the external pressure and P_0 is a standard pressure. By definition, the pressure applied such that $\mu_1 - \mu_1^0 = 0$ is called osmotic pressure, i.e., $\pi = (P^* - P_0)$, and assuming further that the liquid is incompressible, so that V can be taken out from under the integral sign in Equation (1.23),

$$V_1 \pi = -RT \ln X_1 \tag{1.24}$$

or

$$\pi = \frac{RT}{\bar{V}_i} \ln X_1 \tag{1.25}$$

Equation (1.25) is the thermodynamic relationship for osmotic pressure, derived using only two assumptions: ideal solution behavior, which holds true only for very dilute solutions, and the liquid is incompressible, which is valid only at relatively low pressures.

Van't Hoff had independently developed a correlation for osmotic pressure

$$\pi = n_2 RT \tag{1.26}$$

where n_2 is the molar concentration of the solute in moles per liter of the solution. Van't Hoff's Equation (1.26) can be derived from the more rigorous Equation (1.25) by making some rather extreme approximations: since X_1 is mole fraction of water,

$$X_1 + X_2 = 1 \tag{1.27}$$

$$X_1 = 1 - X_2 \tag{1.28}$$

When X_2 is very small, i.e., when $X_2 \ll 1$,

$$\ln(1 - X_2) = -X_2 \tag{1.29}$$

$$\ln X_1 = -X_2 \tag{1.30}$$

By definition,

$$X_2 = \frac{N_2}{N_1 + N_2} \tag{1.31}$$

where N is the number of moles of component 1 or 2. Since $N_2 \ll 1$, Equation (1.31) can also be written to a first approximation as

$$X_2 = \frac{N_2}{N_1} \tag{1.32}$$

Substituting Equations (1.30) and (1.32) into Equation (1.25), we get

$$\pi = \frac{N_2}{V_1 N_1} RT \tag{1.33}$$

By definition, V_1 = molar volume of solvent = volume of solvent/moles of solvent = volume of solvent/N_1. Or

$$V_1 N_1 = \text{volume of solvent} \tag{1.34}$$

When the solution is an ideal, dilute, one, the volumes of the solvent and solution are essentially the same. Therefore, Equation (1.34) can be substituted into Equation (1.33) to get

$$\pi = \frac{N_2}{\text{Volume of solvent}} RT = n_2 RT = i\frac{C}{M} RT \tag{1.35}$$

which is the same as van't Hoff's Equation (1.26), where

C = concentration of solute in g/L of solution
M = molecular weight of solute
i = number of ions for ionized solutes (e.g., $i = 1$ for sugars,
     $i = 2$ for NaCl)
T = temperature of the solution in the absolute scale (e.g., °K or °R)
R = ideal gas constant (e.g., 0.08206 atm-L/gmole · °K, or
     8315 N-m/kgmole · °K, or 1545 ft-lb$_f$/lbmole · °R)

Note that the van't Hoff equation has been modified for ionized solutes to include i.

Table 1.3 shows the relative accuracy of the two models in predicting osmotic pressure. The van't Hoff model deviates significantly even at low solute concentrations because of the several approximations made in its development. The Gibbs (thermodynamic) relationship, Equation (1.25), is more accurate over a wider range of solute concentrations. Higher concentration results in deviations from ideal solution behavior even with the Gibbs equation.

Since the van't Hoff equation resembles the ideal gas law, a common misconception has been to visualize osmotic pressure as being caused by the bombardment of solute molecules against the membrane. Higher concentrations

Table 1.3. Osmotic pressure of aqueous sucrose solutions at 30° C.

Concentration (% w/w)	Molality	Osmotic Pressure (atm)		
		van't Hoff Equation	Gibbs Model	Experimental Data
25.31	0.991	20.3	26.8	27.2
36.01	1.646	30.3	47.3	47.5
44.73	2.366	39.0	72.6	72.5
52.74	3.263	47.8	107.6	105.9
58.42	4.108	54.2	143.3	144.0
64.58	5.332	61.5	199.0	204.3

of solute would then logically result in higher osmotic pressure. This view is incorrect since the presence of the membrane per se is not necessary for the existence of an osmotic pressure.

The physical significance of osmotic pressure in biological and clinical situations is well known: the osmotic pressure difference is what causes germinating seeds to burst open their protective coat, causes the drawing of water from the soil into the root system of plants, and can burst open cells by immersing them in a solution of much lower osmotic pressure. As far as membrane processing is concerned, its major significance lies in the fact that the external pressure that must be applied for significant permeate flux must be higher than the osmotic pressure of the solution. As will be seen later, the basic relationship between applied pressure (e.g., by a pump), osmotic pressure, and flow of solvent through a membrane is, like many transport processes, expressed in terms of the flux (the rate of solvent transport per unit area per unit time) and the driving force and resistances. For an ideal semipermeable membrane:

$$J = A(P_T - \pi_F) \tag{1.36}$$

where J is the flux, A is a membrane permeability coefficient (the reciprocal of resistance to flow), P_T is the transmembrane pressure, and π_F is the osmotic pressure of the feed solution. Thus, there has to be a positive driving force for flux; i.e., P_T must be always greater than π_F.

Even relatively small concentrations of dissolved solutes can develop fairly large osmotic pressures. A concentration difference of 0.1 M across a membrane can result in an osmotic pressure of about 2.5 bar (about 37 psi). Table 1.4 shows some examples of osmotic pressure calculations using the van't Hoff equation. With sodium chloride, a 1% solution results in an osmotic pressure of about 125 psi (860 kPa). Thus, no flux will be obtained unless the pressure is above 860 kPa. On the other hand, a 1% solution of lactose (MW = 342) will have an osmotic pressure of 10 psi (69 kPa) and a 1% solution of casein, a milk protein (MW = 25,000) only 0.28 psi (1.8 kPa). Thus, much lower pressures have to

Table 1.4. Osmotic pressure of 1% solutions at 30°C, calculated using the van't Hoff equation.

Component	Molecular Weight (M)	Number of Ions (i)	Osmotic Pressure (psi)
NaCl	58.50	2	125
Lactose	342	1	10
Casein	25,000	1	0.28

be applied with the protein and sugar solutions than with the salt solution. This is why osmotic pressures are of little or no consequence in UF and MF, but important in RO and NF.

This is shown in Figure 1.12, which shows typical flux during RO of water ($\pi_F = 0$), solutions of 1% NaCl, 1% lactose, and a real liquid food (skim milk of 9.1% total solids with an osmotic pressure of 100 psi). As expected, no permeation was observed until the applied pressure was higher than the osmotic pressure. The slopes of the salt and sugar lines are almost the same as the water line. With milk, however, there is a deviation from linearity. As will be explained in Chapter 4, this is because of "concentration polarization" of

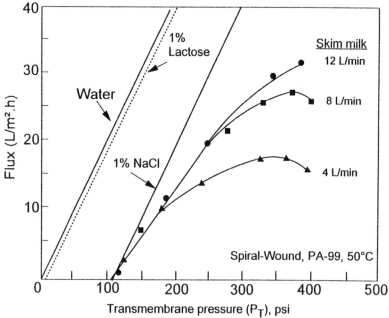

Figure 1.12. *Reverse osmosis of a salt solution, sugar solution, and a complex protein suspension (adapted from Cheryan et al. 1990).*

rejected particles. Flux becomes controlled by the mass transfer characteristics of the system. This explains why turbulence (in the form of higher velocities) has a beneficial effect with skim milk, but not so with salt or lactose where polarization is less important.

The significance of these calculations in MF and UF is that, at the normal concentrations of polymers and macromolecules (e.g., proteins), the osmotic pressure due to the presence of these macromolecules is usually low enough to be negligible. Since MF and UF are designed to retain only the larger dissolved solutes, such as proteins and other colloidal substances, it is assumed that the prevailing osmotic pressures in UF and MF are usually low enough to ignore, and thus the operating strategy will be to maximize mass transfer effects and control viscosity. In RO and NF, osmotic pressure effects are likely to be the dominant resistance.

Osmotic pressure data for macromolecular or colloidal solutes are few, especially as a function of concentration. This is unfortunate, since at sufficiently high concentrations, the osmotic pressure could become significant, especially at the membrane surface due to the polarization phenomenon (see Section 4.E. later). Table 1.5 lists osmotic pressures of food and biological products. Osmotic pressure data obtained from reverse osmosis experiments must be used with caution, since it is frequently obtained by extrapolation of flux data to zero flux.

If the van't Hoff model is used to calculate osmotic pressure, it should be remembered that it assumes that osmotic pressure will increase in a linear fashion with solute concentration. In fact, much of the actual experimental data and the Gibbs osmotic model indicate an exponential increase, as shown in Table 1.3 and also in Figures 1.13–1.15. Figure 1.13 compares the van't Hoff equation with experimental data. For charged molecules such as proteins, the

Table 1.5. Osmotic pressure of foods at room temperature.

Food	Concentration	Osmotic Pressure (psi)
Milk	9% solids-not-fat	100
Whey	6% total solids	100
Orange juice	11% total solids	230
Apple juice	15% total solids	300
Grape juice	16% total solids	300
Coffee extract	28% total solids	500
Lactose	5% w/v	55
Sodium chloride	1% w/v	125
Lactic acid	1% w/v	80
Sweet potato wastewater	22% total solids	870
Perilla anthocyanins	10.6% total solids	330

1 psi = 6.895 kPa = 0.0689 bar

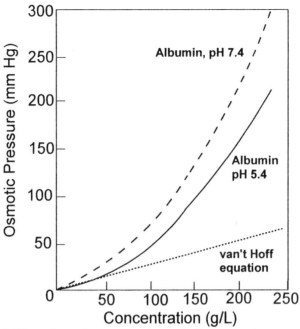

Figure 1.13. *Effect of protein concentration and pH on osmotic pressure of serum albumin. The bottom curve is the calculated curve based on the van't Hoff equation for a solute of 60,000 molecular weight. The other two curves are experimental data. Differences between the van't Hoff curve and the others are due to nonideality of the albumin solutions at higher concentrations. The difference between the two albumin curves is due to the net negative charge at pH 7.4 and the consequent Gibbs-Duheim effect (adapted from Scatchard et al. 1944).*

osmotic pressure also depends on pH and ionic strength of the solution. In general, osmotic pressure of protein solutions is minimum at the isoelectric point and tends to be higher away from the isoelectric point, especially if other charged species and salts are present. This phenomenon is shown in Figures 1.13 and 1.14. This pH effect is usually ascribed to the Gibbs-Donnan effect.

The dextran T10 and whey protein solutions (Figure 1.15) show surprisingly high osmotic pressures at concentrations where the viscosity is relatively low. For example, a 50% w/w dextran solution has a very high osmotic pressure of 25.5 atm, but the viscosity is only 270 cP (Jonsson 1984). Considering the magnitude of the osmotic pressures in Figures 1.13–1.15, it is quite possible that it is the osmotic pressure at the membrane surface that limits the flux, in addition to the resistance of any "gel-polarized" layer (see discussion in Chapter 4). The exponential increase in osmotic pressure with concentration also explains the maxima noticed at high pressures during RO (e.g., with skim milk in Figure 1.12).

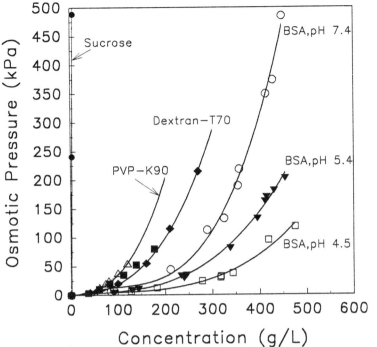

Figure 1.14. *Effect of concentration on osmotic pressure of selected macro-molecules. Sucrose is shown for comparison. Lines are drawn according to the osmotic pressure model [Equation (1.37)] and virial coefficients in Table 1.6. Points are experimental data. The effect of pH on bovine serum albumin (BSA) is shown. Data sources are shown in Table 1.6.*

To account for the curvature in the osmotic pressure-concentration data, the van't Hoff model for osmotic pressure is expressed as

$$\pi = A_1 C + A_2 C^2 + A_3 C^3 + \cdots \tag{1.37}$$

where $A_1, A_2, \ldots$ are known as the "virial coefficients." Table 1.6 lists typical virial coefficients for several solutes. At high concentrations of macromolecules, the second and third virial coefficients may become sufficiently important that osmotic pressure effects may become significant in ultrafiltration.

In summary, the major resistances to be overcome in reverse osmosis are the resistance of the membrane, osmotic pressure of the retained solutes, and possibly mass transfer resistance in the boundary layer. In ultrafiltration and microfiltration, on the other hand, the major resistance is usually due to concentration polarization and the associated boundary layer and, to a lesser extent, the membrane resistance itself, depending on the feed properties and the operating conditions. Under certain conditions and with certain solutes, the osmotic

Table 1.6. Virial coefficients of selected compounds to calculate osmotic pressure (in kPa) with Equation (1.37).

Compound	Valid Range of Concentration	Units of C	A₁	A₂	A₃	Reference
Bovine serum albumin						
pH 7.4	0–450 g/L	g/L	3.787×10^{-1}	-2.98×10^{-3}	1.000×10^{-5}	Vilker et al. (1984)
pH 5.5	0–450 g/L	g/L	5.633×10^{-2}	-2.80×10^{-4}	2.604×10^{-6}	
pH 4.5	0–475 g/L	g/L	7.539×10^{-2}	-4.90×10^{-4}	1.852×10^{-6}	
Cetyl pyridinium chloride	0–200 g/L	g/L	0.39231	1.507×10^{-3}	1.605×10^{-5}	Markels et al. (1995)
Dextran D2	0–16.2%	% w/w	0.05102	0.1047	0.01055	Vink (1971)
Dextran T10	0–50% w/w	% w/w	11.31	−0.49752	0.03	Jonsson (1984)
Dextran T70	0–110 g/L	g/L	0.139	1.1×10^{-3}	3.16×10^{-6}	Nicolas et al. (1995)
MW = 70,400	0–270 g/L	g/L	3.75×10^{-3}	7.52×10^{-4}	7.64×10^{-6}	Wijmans et al. (1985)
MW = 66,300	0–500 g/L	% w/v	0.637	0.0625	7.62×10^{-3}	Clifton et al. (1984)
Dextran T500	0–200 g/L	g/L	8.67×10^{-3}	2.98×10^{-4}	8.98×10^{-6}	Wijmans et al. (1985)
Fibrinogen (bovine)	0–80 g/L	g/L	9.948×10^{-3}	-2.104×10^{-4}	2.833×10^{-6}	Vilker et al. (1984)
Fruit juices	0–0.12	X_c^*	2.675×10^{4}	1.287×10^{4}	1.2715×10^{6}	Sourirajan and Matsuura (1985)
Glycerol	0–35%	% w/w	262.06	2.669	0.0481	Sourirajan (1970)
Hydroxyethyl cellulose	0–8.4%	% w/w	1.232	0.3292	0.0125	Vink (1971)
β-Lactoglobulin	0–250 g/L	g/L	2.699×10^{-2}	1.311×10^{-3}	7.277×10^{-8}	van den Berg et al. (1987)
Low-density lipoprotein	0–300 g/L	g/L	1.001×10^{-3}	2.888×10^{-6}	3.269×10^{-8}	Vilker et al. (1984)
Mushroom blanch water	0–22% TS	% w/w	201.80	7.84	−0.2325	Chiang et al. (1986)
Ovalbumin	0–30% w/w	% w/w	3.55	0	8.34×10^{-3}	Nabetani et al. (1990)

(continued)

(continued)

Table 1.6. Virial coefficients of selected compounds to calculate osmotic pressure (in kPa) with Equation (1.37).

Compound	Valid Range of Concentration	Units of C	A_1	A_2	A_3	Reference
Polyethylene glycol						
PEG 6	14–40%	% w/w	15.72	−0.5738	0.0787	Prouty et al. (1985)
PEG 20	0–60%	% w/w	9.65	−0.177	0.04964	Prouty et al. (1985)
Polyethylene oxide						
MW = 43,500	0–7%	% w/w	3.32506	−0.9779	0.3352	Vink (1971)
MW = 278,000	0–130 g/L	g/L	4.439×10^{-2}	6.18×10^{-3}	3×10^{-5}	Nicolas et al. (1995)
MW = 600,000	0–60 g/L	g/L	1.097×10^{-2}	-9.72×10^{-5}	3.67×10^{-6}	Vilker et al. (1984)
Polystyrene 90 K	Up to 150 g/L	g/L	5.698×10^{-2}	8.3×10^{-4}	2×10^{-5}	Nicolas et al. (1995)
Polyvinylpyrrolidone						
PVP1 (MW = 27,900)	0–12.4%	% w/w	1.1816	0.0438	0.02549	Vink (1971)
PVP40 (MW = 40,000)	0–43%	% w/w	−5.7302	0.11496	0.0323	Prouty et al. (1985)
PVP K90	0–200 g/L	g/L	2.13×10^{-2}	1.626×10^{-3}	1.659×10^{-5}	Nicolas et al. (1995)
Sodium chloride solutions	0–25%	% w/w	869.50	−5.1105	1.0403	Sourirajan (1970)
Sucrose solutions	0–25%	% w/w	163.47	−5.882	0.1324	Sourirajan (1970)
Sweet potato waste water	6–22% TS	% w/w	−78.955	24.845	−0.5724	Chiang and Pan (1986)
Whey protein isolate	0–50%	% w/w	4.4585	-1.723×10^{-2}	8.0×10^{-3}	Jonsson (1984)

*Carbon weight fraction

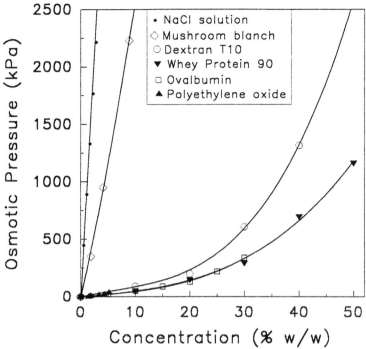

Figure 1.15. *Effect of concentration on osmotic pressure. Lines are drawn according to the osmotic pressure model [Equation (1.37)] and virial coefficients in Table 1.6. Points are experimental data. Data sources are shown in Table 1.6.*

pressure may become the limiting factor in ultrafiltration also. Thus, the operating strategy to maximize the flux will depend on the mechanism of the limiting flux.

REFERENCES

BODDEKER, K. W. 1995. *J. Membrane Science* 100: 65.

CHERYAN, M. and MERIN, U. 1980. *Polymer Sci. Technol.* 13: 619.

CHERYAN, M. and NICHOLS, D. J. 1992. In *Mathematical Modelling of Food Processes,* S. Thorne (ed.), Elsevier, London, p. 49.

CHERYAN, M., VEERANJANEYULU, B. and SCHLICHER, L. R. 1990. *J. Membrane Sci.* 48: 103.

CHIANG, B. H., CHU, C. L. and HWANG, L. S. 1986. *J. Food Sci.* 51: 608.

CHIANG, B. H. and PAN, W. D. 1986. *J. Food Sci.* 51: 971.

CLIFTON, M. J., ABIDINE, N., APTEL, P. and SANCHEZ, V. 1984. *J. Membrane Sci.* 21: 233.

DUTKA, B. J. 1981. *Membrane Filtration. Applications, Techniques, Problems.* Marcel Dekker, New York.

GELMAN, C. 1965. *Anal.Chem.* 87: 29.

GRAHAM, T. 1854. *Phil. Trans., Roy. Soc. (London)* 144: 177.

HWANG, S. T. and KAMMERMEYER, K. 1975. *Membranes in Separations*, Wiley-Interscience, New York.

JONSSON, G. 1984. *Desalination* 51: 61.

Koch Membrane Systems. 1987. Product literature. Wilmington, MA.

LAKSHMINARAYANAIAH, N. 1984. *Equations of Membrane Biophysics.* Academic Press, New York.

LLOYD, D. R. 1985. *Material Science of Synthetic Membranes.* American Chemical Society, Washington, DC.

LOEB, S. 1981. In *Synthetic Membranes. Vol. 1. Desalination*, A. F. Turbak (ed.), American Chemical Society, Washington, DC.

LONSDALE, H. 1982. *J. Membrane Sci.* 10: 81.

MARKELS, J. H., LYNN, S. and RADKE, C. J. 1995. *AIChE J.* 41: 2058.

MERIN, U. 1979. Ph.D. Thesis, University of Illinois, Urbana.

NABETANI, H., NAKAJIMA, M., WATANABE, A. NAKAO, S. and KIMURA, S. 1990. *AIChE J.* 36: 907.

NICOLAS, S., BOULANOUAR, I. and BARCOU, B. 1995. *J. Membrane Sci.* 103: 19.

PROUTY, M. S., SCHECHTER, A. N. and PARSEGIAN, V. A. 1985. *J. Mol. Biol.* 184: 517.

RAMAN, L. P., CHERYAN, M. and RAJAGOPALAN, N. 1994. *Chem. Engr. Progr.* 90 (3): 68.

REID, C. E. and BRETON, E. J. 1959. *J. Applied Polymer Sci.* 1: 133.

SCATCHARD, G., BATCHELDER, C. and BROWN, A. 1944. *J. Clin. Investigation* 23: 459.

SOURIRAJAN, S. 1970. *Reverse Osmosis.* Academic Press, New York.

SOURIRAJAN, S. and MATSUURA, T. 1985. *Reverse Osmosis/Ultrafiltration Process Principles.* National Research Council, Ottawa, Canada.

TOMBS, M. P. and PEACOCKE, A. R. 1974. *The Osmotic Pressure of Biological Macromolecules.* Clarendon Press, Oxford.

VAN DEN BERG, G. B., HANEMAAIJER, J. H. and SMOLDERS, C. A. 1987. *J. Membrane Sci.* 31: 307.

VILKER, V. L., COLTON, C. K., SMITH, K. A. and GREEN, D. L. 1984. *J. Membrane Sci.* 20: 63.

VINK, H. 1971. *Eur. Polym. J.* 7: 1411.

WIJMANS, J. G., NAKAO, S., VAN DEN BERG, J. W. A., TROELSTRA, F. R. and SMOLDERS, C. A. 1985. *J. Membrane Sci.* 22: 117.

YUSTER, S. T., SOURIRAJAN, S. and BERNSTEIN, K. 1958. Report 58-26, University of California-Los Angeles, Department of Engineering.

CHAPTER 2

Membrane Chemistry, Structure, and Function

2.A.
DEFINITIONS AND CLASSIFICATION

2.A.1.
DEPTH VERSUS SCREEN FILTERS

Filters are manufactured from a variety of materials using several methods, but they can all be classified into two general categories: depth filters or screen filters (Figure 2.1). Depth filters derive their name from the fact that filtration or particle removal occurs within the depths of the filter material. Depth filters consist of a matrix of randomly oriented fibers or beads that are bonded together to form a tortuous maze of flow channels (Figure 2.2). Common materials of construction include cotton, fiberglass, asbestos, sintered metals, and diatomaceous earth. Particulates that are insoluble or colloidal in nature are removed from a fluid by entrapment or adsorption to the filter matrix. By utilizing direct interception, inertial impaction, and diffusion, particles as small as 0.01 μ can be retained by these filters. Frequently, several stages of materials are combined in one filter (Figure 2.3), with the feed coming in contact with the most open matrix first. Depth filters are operated in the dead-end mode.

A screen filter, in contrast, separates by retaining particles on its surface, in much the same manner as a sieve (Figure 2.4). The structure is usually more rigid, uniform, and continuous, with pore size more accurately controlled during manufacture. Membrane filters fall into this category. Unlike depth filters, screen filters are rigid, with little danger of material migration, and "grow-through" of microorganisms is not as frequent a problem. Because screen filters have a defined pore size, they can be given an absolute, or quantitative, rating. A further advantage of screen filters is that the retained particles are not lost in its depths, and much higher recoveries of the retained material are possible. This may be important if the objective is to maximize recovery of the retained solids, e.g., in microbial cell harvesting. Figure 2.5 compares the ultrastructures of a

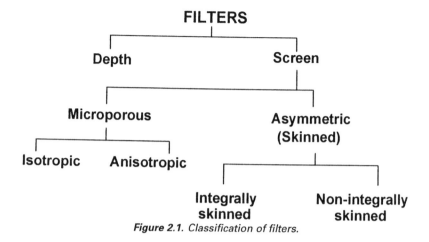

Figure 2.1. Classification of filters.

depth filter and a membrane screen filter, showing how particles are retained on the surface of a screen filter.

2.A.2.
MICROPOROUS VERSUS ASYMMETRIC MEMBRANES

Screen filters can be further classified according to their ultrastructure as either *microporous* or *asymmetric* (the latter are also referred to as "skinned" membranes). Microporous membranes are sometimes further classified as

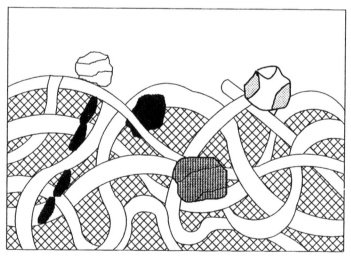

Figure 2.2. Schematic of depth filter, showing the randomly oriented fibers trapping particles on its surface and within its matrix.

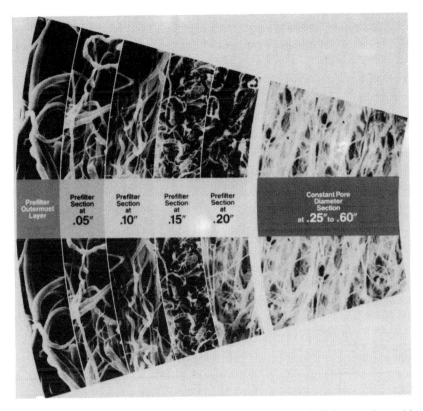

Figure 2.3. Micrograph of a multistage depth filter (Source: Pall Corporation, with permission).

isotropic (with pores of uniform size throughout the body of the membrane) or anisotropic (where the pores change in size from one surface of the membrane to the other). Figure 2.6 is a schematic representation of the microporous type of membrane filters. Frequently, the terms *anisotropy* and *asymmetry* are (incorrectly) used interchangeably. Electron micrographs of microporous membranes (Figure 2.7) show a multilayered screen or sieve of very small mesh size, with passageways in this mesh forming the "pores." Particles are retained on or in the surface mesh-like layer of the membrane.

Microporous membranes are designed to retain all particles above its rating. For example, a 0.45-μ membrane implies that it will not allow particles larger than 0.45 μm to pass through it. This does not mean that the size of the pores is 0.45 μm or less. In fact, it has been frequently observed that pores are much larger than the particles they retain (Zeman and Denault 1992), and there is a distribution of pore sizes on a membrane surface (Chapter 3). Thus, particles

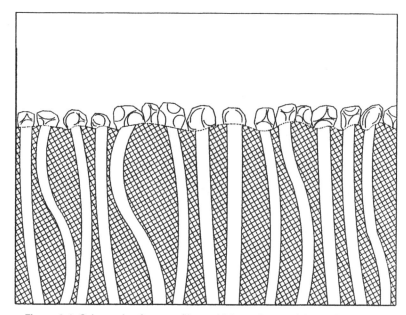

Figure 2.4. Schematic of screen filter, which retains particles on its surface.

Figure 2.5. Comparison of depth filter versus screen filter. Top: Fiberglass depth filter material magnified 2600×. Bottom: A 0.45-μ membrane filter (1000×) showing bacteria trapped on the filter surface (Source: Millipore Corporation, with permission).

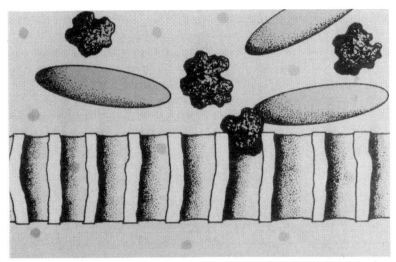

Figure 2.6. Schematic representation of the structure of a microporous membrane filter. Although retention of all particles larger than the rated pore size of the membrane is absolute, particles of the same size as the pore could become lodged in the pores and block them. (Scale is distorted for illustrative purposes. In reality, the total membrane thickness is usually more than several hundred times the pore diameter.) (Source: Millipore Corporation, with permission).

that are approximately the same size as the pores may penetrate partially into the pores and block them. In fact, the larger pores within a particular distribution will initially have the greatest flow through them and will probably be the ones that plug up first. This will result in the classic rapid drop in flux in the first few minutes of operation that we observe frequently with microfiltration (MF) membranes (Chapter 6). If enough pores get blocked, the filter becomes irreversibly plugged. Attempts to produce this type of filter with pore sizes in the ultrafiltration range (1–10 nm) have not been very successful; the few available commercially have low flux and experience rapid plugging.

Asymmetric membranes, on the other hand, whose manufacture and general properties will be discussed later in this chapter, are characterized by a thin "skin" on the surface of the membrane (Figures 2.8 and 2.9; see also Figure 1.7). The layers underneath the skin may consist of voids as shown in Figure 2.9, which serve to support the skin layer. Rejection again occurs only at the surface, but because of its unique ultrastructure, retained particles or macromolecules above the nominal molecular weight cut-off (MWCO) do not enter the main body of the membrane. Thus, these asymmetric membranes rarely get "plugged" in the fashion that microporous membranes do, although, like all filters, they are susceptible to flux-lowering phenomenon such as fouling and concentration polarization. Most ultrafiltration (UF), nanofiltration (NF), and reverse osmosis (RO) membranes have this type of a structure, but not most polymeric MF membranes.

Figure 2.7. *Micrograph of typical screen filter. This is a polyvinylidene fluoride (PVDF) microporous membrane sold under the trade name of Durapore by Millipore Ccrporation. The surface and the underlying internal pore structure are visible. Magnification = 5000× (Source: Millipore Corporation, with permission).*

The method of membrane preparation determines if the skin layer is porous or nonporous. Skin layers resulting from the phase-inversion process discussed later are probably porous, while those skin layers that are deposited from solution or plasma onto a porous support (such as those intended for gas separations) are probably homogenous (Lloyd 1985). The former type of membrane is called "integrally skinned" membranes, while the latter are termed "nonintegrally skinned" structures. The composite and dynamically formed membranes discussed in Section 2.D. fall within the latter category.

Another major difference between asymmetric and microporous membranes is in their definition and characterization of retention limits. Microporous membranes are given an absolute rating in terms of its maximum *equivalent* pore diameter; i.e., they will retain all particles larger than that rating, up to certain specified concentration limits in the feed. Asymmetric ultrafiltration membranes, however, are given "nominal" ratings. This refers to the molecular size

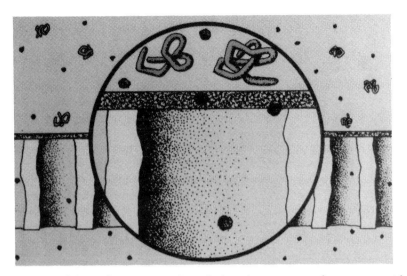

Figure 2.8. Schematic representation of the ultrastructure of an asymmetric ("skinned") membrane. Particles do not pass through the skin and do not enter the filter body. (Scale is distorted for illustrative purposes. In reality, the "skin" layer is only about 0.1–0.5 μm thick, while the rest of the body of the membrane may be 100–200 μm thick.) (Source: Millipore Corporation, with permission).

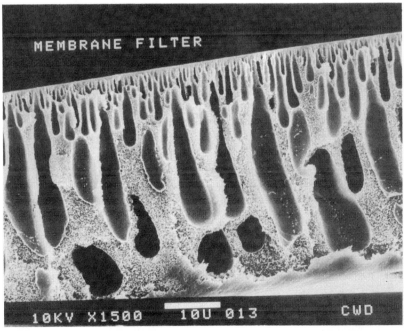

Figure 2.9. Electron micrograph of the cross section of an asymmetric polysulfone ultrafiltration membrane (Source: Cuno Inc., with permission).

or molecular weight above which a certain percentage of the solute in the feed solution (of a specific molecular weight) will be retained by the membrane, under controlled conditions. This is discussed later in Chapter 3.

2.B.
GENERAL METHODS OF MEMBRANE MANUFACTURE

There are several methods for manufacturing membranes (Table 2.1). Some of the methods are applicable to a variety of polymers, and others are material-specific, e.g., the heating-stretching method used to make pores in microporous polytetrafluoroethylene (PTFE) membranes. Each of these methods results in different ultrastructures, porosity, and pore size distribution. For example, track-etched membranes have a narrow pore size distribution, but a low porosity (the number of pores per unit surface area). On the other hand, the phase-inversion process is a good way to form the asymmetric skin structure on a membrane and can result in fairly high porosity in certain cases.

Figure 2.10 is a schematic diagram showing a general procedure for manufacture of synthetic polymeric membranes. The source of the polymers may be naturally available materials such as cellulose or synthetic polymers such as polycarbonate, polyethylene, or polysulfone; modifications of existing polymers such as sulfonated polysulfones; or entirely new polymers developed specifically for membrane applications. The polymer is then combined with a suitable solvent and a swelling agent or nonsolvent to form the casting dope, which is then sent to the casting machine. Prior to actual casting on the machine, a small sample of the casting dope may be cast or poured onto glass plates under controlled conditions as a quality control check of the casting dope. If it gives a satisfactory membrane, the entire batch is then cast on the casting machine. If not, this will be the stage to adjust the composition of the dope to obtain the correct formulation.

Table 2.1. Methods of manufacture of synthetic membranes.

Process	Materials
Phase inversion by	Polymers
Solvent evaporation	Cellulose acetate, polyamide
Temperature change	Polypropylene, polyamide
Precipitant addition	Polysulfone, nitrocellulose
Stretching sheets of partially crystalline polymers	Polymers: PTFE
Irradiation and etching	Polymers: polycarbonate, polyester
Molding and sintering of fine-grain powders	Ceramics, metaloxides, PTFE, polyethylene

Source: Adapted from Ripperger and Schulz, 1986

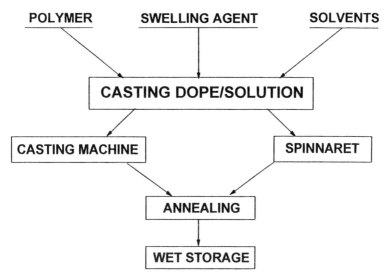

Figure 2.10. *Schematic of general procedure for manufacture of membranes.*

The casting machine is a critical factor in the quality of the membrane. Casting equipment is designed specifically for each type of membrane or module configuration. They are usually designed and fabricated by the membrane manufacturers themselves. Most membranes are manufactured in the "flat sheet" form, although in some cases the membrane may be cast directly onto the module itself (e.g., in tubular modules) or may be extruded or spun into hollow fibers directly. The membrane is subsequently subjected to an annealing process and checked for performance characteristics before being fabricated into its final form, after which it may be subjected to further tests to evaluate the module–membrane integrity and performance.

2.B.1.
PHASE-INVERSION PROCESS OF MEMBRANE MANUFACTURE

This refers to a method of manufacturing asymmetric membranes that results in a "solvent-cast structure which owes its porosity to immobilization of the polymer gel prior to complete solvent evaporation or depletion. This is accomplished by *not* allowing the cast solution to evaporate to dryness before its structure is set; partial solvent loss occurs so that the solution separates into two interspersed liquid phases. One of these phases represents the voids. As evaporation is allowed to continue, the gel structure is set" (Kesting 1971). In practice, the polymer (e.g., cellulose acetate) is dissolved in a suitable volatile solvent (e.g., acetone, dioxane, pyridine, etc., depending on its acetyl content), and a swelling agent is added, which may be magnesium perchlorate or formamide.

When this solution is cast and the solvent allowed to evaporate, it results in an increased concentration of polymer at the solution/air interface, since solvent is lost more rapidly from the surface. The polymer essentially goes out of solution at the surface and forms the so-called skin layer that is characteristic of asymmetric membranes. (This phenomenon is analogous to "case hardening" in products that have been rapidly dried.) After the skin forms, the remaining solvent in the bulk of the mixture evaporates more slowly. Eventually, the swelling agent in the mixture starts separating out as a different phase, resulting in two phases within the substructure: the polymer-solvent as the concentrated phase and the swelling agent as the dispersed phase. As solvent evaporates further, the polymer starts aggregating and forms a "coat" around the swelling agent. These polymer-coated aggregates thicken, approach each other closely, make contact, and deform into polyhedra (Kesting 1971). Eventually, both solvent and swelling agent depart, the polymer coating becomes too thin, and it ruptures, leaving behind an open-celled structure. This sequence of events occurs in the substructure after the dense skin has formed. The open-celled structure is visible in Figures 1.6 and 1.7 as voids in the substructure and is also visible in Figure 2.9.

There are several variations of the phase inversion process. In the wet process, originally developed by Loeb and Sourirajan, the solution is cast at low temperatures (0 to $-11°C$), and the polymer is fairly concentrated. The solvent is allowed to evaporate only partially in open air before the system is immersed in water to complete the phase inversion. The primary membrane, or skin, is thus formed by gelation. The membrane is then heated to 70–90°C to produce the secondary gel membrane. Very small voids are produced by this process. However, because the solvent diffuses out and water diffuses in during the skin formation, these membranes must be kept wet at all times; i.e., the water is an integral physical part of the membrane. If membranes made by the wet process are allowed to dry out, the capillary forces generated during drying may be enough to cause the thin void walls to collapse. Having to keep the membranes wet at all times may be a problem during transportation and storage of the membranes; furthermore, this makes the cellulose backbone of the polymer susceptible to microbial attack.

One of the older techniques is for preparing *dry* membranes. This involves using solvents and nonsolvents of different volatilities. The solvents evaporate, leaving a thin film containing relatively large interconnected voids, which form the pores. This method can be used to produce cellulose nitrate and mixed ester MF membranes. Pore characteristics are determined by a number of factors, such as polymer concentration, molecular weight, relative humidity, and relative volatility of the solvents. For example, varying the ratio of ether and ethyl alcohol allows a wide range of porosities and pore sizes to be produced with cellulose nitrate membranes. The nonsolvent can also be introduced from the vapor phase, e.g., by placing the cast film in a humid atmosphere and allowing the water to diffuse into it (Mulder 1991).

The *thermal inversion* process involves bringing the polymer solution to a high enough temperature to form a single phase and then cooling it so that it splits into a polymer-rich phase and a solvent-rich phase. This results in gelation of the polymer, and its structure can then be set by cooling to below the freezing point of the solvent. A slow rate of cooling will lead to a more homogenous ultrastructure made up of micro-aggregates of polymer. The solvent can then be leached out by a second solvent that acts as a nonsolvent (Lloyd et al. 1988). This method has been used with polypropylene, polyethylene, and polyolefin membranes.

Another process in common use is the *melt spinning* process. This was originally developed by the DuPont Company for producing nylon or polyamide membranes in the hollow fine fiber form. A concentrated polymer solution is used, which is extruded in a spinneret where, under the high temperatures, the solvent is flashed off. Annealing is done at 80–90°C. Because the solvent evaporation is done in a relatively uncontrolled fashion and because of the high polymer concentration used, the voidage is relatively low, which may explain the relatively low flux obtained in hollow fine fibers. It should be noted that there are two types of hollow fiber membranes available commercially: (a) the hollow fine fibers such as those marketed by DuPont, which has an internal diameter of 0.00016″ (about 0.042 mm) and where the feed is pumped from the outer shell side, and (b) the hollow "fat" fibers, where the feed is pumped in through the tube side and permeate flows out through the shell side. These are discussed in greater detail in Chapter 5.

2.C.
POLYMERS USED IN MEMBRANE MANUFACTURE

A survey of the scientific and patent literature indicates that over 130 materials have been used to manufacture membranes; however, only a few have achieved commercial status, and fewer still have obtained regulatory approval for use in food, pharmaceutical, and kindred applications. Table 2.2 is a partial listing of typical materials used in the manufacture of membranes. Not all materials are suitable materials for all pressure-driven membrane processes, with the possible exception of cellulose acetate. Substantial progress has been made since the 1980s in understanding the complex relationships between the variables associated with creating good membranes. A brief summary of the important aspects of membrane chemistry is presented in this section, not so much to provide details of the manufacture of membranes, but more towards understanding the behavior, performance, and limitations of particular membranes in a particular application. Membrane chemistry and manufacture are discussed in

Table 2.2. Materials used for the manufacture of membranes.

Material	MF	UF	RO
Alumina	X		
Carbon–carbon composites	X		
Cellulose esters (mixed)	X		
Cellulose nitrate	X		
Polyamide, aliphatic (e.g., nylon)	X		
Polycarbonate (track-etch)	X		
Polyester (track-etch)	X		
Polypropylene	X		
Polytetrafluoroethylene (PTFE)	X		
Polyvinyl chloride (PVC)	X		
Polyvinylidene fluoride (PVDF)	X		
Sintered stainless steel	X		
Cellulose (regenerated)	X	X	
Ceramic composites (zirconia on alumina)	X	X	
Polyacrylonitrile (PAN)	X	X	
Polyvinyl alcohol (PVA)	X	X	
Polysulfone (PS)	X	X	
Polyethersulfone (PES)	X	X	
Cellulose acetate (CA)	X	X	X
Cellulose triacetate (CTA)	X	X	X
Polyamide, aromatic (PA)	X	X	X
Polyimide (PI)		X	X
CA/CTA blends			X
Composites (e.g., polyacrylic acid on zirconia or stainless steel)			X
Composites, polymeric thin film (e.g., PA or polyetherurea on polysulfone)			X
Polybenzimidazole (PBI)			X
Polyetherimide (PEI)			X

greater detail by Kesting (1985), Klein (1991), Lloyd (1985), Matsuura (1994), Mulder (1991), Sourirajan (1970, 1977), and Sourirajan and Matsuura (1985).

2.C.1.
CELLULOSE ACETATE

Cellulose acetate is the classic membrane material used by the pioneers of modern membrane technology to create skinned membranes. The raw material is cellulose, which is a polymer of β-1,4 linked glucose units (Figure 2.11). One primary and two secondary hydroxyl groups and the β-glucosidic oxygen are in the equatorial position. Cellulose and its derivatives are generally linear, rod-like, and rather inflexible molecules, which are considered as fairly important characteristics for RO and UF applications. The major traditional source of

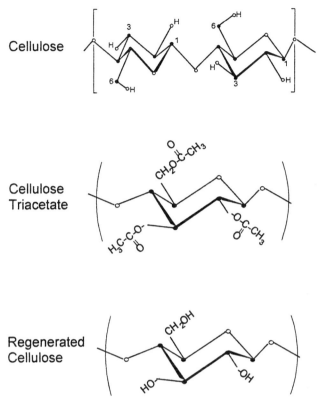

Figure 2.11. *Structures of cellulose, cellulose triacetate (CTA) and regenerated cellulose (RC).*

cellulose is wood pulp or cotton linters, although there has been some recent interest in microcrystalline cellulose, which is chemically modified wood pulp. This may be particularly useful when a narrow molecular weight distribution and a relatively pure source of cellulose is needed.

Cellulose acetate (CA) is prepared from cellulose by acetylation, i.e., reaction with acetic anhydride, acetic acid, and sulfuric acid. Figure 2.11 also shows the typical structure of a completely substituted molecule of cellulose, which would be called cellulose triacetate (CTA). Such a structure with a degree of substitution (DS) of 3.0 is actually quite rare since we seldom get complete acetylation of cellulose. Traditionally, cellulose derivatives with a DS > 2.75, i.e., with an acetyl content greater than 42.3%, have been called cellulose triacetate. Typically, in these structures, more than 92.5% of the available hydroxyl groups have been esterified. It appears to be more common in the industry, however, to use cellulose derivatives with DS values of 2.4–2.5.

Another important physical property that affects membrane properties is the degree of polymerization of the cellulose. The optimum appears to be 100–200 or 100–300, which would result in molecular weights of about 25,000–80,000.

There are several advantages to the use of CA and its derivatives as membrane materials:

1. Hydrophilicity, which is very important in minimizing fouling of the membrane
2. Wide range of pore sizes can be manufactured, from RO to MF, with reasonably high fluxes; this combination has rarely been duplicated with other membrane materials
3. CA membranes are relatively easy to manufacture
4. Low cost

Among the disadvantages of CA membranes are

1. A fairly narrow temperature range: Most manufacturers recommend a maximum temperature of 30°C, which is a disadvantage from the point of view of flux (since higher temperatures lead to higher diffusivity and lower viscosity, both of which lead to higher flux) and sanitation, since this temperature is particularly conducive to microbial growth. Blends of CA and CTA can tolerate temperatures of 35–40°C, although under carefully controlled operating conditions.
2. A rather narrow pH range: Most CA membranes are restricted to pH 2–8, preferably pH 3–6 (see Figure 2.13 later). The polymer hydrolyzes easily under more acidic conditions, since acid tends to attack the β-glucosidic links in the cellulose backbone, which could lead to a loss in molecular weight and a consequent loss in structural integrity. Highly alkaline conditions, on the other hand, cause deacetylation, which will affect selectivity, integrity, and permeability of the membrane. Higher temperatures accelerate the degradation. In water treatment applications, where cleaning is infrequent, CA membranes have a lifetime of about 4 years under normal usage at pH 4–5, 2 years at pH 6, and a few days at pH 1 or 9. This narrow range of pH tolerance is sometimes a problem in developing cleaning procedures with CA membranes, since most cleaning solutions, especially in the food and bioprocessing industries, tend to be alkaline.
3. Another problem is the poor resistance of CA to chlorine. Less than 1 mg/L free chlorine is suggested under continuous exposure and 50 mg/L in a shock dose. Chlorine oxidizes cellulose acetate and weakens the membrane, opening up the pores. This results in a temporary large increase in water flux, but it also leads to poor long-term operating lifetime. Chlorine is almost a universal sanitizer in the process industries, and this poses a special problem in these applications.

4. CA is also reported to undergo the "creep" or compaction phenomenon to a slightly greater extent than other materials, i.e., gradual loss of membrane properties (most notably flux) under high pressure over its operating lifetime.
5. CA is also highly biodegradable; i.e., it is highly susceptible to microbial attack due to the nature of its cellulose backbone. Not being able to use the usual sanitizers such as chlorine adds to the problem, and thus cellulose acetate membranes have relatively poor storage properties.

It is to be noted that these limiting variables (pH, temperature, and pressure) are subject to strong interaction effects; e.g., the effect of pH on membrane properties depends on the prevailing temperature and pressure.

2.C.2.
POLYAMIDE MEMBRANES

This class of materials is characterized by having an amide bond in its structure ($-CONH-$). Typical structures of some polyamide membranes are shown in Figure 2.12. Polyamide (PA) membranes overcame some of the problems associated with CA membranes; e.g., the pH tolerances are wider (Figure 2.13). However, PA membranes are much worse with regard to chlorine tolerance (Figure 2.14) and biofouling tendencies (Figure 2.15). Polyamides form the contact skin layer in many composite membranes (Section 2.D.)

2.C.3.
POLYSULFONE MEMBRANES

The family of polysulfone membranes are widely used in MF and UF. Polysulfone itself (Figure 2.16) is characterized by having in its structure diphenylene sulfone repeating units. The $-SO_2$ group in the polymeric sulfone is quite stable because of electronic attraction of resonating electrons between adjacent aromatic groups. The oxygen molecules projecting from this group each have two pairs of unshared electrons to donate to strong hydrogen bonding of solute or solvent molecules (Leslie et al. 1974). Repeating phenylene rings create both steric hindrance to rotation within the molecule and electronic attraction of resonating electron systems between adjacent molecules; both contribute to a high degree of molecular immobility, producing high rigidity, strength, creep resistance, dimensional stability, and heat deflection temperature. Phenyl ether and phenyl sulfone groups have high thermal and oxidative stability, producing long-term, high-temperature stability during use (Deanin 1972).

Polysulfone (PS) and polyethersulfone (PES)—especially the latter, which is widely used today—are considered breakthroughs for MF and UF applications due principally to the following favorable characteristics:

1. Wide temperature limits (Figure 2.17): Typically, temperatures up to 75°C can be used routinely, although some manufacturers are claiming their PS

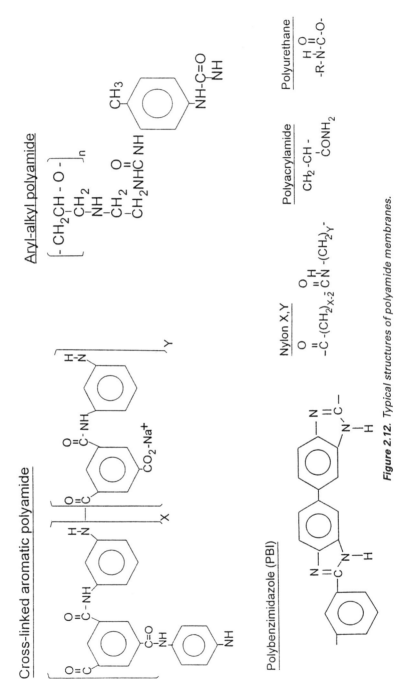

Figure 2.12. Typical structures of polyamide membranes.

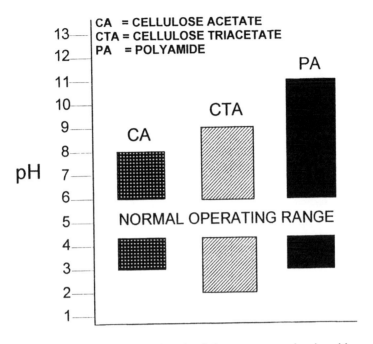

Figure 2.13. Comparison of pH limits of cellulose acetate and polyamide membranes.

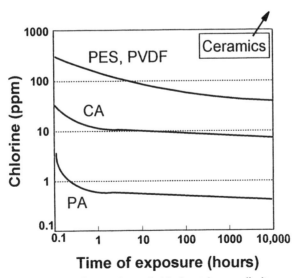

Figure 2.14. Comparison of chlorine tolerance limits.

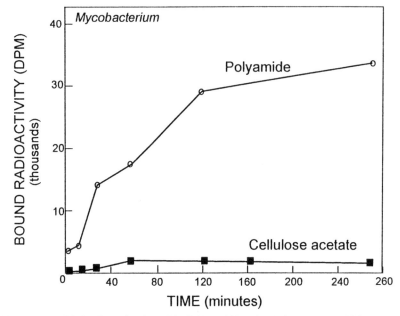

Figure 2.15. *Biofouling of polyamide (PA) and blend cellulose acetate (CA) reverse osmosis membranes by* Mycobacterium. *Membranes were exposed to radiolabeled bacteria in a dilute buffer at neutral pH and 30°C. Initial cell concentration was 5 × 10⁷/mL (adapted from Ridgway 1988).*

and PES membranes can be used up to 125°C. This would be an advantage in fermentation and biotechnology where sterility is maintained by heat treatment at 121°C and in some process applications where the viscosity of the process stream is much lower at high temperatures. It should be remembered that the higher the operating temperature, the more carefully we have to select other operating parameters, such as pH, pressure, and the cleaning regime.

2. Wide pH tolerances: PS/PES can be continuously exposed to pHs from 1 to 13. This is definitely an advantage for cleaning purposes.

3. Fairly good chlorine resistance (Figure 2.14): Most manufacturers permit the use of up to 200 ppm chlorine for cleaning and up to 50 ppm chlorine for short-term storage of the membrane. However, prolonged exposure to high chlorine levels can damage the membrane film.

4. Easy to fabricate membranes in a wide variety of configurations and modules

5. Wide range of pore sizes available for UF and MF applications, ranging from 10Å (1000 MWCO) to 0.2 μ in commercial-size modules.

6. Good chemical resistance to aliphatic hydrocarbons, fully halogenated hydrocarbons, alcohols, and acids. However, it does not offer much resistance to aromatic hydrocarbons, ketones, ethers, and esters.

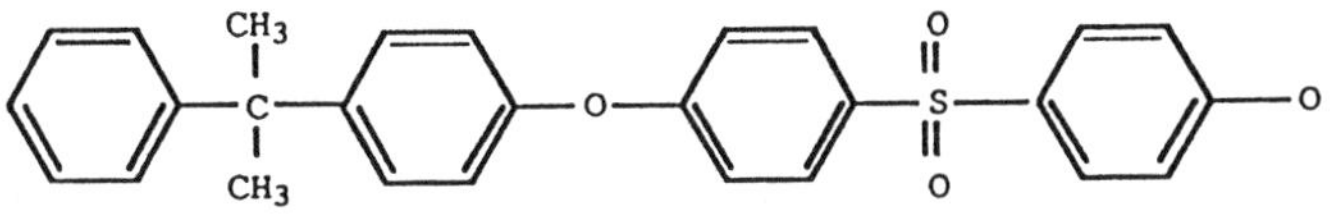

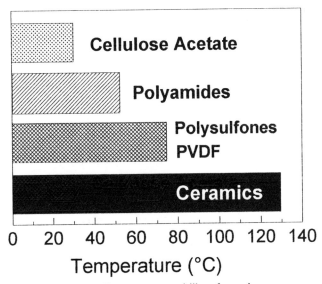

POLYSULFONE   (UDEL, *Union Carbide*)

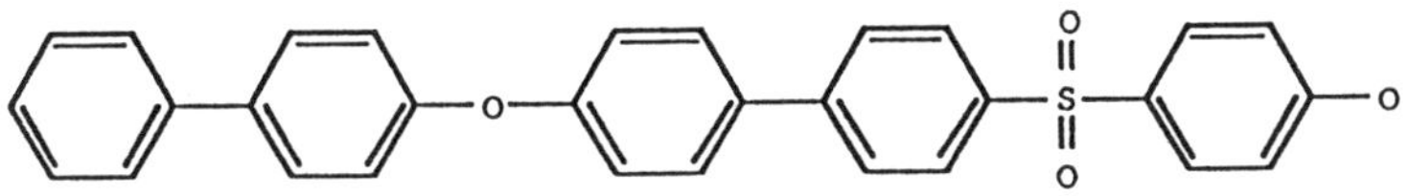

POLYPHENYLENESULFONE (RADEL, *Union Carbide*)

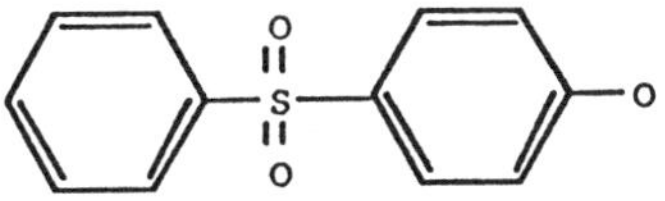

POLYETHERSULFONE (VICTREX, *ICI*)

Figure 2.16. Structure of the polysulfone family of membranes, their trade names, and manufacturers.

Figure 2.17. Temperature stability of membranes.

49

The main disadvantages of polysulfone and polyethersulfone are (a) the apparent low pressure limits (typically 100 psig/7 bar with flat sheet membranes and 25 psig/1.7 bar with polysulfone hollow fibers) and (b) hydrophobicity, which leads to an apparent tendency to interact strongly with a variety of solutes, making it prone to fouling in comparison to the more hydrophilic polymers such as cellulose and regenerated cellulose (more details on the importance of hydrophilicity are provided in Chapter 6).

2.C.4.
OTHER POLYMERIC MATERIALS

Figure 2.18 is a partial listing of polymeric materials that have been used for manufacturing MF and UF membranes. The most common ones available industrially are

- *Nylon*: These membranes (Figure 2.12) are naturally hydrophilic with fluxes of the same order of magnitude as cellulosic membranes. They are also autoclavable; however, they strongly bind biological solutes such as nucleic acids and proteins (see Table 3.14).

POLYACRYLONITRILE (PAN)	$-CH_2- CH - (CN) -$
POLYCARBONATE	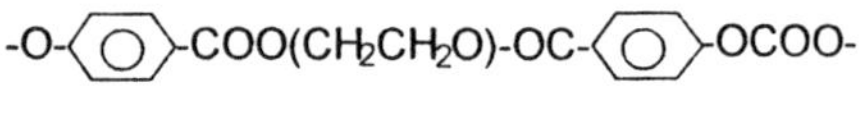 $-O-\langle O \rangle-COO(CH_2CH_2O)-OC-\langle O \rangle-OCOO-$
POLYMETHYLMETHACRYLATE(PMMA)	$-CH_2-C-(CH_3)-(COOCH_3) -$
POLYPROPYLENE (PP)	$-CH_2-CH(CH_3) -$
POLYSTYRENE	$-CH_2-CH(C_6H_5) -$
POLYTETRAFLUOROETHYLENE (PTFE)	$-CF_2-CF_2-$
POLYVINYL ALCOHOL (PVA)	$- CH_2-CH(OH)-$
POLYVINYL CHLORIDE (PVC)	$-CH_2-CH(Cl) -$
POLYVINYLIDENE FLUORIDE (PVDF)	$-CH_2- CF_2-$

Figure 2.18. *Other polymers used in the manufacture of membrane filters.*

- *Polyvinylidene fluoride* (PVDF): It can be autoclaved and its chemical resistance to common solvents is quite good. The membrane is hydrophobic, although some PVDF membranes, such as Durapore from Millipore Corporation (Figure 2.7), has its surface modified to make it more hydrophilic so that its surface can be wetted, an important consideration for analytical (e.g., microbiological) applications. A very popular material for MF and UF, it has better resistance to chlorine than the polysulfone family. It is especially popular for fruit juice clarification because of its resistance to limonene.

- *Polytetrafluoroethylene* (PTFE) is also very stable to strong acids, alkalis, and solvents and can be used at a wide range of temperatures, from −100°C to 260°C. It is made by a combination of heating and stretching melted films, which results in MF-sized pores (Figure 2.19). It is extremely hydrophobic and finds many uses in the treatment of organic feed solutions, vapors, and gases. PTFE membranes are available only in MF pore sizes.

- *Polypropylene* (PP) can be made by the thermal inversion process and also by melt-extruding and stretching. It is widely available in the form of hollow fibers. It is hydrophobic, relatively inert, and can withstand moderately high temperatures (Figure 2.19).

- *Regenerated cellulose* (RC) membranes (such as the YM series from Amicon Corporation) are very hydrophilic and have exceptional nonspecific protein-binding properties. RC also has good resistance to some common solvents such as 70% butanol and 70% ethanol and can tolerate temperatures up to 75°C.

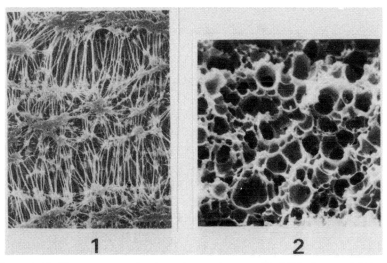

Figure 2.19. *Ultrastructure of polymeric membranes. (1) PTFE (Source: Millipore, with permission). (2) polypropylene (Accurel) from Enka/Microdyn, with permission.*

- *Polycarbonate* is one of two polymers (polyester is the other one) that is used to make track-etch membranes. This describes a process developed in the late 1960s by General Electric Corporation (produced and marketed by Nuclepore Corporation and Poretics Corporation) that was quite unlike other methods. Conventionally manufactured polymeric membranes result in "tortuous-path" or "sponge-like" ultrastructures, which tend to have a wide range or distribution of pore sizes. On the other hand, track-etch membranes are manufactured by a combination of two separate process: the production of "tracks" by nuclear bombardment and the etching of these tracks to form the actual pores.

Figure 2.20 shows a schematic of such a process. A thin sheet of the appropriate polymer (which was polycarbonate in the early days of development of this technology) is bombarded with massive energetic nuclei in a collimated beam of U^{235} fission fragments, which, upon passing through the film, leave narrow trails of radiation-damaged material in its path called *tracks*. Then the film is passed through an etching bath (usually warm caustic) that preferentially attacks the sensitized tracks and thus forms the pores. This method works on any nonconducting material as long as the maximum thickness of the film is about 15 μm. After the etching treatment, the membrane is usually treated to render the surface hydrophilic, antistatic, etc., as required.

Figure 2.21 shows the surface structure of a typical track-etch membrane. Compared to the surface pore distribution on a membrane made by the phase-inversion process, the surface porosity (number of pores per unit surface area)

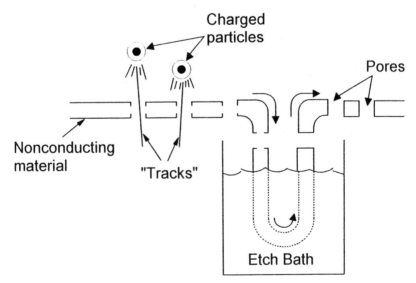

Figure 2.20. *Schematic of process for manufacture of track-etch membranes (adapted from Porter, 1990).*

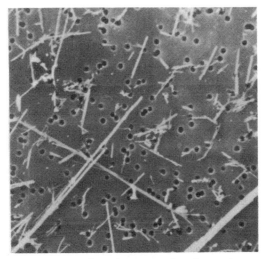

Figure 2.21. *Surface structure of a polyester track-etch membrane. A few overlapping pores can be seen (adapted from Nuclepore catalog).*

is much less; typically, about 10–15% of the surface of track-etch membrane is occupied by the pores, as compared to about 60–90% for other types of MF membranes. The lower porosity would be expected to result in lower flux for track-etch membranes. However, track-etch membranes are also much thinner than normal MF membranes, and the flow path is relatively much straighter and less tortuous, which compensates for the lower porosity, resulting in fluxes that are comparable to membranes made by other processes (see Chapter 3).

In theory, the appearance of the remarkably uniform pores seen in Figure 2.21 should result in a fairly sharp molecular weight cut-off and close to being an "absolute" filter. In practice, there could be overlapping pores due to the random nature of the bombardment of the fission fragments, which could render these membranes less than absolute. However, potential problems caused by overlapping pores are minimized by angling the nuclear beam at a 29° angle, so that, even if a particle does enter the overlapping pore, it will get stuck in the depths of the membrane body and will be effectively retained.

2.D.
COMPOSITE MEMBRANES

The preceding description focused on single-polymer, integrally skinned membranes; i.e., the separating layer/skin of the membrane is developed during the manufacture of the membrane itself. A major milestone in membrane technology occurred in the late 1970s with the development of the second generation of membranes known as "composite" membranes. These are also generically referred to in the trade as thin-film composite (TFC; this is actually a trademark of one particular manufacturer), ultrathin, or thin-layer composite membranes. Primarily developed for RO and NF applications, composites have a thin dense polymer skin formed over a microporous support film. By this definition they are direct descendants of the Sourirajan asymmetric membrane structure but differ from it in the manner in which the dense skin is formed. While integrally skinned membranes are made in a one-step procedure, composite membranes are made in two or more steps (Figure 2.22).

There are four general methods of forming composite membranes (Cadotte 1985; Petersen 1993):

1. Casting the ultrathin barrier layer separately, followed by lamination to the support film
2. Dip-coating of a solution of the polymer onto a microporous support and drying, or dip-coating a reactive monomer or prepolymer solution followed by curing with heat or irradiation
3. Gas-phase deposition of the barrier layer from a glow-discharge plasma
4. Interfacial polymerization of reactive monomers on the surface of the support film

Figure 2.22 shows the principle of manufacture of one of the earliest composite membranes (NS-100), which is formed on a polysulfone backing. The addition of polyethylenimine (PEI) to the polysulfone backing increases its salt

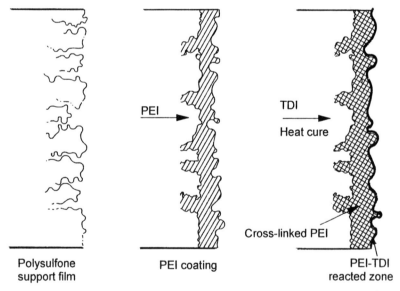

PEI →

TDI →

Heat cure

Cross-linked PEI

Polysulfone
support film

PEI coating

PEI-TDI
reacted zone

Figure 2.22. *Manufacture of composite membranes (adapted from Rozelle et al. 1977). The composite polymeric structure is then placed on a supporting fabric like polyester for reinforcement.*

rejection from essentially zero to about 70% but decreases the flux from about 500 gallons per square foot per day (GFD), or 850 liters per square meter per hour (LMH), to about 50 GFD (85 LMH). Adding toluene 2,4 diisocyanate (TDI) and drying at 110°C further increases salt rejection to 94–99% while reducing flux to 5–20 GFD (8–35 LMH). In a later version, the TDI was replaced with isophthaloyl chloride (IPC). Piperazine (PIP)/IPC composites and PIP/trimesoyl chloride (TMC) combinations also resulted in RO membranes with varying properties. The PIP/TMC membrane is probably the forerunner of today's NF membranes, since it had good flux, low rejection of NaCl, and high rejection of MgSO$_4$.

In 1978, FilmTec Corporation developed an interfacial composite polyamide membrane using m-phenylene diamine (MPD) and TMC as reactants: the FT-30. It consists of a 0.20-μ thick polyamide separation barrier deposited on a 40-μ microporous polysulfone membrane, which is then supported on a 120-μ reinforcing fabric. There are now several clones of this membrane, some made under license from FilmTec (later owned by Dow Chemical) and some with different polyamide structures. Figure 2.12 shows possible structures of the skin layer of some commercially successful composite membranes such as the cross-linked aromatic polyamide and aryl-alkyl polyamide. Figure 2.23 shows an electron micrograph of a commercial TFC® membrane with a polyetheramineurea thin film layer on a polysulfone support. Composite membranes are available in spiral, plate, and tubular module configurations.

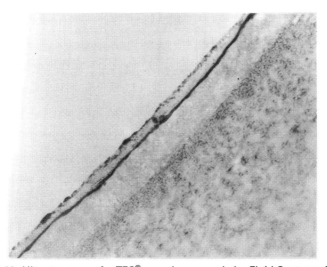

Figure 2.23. *Ultrastructure of a TFC® membrane made by Fluid Systems. The top layer is polyvinyl alcohol, the middle layer is the polyetheramineurea thin-film layer, and the bottom layer is the polysulfone support. Magnification = 126,000× (Source: Fluid Systems, with permission).*

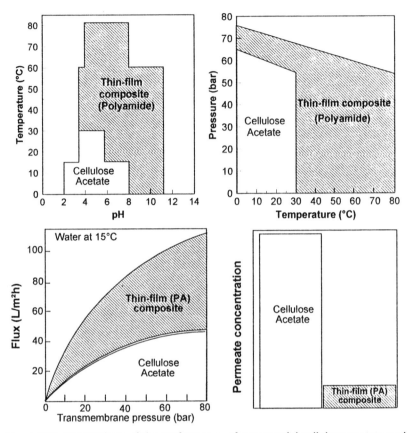

Figure 2.24. Comparison of the performance of commercial cellulose acetate and thin-film composite (polyamide) membranes (data from PCI Membrane Systems).

Composites resulted in a substantial improvement in RO technology since they are superior to CA in terms of pH stability (Figure 2.13), pressure, temperature, flux, and rejection (Figure 2.24). However, they have a greater biofouling tendency than CA membranes (Figure 2.15). In one case of processing coffee effluent, a polyamide composite membrane fouled severely (zero flux in less than 12 hours, probably due to polyphenol interactions with functional groups of the polyamide), while a CA membrane performed successfully (PCI 1994). PA composites are also much less tolerant to chlorine (Figure 2.14). Chlorine tolerance is generally specified by manufacturers in terms of "ppm-hours," which is the number of hours a particular membrane can be exposed to a certain concentration of chlorine before seriously affecting membrane properties.

Composite UF membranes are rare. One of the few available commercially is the G-series manufactured by DESAL (Desalination Systems, Inc.). They are

**Table 2.3. Methods for manufacture
of inorganic membranes.**

- Particle dispersion and slip casting
- Phase separation and leaching
- Anodic oxidation
- Thin-film deposition
- Track-etching

available in MWCOs of 1000–15,000 (as measured by polyethylene glycols). Chlorine tolerance can range up to 24,000 ppm-hours, and they can be operated at temperatures of 90°C. pH tolerance is good with continuous operation possible at pH 2–11 below 50°C.

2.E.
INORGANIC MEMBRANES

The commercialization of inorganic membranes (also generically referred to in the trade as "ceramic" or "mineral" membranes) in the early 1980s was the next major advancement in the science and art of membrane technology. Because of its distinct properties, it opened up many new areas of applications that formerly could not be supported by polymeric membranes. Generally, manufacturing of inorganic membranes starts with selection of the appropriate material in the form of a powder with a narrow particle distribution. A macroporous substrate (e.g., as shown in Figure 2.25) is first formed, perhaps by thermal sintering of an extruded paste of the powder. If a tubular geometry is used, pastes from two powders of different grain sizes may be coextruded, with the finer grain being closer to the axis. After baking at high temperatures (>1000°C), the inside may be coated by slip casting with the final fine-grain powder. A series of such layers may be necessary to obtain the asymmetric-type ultrastructure seen in Figure 2.25. The membrane is finally set by a series of pressurizing, drying, and baking steps. Table 2.3 lists several methods of forming the active layer in inorganic membranes.

With a couple of exceptions, inorganic and ceramic membranes are available in tubular form, either as a single-channel tube or multichannel element (Figure 2.26). The multichannel elements may each contain 7 to 37 individual circular channels, or "lumens," depending on the relative diameters of the channel and the element. The inner diameter of individual channels varies from 2–6 mm and lengths from 0.8–1.2 m (36″ to 47″). For example, the Kerasep monolithic element (marketed by TechSep) has an alumina/titania support and a titania or zirconia membrane. They are available with pore sizes of 15,000–300,000 MWCO and 0.1–1.0 μ. The element is 2 cm in diameter and 85.6 cm in length, containing either seven channels of 4.5 mm diameter (Figure 2.26) or 19

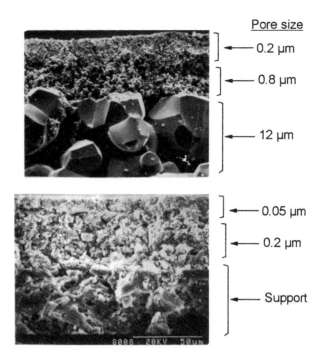

Pore size

]◄— 0.2 µm

]◄— 0.8 µm

]◄— 12 µm

]◄— 0.05 µm

]◄— 0.2 µm

]◄— Support

Figure 2.25. *Ultrastructure of the cross section of ceramic membranes. The pore sizes are distances between nonporous alumina crystals (Sources: top: USFilter; bottom: CeraMem, with permission).*

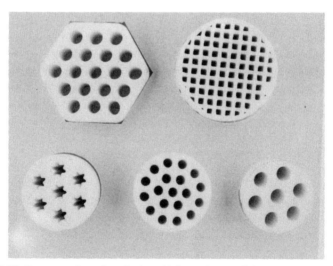

Figure 2.26. *Cross section of typical multichannel monolith geometries available with inorganic membranes: top row, left to right: Membralox (USFilter), CeraMem; bottom row, left to right: Tecramics (Fairey Industrial Ceramics Ltd., UK), Ceraflo (originally developed by Norton, now marketed by USFilter), Kerasep (TechSep).*

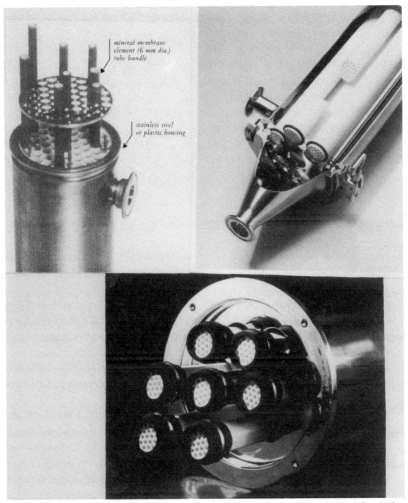

Figure 2.27. Ceramic membrane modules in their housing: top left: TechSep's Carbosep module with individual tubes; top right: NGK's Cefilt module; bottom: USFilter's Membralox.

channels of 2.5 mm diameter, resulting in 0.08 m^2 or 0.12 m^2 per element. As many as 99 of these elements may be put together in a single housing, resulting in 8 or 12 m^2 per module. Typical housings are shown in Figure 2.27. Normal process ratings are 15 bar and 150°C.

The USFilter membrane elements are recognizable by their hexagonal shape (Figures 2.26–2.28). A single element of 1.02 m in length will have 37 of the 3-mm channels for an area of 0.36 m^2 per element or 19 of the 4-mm or 6-mm channels, resulting in areas of 0.24 m^2 and 0.35 m^2, respectively. For the 0.85 m

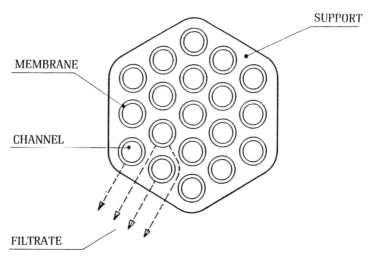

Figure 2.28. *Flow path for the feed and permeate in a USFilter/Membralox ceramic element.*

length elements, the membrane areas are 0.2 m^2 and 0.3 m^2, respectively. Each USFilter module will contain 1, 3, 7, 19, 37, or 60 such elements in the same housing. In contrast, the CeraMem monolith, which has flow channels with a square cross section (Figure 2.26), may contain about 1800 of the 2 mm × 2 mm channels (11 m^2) or 480 of the 4 mm × 4 mm channels (6 m^2) in their industrial size modules (Figure 2.29). One such monolith element is fitted per housing, forming one module.

In all inorganic modules, the feed flows through the inside of the channels, while the permeate flows through the support layer around the lumens in the monolith and to the outside of the element (Figure 2.28). Several individual elements are assembled in one housing (Figure 2.27), and two to four housings are placed in series in a stack, with several stacks in parallel (Figure 2.30).

Examples of single-channel tubes are the Carbosep (TechSep) carbon-zirconia and carbon-titania composites, GFT carbon–carbon composite, and the stainless steel-based composites, originally developed by CARRE and marketed by DuPont, but now marketed under the SCEPTER trademark by Graver Separations. Single tubes of carbon–zirconia or carbon–titania composites have been available for a long time under the "Carbosep" trade name. They are 6 mm (1/4″) inner diameter, 10 mm outer diameter, and 120 cm (47″) in length (Figure 2.27). These membranes are available in a wide variety of pore sizes, from the M5 model of 10,000 MWCO to the M9 with 300,000 MWCO. MF available pore sizes range from 0.08 to 0.45 μ. These single tubes are bundled together in a shell-and-tube arrangement in a stainless steel housing, as shown in Figure 2.27, resulting in modules with membrane areas of 0.023 m^2 (Type S1 module) to 5.73 m^2 (S252). Normal process ratings are up to 15 bar and 150°C.

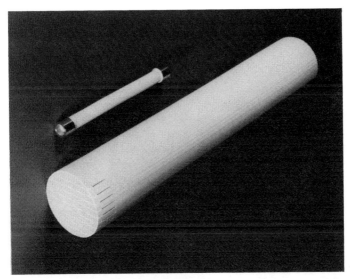

Figure 2.29. *CeraMem ceramic membranes. Shown are the commercial module (length = 0.9 m, diameter = 15 cm) and the laboratory module (30 cm length, 2.5 cm diameter). The flow channels in these modules have a square cross section. PMA industrial modules (shown) have 1800 channels with 2-mm sides and 11 m² of area. PMC modules (not shown) have 480 channels with 4-mm sides and 6 m² of area. The LMA laboratory module (shown above) has 60 channels of 2 mm (0.13 m²), and the LMC module has 12 channels of 4 mm and 0.06 m² (shown in Figure 5.7).*

Figure 2.31 shows the structure of the Scepter stainless steel membrane. Powder metallurgy is used to produce porous stainless steel (SS), large-diameter (12–75 mm, 0.5–3″) tubes of 5–15 ft (1.5–6.1 m) in length. Titania (TiO_2) is sintered on this porous material on the inside of the tubes to form a permanent MF membrane of about 0.1 μ pore size. If necessary, the TiO_2-SS composite can serve as the base for the manufacture of UF or NF membranes using the company's formed-in-place (FIP) technique. Organic or inorganic chemicals are added at the final stages of the clean-in-place (CIP) sequence to allow the surface to be coated with these chemicals/polymers and thus produce the required semidynamic membrane. For example, zirconium oxide or other food-grade polymers result in UF membranes with an MWCO of 20,000–80,000, depending on the chemical. Zirconia-polyacrylate and zirconia-polyacrylate-polyethylene imine result in NF membranes of 250–500 MWCO. Aggressive processing and cleaning conditions can be used, since recoating is apparently simple and inexpensive.

The carbon composite membrane from GFT also uses several layers of material, ending with a 0.5- to 1-μ thick layer of the membrane on the inside of 6-mm diameter tubes. The thickness of the wall is 1 mm.

Inorganic membranes are also available in the flat-sheet form. A silver membrane is shown in Figure 2.32. It is manufactured by Osmonics from amorphous

Figure 2.30. Ceramic membranes being used as a prefilter in a water application (Source: USFilter, with permission).

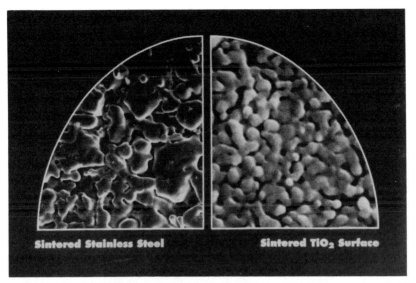

Figure 2.31. Sintered stainless steel membrane (Source: Graver Separations, with permission).

Figure 2.32. Ultrastructure of a silver flat-sheet membrane (Source: Millipore Corporation).

silver into a homogenous crystalline network of porous silver in pore sizes of 0.2–5 μ. It is probably the only bonded metal membrane available (others are sintered) and is also available from several other suppliers for specialized applications. Because of their excellent chemical and high-temperature resistance, they are especially useful in chemically and thermally aggressive environments, such as high-temperature vents, critical filtration of lubricating oils, cleaning liquids, coolants, dopant gases, and photoresists. They are also used in analytical applications. Silver membranes are attacked by nitric and sulfuric acids. Like most inorganic membranes, they are expensive: a 29.3-cm diameter silver membrane costs $210.

Anodized aluminum membranes are available through Anotec Separations. Technically, these are not "ceramic" membranes, even though the membrane material is alumina. The starting material is aluminum, which is subjected to electrolysis using oxalic or sulfuric acid. This results in the formation of a porous aluminum oxide film on the anode. The pore size and structure are controlled by adjusting the voltage. When a suitable film thickness has been reached (usually 60 μm), the oxide film is then removed from the anode, and after other proprietary treatments, the remarkably uniform structure shown in Figure 2.33 is obtained. At present, these membranes are available only in small disc form of 0.02-μ and 0.2-μ pore sizes.

One step up the evolutionary ladder of flat sheet inorganic membranes is the Ceramesh (U.K.) spiral-wound element. It is made with an Inconel 600 metal mesh consisting of 100-μm diameter wires woven to form square apertures 150 μm across. A slurry of zirconia is then suspended within the apertures to form a meniscus. After drying and firing, the result is a ceramic "membrane"

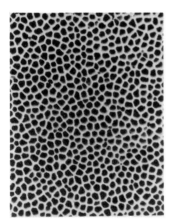 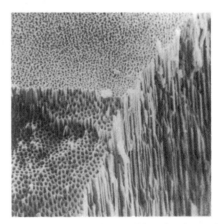

Figure 2.33. Anodized aluminum membrane (0.2-μ pore size), sold under the trade name of ANOPORE: top view and angular cross section (adapted from Anotec Separations catalog).

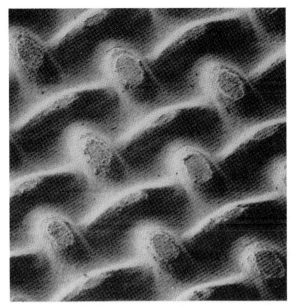

Figure 2.34. *Top surface of the Ceramesh flat sheet ceramic membrane. Zirconia is suspended within a metal mesh (Inconel 600), fired, and then baked to form this composite. The thickness of the membrane at the center of the meniscus is about 20 μ (Source: Acumem Corporation, with permission).*

in the menisci as shown in Figure 2.34. At the center of the meniscus, the membrane is about 20 μm thick. The manufacturer claims a narrow pore size distribution. These flat sheets can be placed in a plate-and-frame type module, or even wound as a spiral-wound module with a curvature radius of 12.5 mm without inducing cracks. Feed spacers of polypropylene mesh (2 mm, 90-mil) and 316 SS corrugated sheet have been used to make modules with areas of 0.4, 0.8, and 2.5 m^2.

2.E.1.
PROPERTIES OF INORGANIC MEMBRANES

   Mineral or ceramic membranes should be extremely versatile, since they are made of inorganic materials and thus should have few of the disadvantages associated with polymeric membranes. Sintered alumina layers will not peel off under high temperatures, pressures, or backflush conditions. However, keep in mind that, even if the membrane itself is very resistant to extremes of operating parameters, it is the noninorganic materials in the module that limit its performance. Special care should be taken about the seal, gaskets, and O-rings that are used to hold the membrane elements in the housing. These are probably made of synthetic polymers, which may themselves have limitations. Another important point to remember is the interaction effect, e.g., very strong doses of certain chemicals, while harmless at room temperature, may be very aggressive at very high temperatures.

- *Inert to common chemicals and solvents:* Very few chemicals will bother these membranes, except perhaps for hydrofluoric acid and, in the case of alumina membranes, phosphoric acid. Cordierite supports (e.g., as with CeraMem membranes) should not be exposed to strong alkalis for long periods of time. The alumina (Al_2O_3) monolith of the Kerasep membrane module is protected by titania (TiO_2), reportedly to make it less susceptible to attack by high-temperature (>70°C) alkaline (>pH 12) cleaning solutions. Keep in mind that acids and alkalis, commonly used for cleaning membranes, have different effects on water flux of these inorganic membranes. This is discussed later in the chapter on cleaning (Chapter 6). Perhaps the inorganic membrane's most useful feature is its ability to tolerate strong doses of chlorine (up to 2000 ppm in certain cases). Chlorine at alkaline pH is an extremely effective cleaning agent.
- *Wide temperature limits:* Depending on the seals and type of housing, some inorganic membranes can be operated as high as 350°C. Routine sterilization by steam or hot water (e.g., at 125°C) is to be expected with all inorganic membranes. Care must be taken during high-temperature sterilization to carefully control the rates of temperature increase during heating and temperature decrease during cooldown. The metal housing and the ceramic module expand and contract at different rates, and uncontrolled heating and cooling will result in thermal shocks and possible crack damage to the element. Again, due to its brittleness, liquid should not be allowed to freeze in the pores of an inorganic membrane due to the possibility of rupture.
- *Wide pH limits:* Most manufacturers will specify pH ranges of 0.5 to 13, although some advertise pH 0–14. With some inorganics, the type of acid or alkali may be a limitation, as mentioned earlier.
- *Pressure limits:* Considering the nature of the membrane, it may seem surprising that many manufacturers limit applied pressures to about 10 bar (150 psi). However, this is mainly a function of seal limitations and the type of housing. In addition, since most of the inorganic membranes are designed for MF or UF applications, there is really little need to design modules and housings to operate at much higher pressures. It is more important to avoid sudden surges in pressures, as described below.
- *Extended operating lifetimes:* This is another big advantage. Many ceramic systems are still operating 10–14 years after installation with the first set of membranes. Unlike polymeric membranes, whose membrane life is most affected by the frequency and nature of cleaning, inorganic

membranes appear to be able to tolerate frequent aggressive cleaning regimes.
- *Backflushing capability:* This is discussed in greater detail in Section 6.E.2. It essentially means the application of pressure from the permeate side of the membrane to the feed/retentate side. The brief periodic application of a pulse or backpressure lifts off some of the solids that have accumulated on the membrane surface, thus improving the flux. To be effective, the backpulse pressure should be higher than the normal feed pressure. Most tubular inorganic membranes are well suited for backflushing and backpulsing, unlike spiral-wound and polymeric tubular membranes (polymeric hollow fibers can also be backflushed).

Some limitations of ceramic membranes are also mentioned below:

- *Brittle* is the term best used to describe these membranes (perhaps the stainless steel membrane is an exception). If dropped or subjected to undue vibrations, it could be damaged. Cavitation and pressure surges in the pumps should be avoided.
- *Pore sizes* available today are mostly UF and MF. Some NF membranes are currently being manufactured, although in some cases, NF size pores are being obtained by filling UF-size pores with silicone-type materials or other organic compounds, thus not making them truly all-inorganic (e.g., the Scepter stainless steel ZOPA-PEI composite).

  Also, because of the nature of the material, care should be taken to avoid pumping abrasive materials through the membrane system. These include carbon particles, bleaching earths, filter aids, rigid fibrous materials, etc. For example, a suspension of calcium carbonate reportedly caused substantial erosion of the zirconia membrane layer of a Carbosep membrane element, leaving behind only the macroporous carbon support (Colomban et al. 1993).
- *A large pumping capacity* is required for almost all inorganic membranes in order to operate them at the recommended velocities of 2–6 m/sec. A single USFilter 37P19-40 module containing 37 Membralox elements, each 102 cm long with 19 of the 4-mm diameter channels, would have a membrane area of about 8.9 m^2 (95 ft^2) and require a cross-flow rate of 700 gallons per minute (gpm) or 160 m^3 per hour. The pressure drop with water would be 0.5–1.5 bar. In general, one should choose the smallest channel size available that can handle the feed. Not only will it require less pumping, but it will also lower the unit cost of the membrane, since more area will be available in the element with smaller channels.
- *Expense* is probably the major limitation of ceramic membranes. For example, in 1996, polymeric spiral-wound membrane elements cost $50–100/m^2 in the United States (Chapter 5). Inorganic membranes cost

$500–3000/m^2 for just the membrane elements alone. On a system basis, polymeric spiral-wound UF and MF membrane plants could be purchased for $225–$350/m^2, including controls, pumps, fittings, and the first set of membranes (less tankage). In contrast, ceramic systems would cost $2200–$6000/m^2 on the same basis. However, the initial purchase price alone should not be the sole criteria if inorganic membranes are being considered. Inorganic membranes should have considerably longer lifetimes than polymeric membranes. For most applications, polymeric membranes are assumed to have a life of 12–18 months if daily chemical cleaning is done. (In water treatment applications, where cleaning may be done on a monthly or half-yearly basis, lifetimes of 3–6 years or more are not uncommon). Inorganic membranes have been on the job for more than 10 years, although it may be difficult to obtain such a lengthy warranty from the manufacturer. This long lifetime, coupled with savings in labor (no need to replace the membrane annually), a generally higher flux, and a wider range of operating parameters could tip the balance in favor of inorganic membranes, despite the higher initial cost.

In summary, considerable progress has been made in membrane science and manufacture since the mid-1980s. There continues to be a vast amount of research underway on membrane materials. The objectives still remain the same: precise control of the pore size distribution and the skinniest possible surface/contact layer. The chemical nature of the membrane governs compatibility and physicochemical properties to a large extent, while the method of preparation primarily governs the physical structure of the membrane. A better understanding of solute-solvent-membrane interactions has enabled membranes to be used in an ever-increasing variety of applications, from chemical processing to medicine and biotechnology. It is also possible to modify the surface of the membrane to change its properties. Cross-linking, copolymerization, and coating hydrophobic membranes such as PVDF and PES with nonionic polymers can make it more hydrophilic and less prone to fouling (see Chapter 6).

However, despite the vast number of materials that have been studied as possible membrane materials, there are still only a few that have succeeded commercially. For ultrafiltration and microfiltration applications, polyethersulfone is practically in a class by itself. Other popular materials include PVDF, polyacrylonitrile (PAN), polysulfone, polypropylene, and the ceramics. Interestingly, despite its apparent limitations, cellulose acetate is still holding on to a respectable share of the market, perhaps due to its relatively low-fouling tendencies (Chapter 6) and low cost. "Tailoring" membranes for specific applications is still in the future. Thus, even though the state of knowledge has improved considerably vis-à-vis the material science of membrane technology, there is still no substitute for experimental trials to determine if a particular membrane material/module configuration is suitable for a particular application.

REFERENCES

CADOTTE, J. E. 1985. In *Materials Science of Synthetic Membranes*, D. R. Lloyd (ed.), American Chemical Society, Washington, DC.

COLOMBAN, A., ROGER, L. and BOYAVAL, P. 1993. *Biotechnol. Bioeng.* 42: 1091.

DEANIN, R. D. 1972. *Polymer Structure, Properties and Applications.* Chaners Books, Boston, MA.

KESTING, R. E. 1971. *Synthetic Polymeric Membranes.* McGraw-Hill, New York.

KESTING, R. E. 1985. *Synthetic Polymeric Membranes. A Structural Perspective.* 2nd edition. Wiley-Interscience, New York.

KLEIN, E. 1991. *Affinity Membranes.* John Wiley, New York.

LESLIE, V. L., ROSE, J. B., RUDKIN, G. O. and FELTZIN, J. 1974. In *New Industrial Polymers*, R.D. Deanin (ed.), Symposium Series No. 4. American Chem. Society, Washington, DC.

LLOYD, D. R. 1985. *Materials Science of Synthetic Membranes.* American Chemical Society, Washington, DC.

LLOYD, D. R., BARLOW, J. and KINZER, K. 1988. *AIChE Symp. Series.* 84 (No. 261): 28.

MATSUURA, T. 1994. *Synthetic Membranes and Membrane Separation Processes.* CRC Press, Boca Raton, FL.

MERIN, U. 1979. Ph.D. Thesis, University of Illinois, Urbana.

MULDER, M. 1991. *Basic Principles of Membrane Technology.* Kluwer Academic Publishers, Dordrecht, The Netherlands and Norwell, MA.

PCI. 1994. Personal communication. PCI Membrane Systems, UK.

PETERSEN, R. J. 1993. *J. Membrane Sci.* 83: 81.

PORTER, M. C. (Ed). 1990. *Handbook of Industrial Membrane Technology*, Noyes Publications, Park Ridge, NJ.

RIDGWAY, H. F. 1988. In *Reverse Osmosis Technology*, B. S. Parekh (ed.), Marcel Dekker, New York.

RIPPERGER, S. and SCHULZ, G. 1986. *Bioprocess Engr.* 1: 43.

ROZELLE, L. T., CADOTTE, J. E., COBIAN, K. E. and KOPP, C. V. 1977. In *Reverse Osmosis and Synthetic Membranes. Theory—Technology—Engineering*, S. Sourirajan (ed.), National Research Council Canada, Ottawa, Canada.

SOURIRAJAN, S. 1970. *Reverse Osmosis*, Academic Press, New York.

SOURIRAJAN, S. (Ed.). 1977. *Reverse Osmosis and Synthetic Membranes. Theory—Technology—Engineering.* National Research Council Canada, Ottawa, Canada.

SOURIRAJAN, S. and MATSUURA, T. (eds.). 1985. *Reverse Osmosis and Ultrafiltration.* American Chemical Society, Washington, DC.

ZEMAN, L. and DENAULT, L. 1992. *J. Membrane Sci.* 71: 233.

CHAPTER 3

Membrane Properties

This chapter covers properties of membranes such as those listed in Table 3.1, as well as how the two key parameters of membrane performance—flux and rejection—are affected by those properties and operating factors. Of special importance for ultrafiltration (UF) and microfiltration (MF) membranes are the pore statistics, e.g., pore "size" (usually expressed as pore diameter), pore density (number of pores per unit membrane surface area), and bulk porosity (or void volume), which is the fraction of the membrane volume occupied by the pores. However, not all membrane manufacturers use the same methods for determining these properties. Thus, the identical numerical designation of pore size by different manufacturers does not necessarily mean that the membrane pores are identically sized. Indeed, it is highly unlikely that this is so, considering the vast combinations of materials, manufacturing methods, and measuring methods. This lack of standardization may be more noticeable with UF membranes than with MF membranes, due partly to the manner in which each membrane is characterized. Despite this, however, it has been the practice among users to interchange similarly rated membranes from different manufacturers and to obtain quite similar results, especially with MF membranes. Nevertheless, there is the possibility for some ambiguity in the specifications provided by membrane manufacturers. The guidelines presented in this chapter should provide some understanding about these methods of characterization, especially regarding how it affects separation properties.

3.A.
PORE SIZE

The most common methods of determining pore size are (1) bubble point (Blasendruck technique); (2) direct observation, e.g., with electron microscopes; and (3) function measurement, or the challenge test. It should be recognized, of course, that polymeric membranes used for UF and MF are essentially screen or surface filters (as opposed to depth filters) and that very rarely do we observe pores of just one diameter, but rather a distribution of pore sizes.

71

Table 3.1. Properties of membrane filters requiring standardization.

Property	Method of Measurement	Significance
Pore size and pore size distribution	Hg intrusion, bubble point, electron microscopy	Most critical property of the membrane
Surface porosity	Pore size and number of pores	Flux
Bulk porosity	Thickness/weight measurement	Filter flow, life
Retention/rejection	Passage testing	Yield, sterile filtration
Flux	Permeate flow rate	Economics
Temperature stability	Exposure to higher temperatures temperatures	Flux, sterilization
Solvent resistance	Compatability testing	Stability, cleanability, life
Wettability	Water-wetting pattern	Flux
Sterility	U.S. Pharmacopoeia sterility test	Sterile filtration
Bacteria growth	Bacterial culture	Bacterial recovery
Bacteria inhibition	Bacterial culture	Inhibition or inactivation
Thickness	Thickness gauge	Strength
Strength	Microtensile tester	Physical stability
Refractive index	Refractometer	Rendering filter transparent
Extractable components	Extraction and gravimetric analysis	Purity of products
Residual ash	Combustion and gravimetric analysis	Analytical methods
Grid marking	Microscopy and coliform growth	Counting colonies

Source: Adapted from Brock (1983)

3.A.1.
BUBBLE POINT AND PRESSURE TECHNIQUES

This method of measuring pore diameter was probably first used by Bechhold and is the most widely used method for measuring pore sizes of membrane filters and for testing integrity of membranes. The principle of the method is shown in Figure 3.1. Due to capillary forces, liquid inside that capillary will rise to a greater height than the surrounding liquid, assuming both the capillary and the surrounding liquid are exposed to the same ambient pressure. The liquid in the capillary can be pushed out by the application of pressure, as shown by Cantor's equation:

$$P = \frac{4\,\sigma\,\cos\theta}{D} \qquad (3.1)$$

where P is the bubble point pressure, σ is the surface tension at the solvent/air interphase, θ is the liquid–solid contact angle, and D is the diameter of the capillary. Since D is in the denominator, the liquid will rise to a greater height in

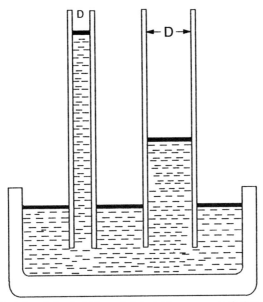

Figure 3.1. *Principle of the bubble point test. Two open-ended capillary tubes of different diameters (D) are immersed in a liquid in an open container. The surface of the liquids in the container and in the capillaries are exposed to atmospheric pressure.*

the narrow diameter tube than in the larger diameter tube; i.e., a greater pressure will be needed to push out the liquid from a smaller sized capillary. Extending this concept to membrane pores (which can be considered as capillaries), if the membrane is dry, it takes very little pressure for air to pass through the pores. However, if wet and if all the pores are filled with liquid, then it will be impermeable to air, unless the wetting liquid has been forced out of the pores.

To conduct the bubble point test, the membrane is thoroughly wetted with a liquid of known surface tension and contact angle (for routine quality control tests, the precise values are not necessary). The wetted membrane is then placed in a holder, or cell, as shown in Figure 3.2. Air is then passed through the cell from the bottom. As the pressure is increased, the liquid is forced out of the pores. When all of the liquid has been forced out of the pores, a steady stream of bubbles will be observed. This is the bubble point of the membrane, which is a reproducible number if the integrity of the membrane filter has not been breached. If there is a leak or there are pores larger than the rated size, the pressure at which the bubble point occurs will be lower, as seen from Equation (3.1).

As an example, Equation (3.1) predicts that, if water at 20°C (surface tension of 73 dynes/cm, contact angle of zero) were used, then the Cantor equation can

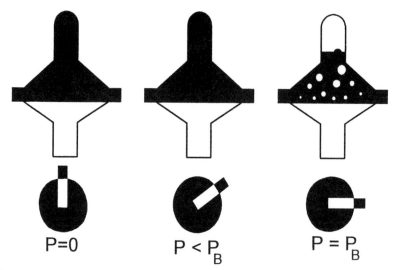

Figure 3.2. Schematic of method for measuring bubble point. P_B is the bubble point.

be simplified to (Brock 1983)

$$D(\mu m) = \frac{41.6}{P \text{ (psig)}} \tag{3.2}$$

Theoretical bubble points for typical MF membranes are given in Table 3.2. In practice, actual bubble point pressures are frequently much lower. This could be partly caused by uncertainties in contact angle values, the irregular shapes of the pores (the Cantor equation assumes capillaries are right circular cylinders), and other physical properties that are not accurately known. In all likelihood, however, the major reason is that the pores are actually larger than their rated size. As far back as the 1930s, Ferry (1936) predicted that the diameter of a capillary in a membrane would be 1.4 times larger than the diameter of a spherical particle rejected at the 90% level, and Elsford suggested that the

Table 3.2. Theoretical and actual bubble points for typical microfiltration membranes.

Pore Size (μm)	Theoretical Bubble Point (psi)[*]	Actual Bubble Point (psi)		
		Millipore	Gelman	Sartorius
0.1	416	250	—	125
0.22	190	55	45	56.6
0.8	52	16	14	18.4

[*]From Cantor equation
Source: Brock (1983)

Table 3.3. Morphological parameters and bubble points for selected MF membranes.

Pore Size Rating (μm)	Water Bubble Point (psi)	IPA Bubble Point (psi)	Bulk Porosity (%)	Measured Pore Size (μm)	Surface Porosity (%)
0.025	—	81.6	49	0.18	4
0.05	—	81.6	62	0.20	6
0.10	—	73.0	62	0.26	10
0.22	—	24.0	70	0.77	32
0.45	—	11.0	76	1.00	34
0.20	51	19.2	69	0.60	14
0.45	26	10.0	74	0.79	22
0.65	20	9.7	74	0.62	11
0.80	15	7.3	75	1.14	38
1.20	10	5.7	77	1.14	32

The first five membranes are Millipore mixed cellulose ester membranes. The bottom five are Sartorius cellulose acetate membranes with side 1 values. IPA = isopropyl alcohol
Source: Data adapted from Zeman (1992)

pores could be three times larger than the particles retained for UF membranes of 10–100 nm, 1.7–2 × larger for 0.1- to 0.5-μm membranes and 1–1.4 × larger for 0.5- to 1-μm MF membranes. This has been confirmed by electron microscopy (Cherkasov and Polotsky 1995; Wu and Wu 1995; Zeman 1992). Image analysis of electron micrographs showed that the actual pore size of MF membranes is often much larger than the rated pore size (Table 3.3). Even polycarbonate track-etch membranes rated at 10, 50, and 100 nm were measured by atomic force microscopy (AFM) to be closer to 18, 65, and 113 nm (Dietz et al. 1992). There also appears to be (at least) two groups of pores (each group with a different mean pore size) on each membrane, perhaps caused by the manner in which phase inversion membranes are made (Persson et al. 1994; Sourirajan and Matsuura 1985; Wu and Wu 1995). Nevertheless, these membranes still manage to retain particles according to their ratings, if based on challenge tests.

Pressures needed to obtain a bubble point with water may be too high with tight MF and UF membranes to conduct the test properly without artifacts induced by deformation of the membrane due to the high pressures. UF membranes have pore sizes of the order of 1 to 10 nm (0.001–0.01 μm) and thus should, in theory, require 4000–40,000 psig (275–2750 bar) with water as the test liquid. Of course, solvents with lower surface tensions, such as alcohols, could be used in place of water, provided they are chemically compatible with the membrane; e.g., ethyl alcohol will result in a bubble point of only 1.5–1.7 bar compared to 3.5–4 bar with water for a 0.2-μm membrane. This is also shown in Table 3.3 where both water bubble points and bubble points obtained with isopropanol are listed. It should be remembered that it is the largest pore on the membrane surface that will allow air to pass through first, and so the bubble point test is really a measure of the diameter of the largest pore.

Integrity testing of membrane filters, especially for "sterilizing filters," involves more than just ensuring that pore sizes are appropriate. It also means making sure that the membrane is intact and properly placed in the filter housing to avoid the feed bypassing the membrane and leaking into the permeate side. The usual practice is to conduct *destructive* integrity tests on samples from each manufacturing lot; these tests are described in Section 3.C. *Nondestructive* integrity tests should be conducted on each sterilizing-grade filter prior to sale and/or use to ensure integrity. These latter tests include the bubble point test, the diffusion test, and the "pressure hold" (decay) test. Integrity test kits are available from membrane manufacturers (e.g., Sartocheck from Sartorius, Compact 100 from Pall, Integritest II Plus and HydroCorr from Millipore) that automate the procedure and provide a written record of the test, as shown in Figure 3.3.

In the diffusion and pressure hold/decay tests, compressed gas is fed into the inlet side of a fully assembled membrane system in which the filter has been prewetted (A–B in Figure 3.3). The pressure applied should be about 80% of the expected bubble point value. After a stabilizing period of 3–5 minutes, the gas supply is shut off. Since the pressure is below the bubble point, the only way air can flow through the membrane in this state is by diffusion through the liquid in the pores. This is very slow: theoretically, the maximum amount of gas that will diffuse through water-wetted pores is 6–15 ml/min · m^2 · bar. Thus,

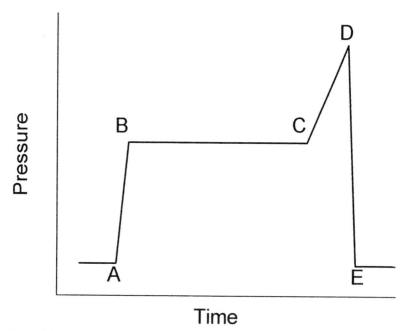

Figure 3.3. *Typical data obtained with pressure testing of microfiltration membranes.*

there should be less than a 10% decrease in pressure (or < 1 psi in 10 minutes) during the measuring period (B–C) for the device to pass the test. If there is a defect, there will usually be a sharp drop in pressure during the measurement phase or a large increase in flow rate of the air in the permeate tubing (Adham et al. 1995).

For the bubble point test, the pressure is increased in 0.05 bar (0.75 psi) increments (C–D) until a continuous stream of bubbles is observed in the permeate tubing (which can be immersed in a liquid to make it easier to observe the bubbles). The onset of bulk gas flow indicates that the bubble point has been exceeded.

Pore size distribution, rather than just the average or maximum pore size, can be determined by liquid permeation for hydrophobic membranes. The membrane pores are initially completely dry, and a nonwetting liquid (with a contact angle >90°) is brought into contact with it. At low pressures, no flow occurs through the membrane. As pressure is increased, the largest pores become flooded with the liquid and the liquid starts to flow through these pores. Higher pressures cause the successively smaller pores to become flooded until the entire membrane is flooded out. Any further increase in pressure causes a corresponding increase in flux. The flux–pressure curve is usually S-shaped, and the data can be analyzed by graphical or statistical methods (McGuire et al. 1995).

It is possible to obtain pore size distribution data of membranes with very small pores (0.5–1 nm) with no need of high pressures using solvent permeability tests. The membrane is allowed to swell in one of two immiscible liquids (this liquid should wet the membrane much better than the other). The other liquid (with a lower interfacial tension) is brought into contact with the surface of the membrane and pressure applied. The Hagen-Poiseuille equation is then used to estimate the pore diameter and pore area.

Other methods use gas permeability, air flow, permporometry and thermoporometry (Abaticchio et al. 1990; Cao et al. 1993; Cuperas et al. 1992; Muldur 1991; Nakao 1994). However, there appears to be little correlation between these methods: biliquid permporometry and thermoporometry gave higher pore sizes than electron microscopy or passage testing, sometimes by as much as one order of magnitude (Kim et al. 1994).

For ultrafiltration membranes, bubble point testing is impractical. This leads to an interesting situation: we know intuitively that a filter that does not allow macromolecules such as proteins or even pyrogens to pass through should also not allow much bigger particles such as microorganisms to pass through. However, the lack of a simple validation test such as the bubble point method prevents most UF membranes from being classified as sterilizing filters. It is possible to correlate challenge tests (as described in Section 3.C.) to water pressure hold tests and to use them as gross indications of module failure. Characterization of pore size and pore size distribution of UF membranes is more commonly done using other techniques described in this section.

For more robust membranes such as inorganic membranes, the mercury intrusion test can be used. Mercury is forced into the pores, and the change in volume of the mercury is measured. Mercury does not wet the membrane; thus, a plot of cumulative volume versus pressure provides an idea of the pore size distribution.

While these tests are easy to perform and are reliable for small flat sheet sections of a membrane done in a controlled laboratory environment, it becomes more difficult when done in situ on a module, cartridge, or an entire system. Not only is it difficult to detect the small increase in gas flow caused by the presence of a few outsize pores among several trillion in a module, but the temperature and system volume are more difficult to control in a system test. A variation of 0.7 mL/min in the gas flow rate can occur per °C change in temperature. With hydrophobic membranes that resist wetting by water, organic solvents could be used for batch tests of flat sheets, as mentioned earlier. However, it is not feasible for in situ integrity tests of membrane modules and cartridges. Instead, the water pressure hold test could be used. It is analogous to the gas pressure test for hydrophilic membranes. The upstream volume of the filter assembly is filled with pure water at a controlled temperature. Gas pressure is applied upstream of the water for a few minutes, and pressure decay is observed for another few minutes. A decay in pressure of more than 0.5 psi in 5 minutes is indicative of a nonintegral cartridge.

Whatever method is used to ensure integrity of the filtration system, consistency of testing conditions is just as important as adhering to standard procedures. As will be shown later, the test conditions can have a marked influence on the results. Thus, for a user, it is important to conduct tests under conditions simulating actual process conditions for a particular application and to compare results to the manufacturer's ratings for the same filter.

3.A.2.
DIRECT MICROSCOPIC OBSERVATION

Electron microscopy has been particularly useful for studying the surface structure of MF membranes, where equivalent pore diameters are 0.05–2 μm, which are well within the resolution capabilities of modern electron microscopes (Nakao 1994). UF membranes, however, have pore sizes ranging from 1 nm to 30 nm. This poses some difficulty for a conventional scanning electron microscope (SEM), which has a resolution limited to 5–10 nm (for a perfect specimen with excellent instrument conditions). An example was shown in Figure 1.7, which is a typical SEM micrograph of a UF membrane with a nominal molecular weight cut-off (MWCO) of 10,000. The surface of the membrane appears to be relatively smooth with no "pores" visible. An SEM was particularly useful in demonstrating the anisotropic and asymmetric nature of membranes (Figures 1.6 and 1.7): the voids and void walls are clearly visible in the supporting substructure of the membrane.

The transmission electron microscope (TEM), on the other hand, has a much higher resolution of 3–4 Å, but much greater care is required during sample preparation. Void volumes and bulk porosities of 50–90% are common with UF and MF membranes (Table 3.3). Very thin sections of the membrane would tend to evaporate under the beam, yet thick sections would not reveal any pores. The replica technique is quite useful under these circumstances. Pieces of the membrane are air-dried and coated with platinum at 300 Hz, followed by carbon, using a freeze-etch unit. The replica can then be peeled off the membrane using a series of acetone-dimethylformamide (DMF) solutions in varying ratios to completely dissolve the membrane. The replica is washed in concentrated sulfuric acid for a few hours and viewed with a TEM.

Figure 3.4 shows the surface structure of a polysulfone UF membrane using the replica technique. The pores appear to be quite uniform with regard to shape and range in size from 1–15 nm in diameter. The pore density was about 4×10^{11} pores per cm^2, as calculated from the micrographs. Figure 3.5 shows

Figure 3.4. *Surface structure of PTGC (Millipore) polysulfone UF membrane with MWCO of 10,000. Replica technique was used for sample preparation: see text for details (adapted from Merin and Cheryan 1980).*

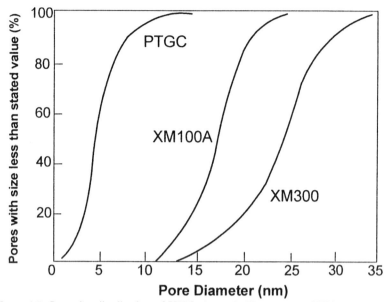

Figure 3.5. *Pore size distribution of PTGC shown in Figure 3.4 and XM membranes shown in Figures 3.6 and 3.7 (adapted from data provided by Merin and Cheryan 1980 and Fane et al., 1981).*

the pore size distribution of the PTGC membrane (from Millipore Corporation), determined from the micrographs. A mean pore diameter of 5.9 nm was reported, and it was estimated that the pores occupied 7–12% of the membrane surface. However, Dietz et al. (1992) reported much larger pore sizes of 14.1 nm (with a range of 12.2 to 17.1 nm) using atomic force microscopy (AFM); the pore density was 10 times lower, but the surface porosity was about the same (6.8%). Once again, the pores appear to be larger than the molecules they reject, considering that this membrane should have at least 90% rejection of 10,000 molecular weight proteins, which are about 1–3 nm in effective diameter.

Figures 3.6 and 3.7 show micrographs of UF membranes with fairly large pores, and the corresponding pore size distributions are shown in Figure 3.5. Figure 3.8 shows the surface structure and pores of membranes as visualized by AFM.

Surface porosity (also termed *pore density*) is less than 10% for many UF membranes (Tables 3.3 and 3.4). This is especially true of those membranes made by phase inversion and track-etch methods. For example, track-etch membranes with pores of 18–113 nm had surface porosities of 0.6–4.5%, while porosities were 2.2–6.8% with UF membranes of 5 K–100 K MWCO (Dietz et al. 1992), 7–12% for the PTGC membrane of 10 K MWCO (Cheryan and Merin 1981; Dietz et al. 1992), and 4.4–15.1% for 30 K–300 K MWCO UF membranes as measured by field emission SEM (Kim et al. 1990). However,

Figure 3.6. Surface structure of Amicon XM100A ultrafiltration membrane (adapted from Fane et al. 1981).

Figure 3.7. Surface structure of Amicon XM300 ultrafiltration membrane (adapted from Fane et al. 1981).

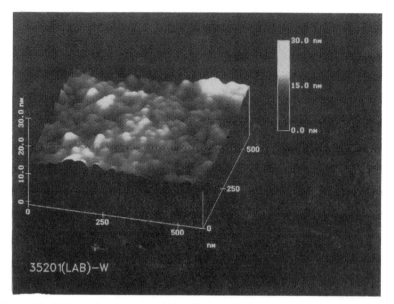

Figure 3.8. *A 3D surface image obtained by atomic force microscopy. This is a 30° view angle of a 30,000 MWCO PAN membrane immersed in water. [Photo courtesy of A.K. Fritzsche, Parker Hannifin Corporation; see Fritzsche et al. (1992) for details].*

these numbers should be interpreted with caution, since the technique used may affect the values. Kim et al. (1990) report that those same membranes cited above measured by TEM with unidirectional shadowing showed porosities of only 0.3–2%. [They also reported much smaller pore sizes with field emission scanning electron microscopy (FESEM) compared to TEM measurements: see

Table 3.4. Pore size and surface porosity of ultrafiltration membranes.

Manufacturer	Membrane[*]	Pore Size (nm)	Number of Pores per cm^2	Surface Porosity (%)
Millipore[**]	PTTK (30,000)	4.8	1.6×10^7	5.5
	PTHK (100,000)	9.2	7.5×10^6	9.9
	SKIP (300,000)	15.1	6.8×10^6	15.1
Amicon[**]	PM30 (30,000)	4.0	1.8×10^7	5.9
	XM100A(100,000)	6.1	1.3×10^7	4.4
	XM300 (300,000)	8.6	6.8×10^6	6.4
Filtron[†]	FNS-10 (10,000)	4.5	—	2.8
	FNS-30 (30,000)	—	—	3.0
	FNS-50 (50,000)	—	—	4.2
	C-10 (10,000)	—	—	2.3

[*]All membranes are polysulfone, except Amicon's XM membranes, which are Dynel (PAN-co-PVC), and Filtron's C-10, which is cellulosic. Numbers in parentheses are MWCO
[**]Data from Kim et al. (1990)
[†]Data from Gekas and Zhang (1989)

Table 3.7 later.] The XM100A Dynel membrane of Amicon's has been reported to have average pore sizes as small as 6.1 nm to as large as 17.5 nm, depending on the technique used. The PTHK membrane (from Millipore Corporation) measured by AFM was reported to have pore sizes of 22.1 nm (range of 16.8–33.6 nm) and a pore density of $8.8 \times 10^9/\text{cm}^2$ (Dietz et al. 1992), which are much larger numbers than those measured by FESEM (Table 3.4). It is interesting to note that, despite large differences in these particular pore statistics, the surface porosities reported by the different techniques are in the same general region. If there is a trend, it is a decrease in surface porosity with a decrease in pore size (Tables 3.3 and 3.4).

In contrast, MF membranes, especially those made by techniques other than classical phase-inversion methods, show higher surface porosities. Values as high as 80% have been reported for the Enka polypropylene membrane (shown earlier in Figure 2.19); 25–38% for the polytetafluoroethylene (PTFE) membrane (Figure 2.19), as reported by Vivier et al. (1989); and 32–35% for the larger MF membranes listed in Table 3.3. The very tight MF membranes in Table 3.3 approached surface porosities (and pore sizes) of UF membranes. Such trends can be seen with many other series of membranes, provided the pore measurements were made by the same technique and the same investigators and membranes were made by the same company using similar methods. For example, in Table 3.4, the 30,000 MWCO polysulfone membranes of Millipore's and Amicon's are comparable, and all Millipore's polysulfone membranes show consistent trends. However, the 100 K and 300 K Dynel membranes of Amicon show different pore statistics than Millipore's 100 K and 300 K polysulfone membranes, even though the data were obtained by the same investigator.

3.B.
PREDICTING FLUX FROM PORE STATISTICS

Tables 3.5 and 3.6 list UF and MF membranes that were available from commercial sources. As expected, there is a fairly good correlation between pore size and experimental water flux. Solvent flow through pores of a membrane will be a function of pore diameter (d_p), number of pores (N), porosity (ε), applied pressure (P_T), viscosity of the solvent (μ), and thickness of the membrane (Δx). The model most frequently used for describing flow through pores is based on the Hagen-Poiseuille model for streamline (laminar) flow through channels (see also Section 4.C.):

$$J = \frac{\varepsilon d_p^2 P_T}{32 \Delta x \mu} \tag{3.3}$$

In this equation, J is given in terms of velocity (e.g., cm/sec).

***Table 3.5. Characteristics of Amicon
ultrafiltration membranes.***

Membrane Name	MWCO	Pore Diameter Å	Pore Diameter nm	Water Flux at 3.8 bar, 24°C (LMH)
UM 05	500	21	2.1	17
UM 2	1,000	24	2.4	34
UM 10	10,000	30	3.0	102
PM 10	10,000	38	3.8	935
PM 30	30,000	47	4.7	850
XM 50	50,000	66	6.6	425
XM 100A	100,000	110	11.0	1105
XM 300	300,000	480	48.0	2215

Source: Amicon Corporation; Porter (1979)

3.B.1.
EXAMPLE

Consider the case of the XM100A UF membrane, shown in Figures 3.5 and 3.6. The following pore statistics were obtained from the original data (Fane et al. 1981):

$$\text{Mean pore diameter} = 17.5 \text{ nm} = 17.5 \times 10^{-7} \text{ cm}$$

$$N = \text{pore density} = 3 \times 10^9 \text{ pores/cm}^2$$

$$\Delta x = \text{"skin" thickness} = 0.2 \text{ } \mu\text{m} = 2 \times 10^{-5} \text{ cm}$$

Thus,

$$\text{Porosity } (\varepsilon) = N \cdot \frac{\pi}{4} \cdot d_p^2 = (3 \times 10^9) \frac{(3.142)}{4} (17.5 \times 10^{-7})^2$$

$$= 7.216 \times 10^{-3}$$

Table 3.6. Standard Nuclepore membrane specifications.

Pore Size (μm)	Pore Size Range (μm)	Pore Density (pores/cm²)	Nominal Thickness (μm)	Water Flux at 0.72 bar and 30°C (LMH)
0.03	0.024–0.03	6×10^8	5.4	14
0.05	0.04–0.05	6×10^8	5.4	267
0.08	0.064–0.08	3×10^8	5.4	886
0.10	0.08–0.10	3×10^8	5.3	4,550
0.20	0.16–0.20	3×10^8	12.0	7,424
1.00	0.8–1.00	2×10^7	11.5	160,000

Source: Nuclepore Corporation

i.e., 0.72% of the membrane surface is occupied by pores. To calculate the water flux at 20°C at an applied pressure of 100 kPa (which is 10^6 g/cm · sec²), viscosity (μ) is 1 cp = 0.01 g/cm · sec. Inserting all these quantities into Equation (3.3):

$$J = \frac{(7.216 \times 10^{-3})(17.5 \times 10^{-7})^2(10^6)}{32(0.2 \times 10^{-4})(0.01)}$$

$$= 3.45 \times 10^{-3} \text{ cm/sec}$$

$$= 124.2 \text{ L/m}^2 \cdot \text{hour}$$

Table 3.7 lists membrane pore statistics and water flux for several UF membranes. The calculated values are fairly close to the experimental values for the larger pore membranes (depending on what pore statistics are used). For the smaller pore membranes, the experimental fluxes show the expected trend of decreasing with smaller pore size. However, the calculated flux is *higher* for the smallest pore size. Further, the discrepancy between calculated and experimental flux appears to be larger as the pore size of the membrane gets smaller.

Several factors could be involved: (a) The Hagen-Poiseuille equation assumes that all pores are right-circular cylinders, which is highly unlikely with these membranes. A "tortuosity" factor should have been included in the equations. (b) As pores get smaller in diameter, the number of pores of various sizes increases, which makes estimation of pore size distribution and pore density more difficult. (c) The number of "dead-end" pores cannot be accounted for. (d) Smaller pores could be more "tortuous." (e) The contribution of the chemical nature of the membrane material is not taken into account. (f) The methods used for measuring pore size and pore density provide contradictory data, as suggested by the data of Kim et al. (1994).

3.C.
PASSAGE (CHALLENGE) TESTS

The only practical measure of a membrane's separation capability, i.e., its rejection properties, is the function measurement, or the "challenge test." In this method, the permeability of selected particles or solutes of different sizes is measured under controlled conditions. The nature of the separated particles determines the type of membranes to be used and the operating strategy. Thus, the protocol for testing MF membranes is different than for UF membranes.

3.C.1.
MICROFILTRATION MEMBRANES

MF membrane ratings were established based on their ability to retain microorganisms. Thus, the microbial challenge test is the most common method of evaluating MF membranes. Figure 3.9 shows a typical procedure for conducting

Table 3.7. Membrane pore statistics and water flux calculations.

	PTGC*		PM30**		XM100A[†]		XM300[†]	
Material	Polysulfone		Polysulfone		PAN-co-PVC		PAN-co-PVC	
MWCO	10,000		30,000		100,000		300,000	
Mean pore diameter (nm)	5.9	14.1	7.5	4.0	17.5	6.1	24.5	8.6
Pore density (pores/cm^2)	4.0×10^{11}	4.35×10^{10}	4.5×10^{10}	1.8×10^{11}	3.0×10^9	1.3×10^{11}	6.7×10^8	6.8×10^{10}
Surface porosity (%)	10.9	6.8	2.0[§]	5.5	0.75	4.4	0.30	6.4
Water flux (LMH) Calculated[‡]	214	759	—[§]	20	124	79	104	164
Experimental	18		63		80		98	

*Data in left column from Merin and Cheryan (1980). Data in right column from Dietz et al. (1992)
**Data in left column from Porter (1979). Data in right column from Kim et al. (1990)
[†]Data in left column from Fane et al. (1981); data in right column from Kim et al. (1990)
[‡]Calculated from the Hagen-Poiseuille equation with $P_T = 100$ kPa, 20°C, $\Delta x = 0.2$ μm
[§]Porosity was back-calculated from experimental water flux data

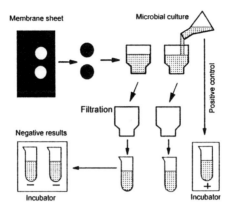

Figure 3.9. *Typical procedure for challenge testing microfiltration membranes.*

such tests on flat sheet membranes. Pieces of the membrane are cut and used to filter a certain volume of the culture medium containing actively growing cells. Pressures used should be typical of what the process will use, although membrane manufacturers could use elevated pressures, possibly pulsed. The filtrate is then cultured to test for the growth of the organism. Simultaneously, the original culture itself is incubated to provide a positive control.

Depending on the membrane pore size, the appropriate microorganism is selected (Table 3.8) and grown in the appropriate culture medium to a cell concentration that would result in a monolayer coverage of the membrane with cells. This is because it is assumed that sieving and, to a lesser extent, adsorption are probably the mechanisms by which these particles are rejected by the membrane. For an organism of 1-μm^2 cross-sectional area, this implies a concentration of 10^8 cells/cm^2 of membrane area. Any higher concentration would measure the retention by the cake of microbes rather than the membrane

**Table 3.8. Microorganisms used for challenge tests
of MF membranes.**

Pore Size (μ)	Organism	Culture Medium
0.1*	*Acholeplasma laidlawii*	Hayflick's media
0.2	*Pseudomonas diminuta*	Saline lactose broth
0.45	*Serratia marcescens*	Wilson's peptone agar
0.8	*Bacillus subtilis* or *Saccharomyces cerevisiae*	Tryptone glucose extract
1.2	*Saccharomyces cerevisiae*	Tryptone glucose extract

*As of this writing, the test for 0.1-μm membranes had not been officially accepted by all regulatory agencies

filter itself. Some manufacturers use the criteria of one organism per pore opening as the true challenge test. Since pore density is of the order of 10^6–10^8 pores per cm^2 for MF membranes, cell concentrations in the test media would need to be about $10^7/cm^2$. For most sterilizing applications involving water or pharmaceutical liquids, this concentration of microorganisms is probably much greater than real-world process streams. However, it does provide for a rigorous test of a membrane's capabilities. In any case, it is important to know the total challenge in terms of colony forming units (cfu) per volume per filter, as well as the total challenge per unit area. Otherwise, it becomes difficult to translate results from a small test disc to an industrial size cartridge or module.

The data is expressed in terms of the log reduction value (LRV):

$$LRV = \log_{10} \left(\frac{\text{Concentration of cells in the feed}}{\text{Concentration of cells in the permeate}} \right) \quad (3.4)$$

Table 3.9 shows LRV data obtained for one set of commercial membranes. Note that the value of zero for the permeate really means that no growth was observed in the culture tubes or no colonies were formed on the agar plates, when the filter was challenged at that particular level. Higher challenge levels may show some passage of microorganisms through the membrane filter, possibly due to a "grow-through" mechanism. However, the LRV for that membrane is still valid even with higher microbial loads, e.g., a membrane with an LRV of 7, if challenged with 10^9 cells/mL, could show about 10^2 cells/mL in the permeate.

There are many organizations involved in setting standards for sterilizing filters, among them American Society for Testing and Materials (ASTM 1992), U.S. Pharmacopoeia (USP), Food and Drug Administration (FDA, USA), DIN (Germany), and Health Industry Manufacturers Association (HIMA). Most agree that a 0.2-μm membrane used for sterilizing purposes should retain *P. dimunita* at a level of at least 10^7 cfu per cm^2 of membrane area (Meltzer

Table 3.9. Retention of microorganisms by MF hollow fiber membranes.

Pore Size (μ)	Organism	Challenge Level in Feed (cells/mL)	Permeate Concentration	Minimum Log Reduction Value
0.02*	E. coli	6.0×10^9	0	9
0.1	E. coli	6.0×10^9	0	9
0.2	E. coli	6.0×10^9	0	9
	P. dimunita	2.5×10^7	0	7
	S. marcescens	3.1×10^7	0	7
0.45	E. coli	6.0×10^9	0	9
	S. cerevisiae	5.0×10^7	0	7

*Nominally rated as 500,000 MWCO
Source: Data from A/G Technology Corporation

1987). The appropriate culture medium is important to ensure that the smaller individual cells predominate (Waterhouse and Hall 1995). Although there is a good correlation between bubble point pressure and retention of test microorganisms (Roche et al. 1993), it should be remembered that, while the former test gives valuable information about the potential ability of the membrane to retain organisms, only the latter test can confirm that ability.

At present, there are no standards for testing membranes with pore size ratings less than 0.2 μm. It is known that microorganisms smaller than *P. dimunita* (e.g., *P. picketti, P. cepacia, Leptospira, Acholeplasma, Mycoplasma*) can penetrate 0.2-μm sterilizing filters (Meeker et al. 1992; Roche and Levy, 1992), probably because *Mycoplasma* lack a cell wall and are thus deformable, allowing them to squeeze through the pores. They are known to contaminate animal and plant cell tissue cultures. There have been suggestions from membrane manufacturers that *A. laidlawii* be adopted as the standard test organism, using the worst-case process conditions.

Note that membrane filters can remove much smaller particles if the fluid is a gas rather than a liquid. Hydrophobic membrane filters [polyvinylidene fluoride (PVDF), PTFE] are commonly used in gas filtration for fermentations. A 0.22-μm PVDF membrane can quantitatively remove 0.03-μm particles from a gas stream. The organism for retention testing is usually a bacteriophage such as ϕX174, which is small (28 nm) and spherical, can withstand high shear forces, and can be easily cultured and grown. Grow-through tests are conducted with *P. dimunita*. The tests should properly simulate the mechanical, hydraulic, hydrolytic, thermal, oxidative, and chemical stresses that such filters are normally subjected to in use (Keating et al. 1992).

3.C.2.
ULTRAFILTRATION MEMBRANES

Since the early applications of UF membranes were targeted at purification of biological solutions, rejected solutes were usually macromolecules such as proteins, and permeable solutes were sugars and salts. Since the size of proteins are commonly characterized in terms of their molecular weight, it became customary to characterize UF membranes in terms of their ability to retain proteins of a particular molecular weight (MW). The term *molecular weight cut-off* came into being for UF membranes, to define the size of the protein that would be (almost completely) retained by a particular membrane.

To determine complete rejection characteristics of a membrane, globular molecules of known sizes should be selected, which are soluble in water or mildly buffered solutions, cover the entire expected rejection range from zero to 100% rejection, and do not interact with, foul, or be adsorbed by the membrane. In practice, it is quite difficult to find macromolecules that meet this criteria. At the low end, solutes such as sodium chloride (MW 58.5) and glucose (MW 180) are used, since they are expected to have zero rejection by UF membranes. On

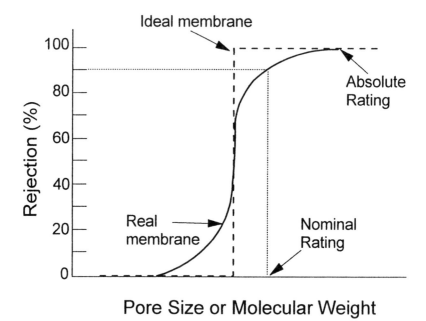

Figure 3.10. *Relationship between pore size, molecular weight of ideal solutes, and ratings of ideal and real membranes.*

the high end of the rejection scale, large proteins (e.g., immunoglobulins, MW >900,000) or blue dextran are used, since they are almost completely rejected by most commonly used UF membranes. Several other solutes of intermediate molecular weights should be used, such as those listed in Tables 3.10 and 3.11. About five to eight solutes, with sizes around the expected rejection of the membrane, are required to adequately characterize a membrane's rejection profile.

Figure 3.10 is a graphical representation of solute rejection data for ideal and real membranes. The sharp pore size distribution shown for the ideal membrane is never seen in practice. MF membranes are given "absolute ratings" based on the largest pore on the membrane surface, as discussed earlier in this chapter. In contrast, UF membranes are given "nominal" ratings that refer to the MW of a test solute (ideally, it should be a globular protein) that is 90% rejected by the membrane under standard conditions. Since this does not imply an absolute retention by the membrane, these designations are sometimes referred to as *nominal* molecular weight cut-off (NMWCO) or nominal molecular weight *limit* (NMWL) of UF membranes. Figures 3.11 and 3.12 show typical rejection curves for Amicon membranes that were obtained from data such as those shown in Tables 3.10 and 3.11. These nominal ratings may appear to be arbitrarily given; e.g., in Figure 3.12, it appears to be at 90% rejection for UM05 and UM10 membranes and 80–85% rejection for the others. As expected, there will be some lot-to-lot variations in the manufacture of membranes.

Table 3.10. Typical solute rejections for Amicon membranes (HF = hollow fibers; all others flat sheets).

Solute	Molecular Weight	YC05	YM1	P3* (HF)	YM3	YM10	PM10	P10* (HF)	YM30	PM30	P30* (HF)
Sodium chloride	58.5	<20	0	—	0	—	0	0	0	0	—
Glucose	180	>70	0	—	0	—	0	0	0	0	—
Sucrose	342	>85	45	—	20	—	0	0	0	0	—
Raffinose	504	95	65	0	25	—	0	0	0	0	—
Bacitracin	1,400	98	92	50	>80	20	—	35	0	—	—
Inulin	5,000	>98	95	>98	—	25	—	—	—	0	—
PVP K15	10,000	—	—	85	—	55	35	70	—	—	45
Cytochrome C	12,400	>98	>98	98	>98	>95	>90	95	<15	0	45
Myoglobin	17,800	>98	>98	—	>98	>98	94	—	—	60	—
α-Chymotrypsinogen	24,500	>98	>98	>98	>98	>98	>98	95	>80	76	75
PVP K30	40,000	—	—	>98	—	>80	—	>98	60	—	75
Albumin	67,000	>98	>98	>98	>98	>98	>98	>98	>98	90	95
PVP K60	160,000	—	—	>98	—	—	—	>98	>90	—	95
Immunoglobulin G	160,000	>98	>98	—	>98	>98	>98	—	>98	>98	>98
Apoferritin	480,000	>98	>98	>98	>98	>98	>98	>98	>98	>98	>98
Immunoglobulin M	960,000	>98	>98	>98	>98	>98	>98	>98	>98	>98	>98

*Data measured at 0.7 bar (10 psi), 30°C. All others measured at 3.8 bar (55 psi)
Source: Data from Amicon Corporation catalogs

Table 3.11. Typical solute rejections for Amicon membranes (HF = hollow fibers; all others flat sheets).

Solute	Molecular Weight	XM50	YM100[1]	P100[1] (HF)	XM100A	XM300[1]	ZM500[1]	MP01[1]
Sodium chloride	58.5	0	0	—	0	0	—	—
Glucose	180	0	0	—	0	0	—	—
Sucrose	342	0	0	—	0	0	—	—
Raffinose	504	0	0	—	0	0	—	—
Bacitracin	1,400	0	—	—	0	—	—	—
Inulin	5,000	0	0	0	0	0	—	—
Cytochrome C	12,400	38	7	—	—	—	—	—
Myoglobin	17,800	20	12	—	—	—	—	—
α-Chymotrypsinogen	24,500	70	10	20	—	0	—	—
PVP K30	40,000	85	20	15	25	0	—	—
Albumin	67,000	95	20	30	—	20	10	0
PVP K60	160,000	>90	65	60	45	10	—	0
PVP K90	360,000	—	>85	>98	—	—	—	35
Immunoglobulin G	160,000	>98	>95	—	—	65	20	—
Apoferritin	480,000	>98	>98	>98	—	85	20	—
Thyroglobulin	677,000	—	>95	—	—	85	90	—
Immunoglobulin M	960,000	>98	>98	>98	90	98	—	—

[1]Data measured at 0.7 bar (10 psi), 30°C. All others measured at 3.8 bar (55 psi)
Source: Data from Amicon Corporation catalogs

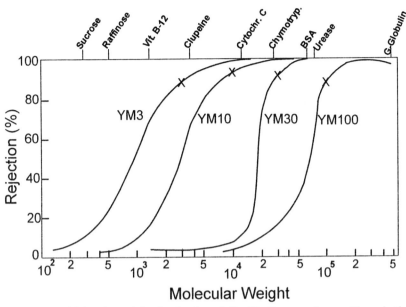

Figure 3.11. *Molecular weight distribution of ultrafiltration membranes. X marks the nominal MWCO rating of the membranes. The solute markers used for this series of membranes is shown on the top axis (adapted from Amicon Corporation literature).*

Unfortunately, at the present time, there does not appear to be any standardization of the procedure, the solute markers, or the test conditions for passage testing of UF membranes. Dead-end stirred cells (such as those shown in Figure 5.3) are frequently used for this testing. Conditions to be standardized are the transmembrane pressure, temperature, stirring rate (or cross-flow velocity if cross-flow cells are used), concentration of the solute, buffer system, ratio of volume of test solution to membrane surface area, the permeate-to-feed ratio, whether solutes should be tested individually or mixed together, and the pretreatment of the membrane. The author's personal recommendation is to use the following conditions: (1) pressure of 100 kPa; (2) temperature of 25°C; (3) maximum possible agitation in the test cell: every effort should be made to minimize concentration polarization to obtain as true a rejection curve as possible; (4) low concentrations of solute (e.g., 0.1%) made up individually in water or mildly buffered solution; (5) 200 mL of solution in the UF dead-end cell using 47–62 mm diameter membranes (18–29 cm² surface area), or proportionally larger amounts for larger diameter flat sheets; (6) removal of not more than 10% of the solution as permeate to avoid concentration effects. If a cross-flow device is used, the permeate and retentate should be returned to the feed reservoir to minimize concentration effects during the test; and (7) the membrane should be new, washed to remove any preservative or glycerin as recommended by the manufacturer, and tempered using a series of pressurization,

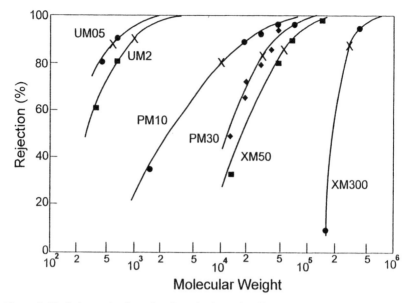

Figure 3.12. *Solute rejection of various Amicon ultrafiltration membrances. X marks the nominal MWCO rating of the membranes (adapted from Porter 1979).*

soaking, and cleaning steps until a consistent and reproducible pure water flux is obtained prior to passage testing.

When comparing MWCO measured with proteins to pore sizes measured by other techniques, it should be remembered that proteins that differ by 10 times in MW may only differ by three times in size in their globular form. In addition, their molecular size can be affected by pH, ionic strength, and interactions with buffer components. Proteins can have different isoelectric points, solubility, and hydrophobicity, thus causing them to interact with and foul the membrane to different extents, which affects measured rejections. Owing to the difficulty of finding proteins that are sufficiently pure (and inexpensive) to conduct MWCO evaluations, other compounds such as polyethylene glycols (PEGs) have been used because they are water-soluble and can be readily obtained with well-defined and narrow size distributions. In addition, they show little or no tendency to foul or adsorb onto most polymeric membranes. However, they are not readily available in very large MW sizes and tend to provide lower estimates of MWCO than with proteins. Polyvinylpyrrolidone (PVP) has also been used.

Dextrans have the advantage of being available in a wide variety of molecular sizes, tend not to be adsorbed by most membranes, and are relatively unaffected by changes in the micro-environment. One test can generate a complete rejection curve. However, analysis of the dextrans is not simple; it requires techniques such as size exclusion chromatography (SEC). Millipore Company's recommendations are to use 0.04M phosphate buffer (with 10 ppm sodium azide as a bacteriostat, if necessary) for membranes in the range of 10,000–1,000,000

MWCO. With smaller pore membranes (1000–10,000 MWCO), a mixture of maltodextrins and dextrans can be used and dissolved in deionized water to avoid interference of buffer salts with small dextran fractions in the SEC analysis (Tkacik and Michaels 1991).

Figure 3.13(a) compares the MWCO profiles obtained with dextrans to that obtained with proteins for a particular Millipore membrane. Dextran and PEG profiles tend to provide lower MWCO estimates than proteins, a phenomenon also observed by Meireles et al. (1995) and Kim et al. (1994). The latter obtained MWCO estimates for Amicon YM30 membranes as low as 9000–10,000 instead of 30,000, and 13,800–26,000 for Millipore's PTHK membranes, which are rated at 100,000. One reason could be that the hydrodynamic diameter of proteins is different than PEGs or dextrans of the same molecular weight. In fact, much better agreement is obtained when the rejection data is plotted in terms of equivalent Stokes radius [Figure 3.13(b)]. Similarly, very good

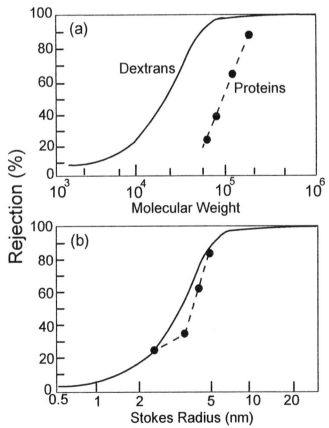

Figure 3.13. Comparison of dextran and protein rejections in terms of (a) molecular weight, and (b) Stokes radius (data adapted from Tkacik and Michaels 1991).

agreement was obtained with PEG, proteins, and dextrans when the sieving coefficient was plotted against the logarithm of the hydrodynamic radius instead of against the molecular weight of the marker (Meireles et al. 1995). In fact, Michaels (1980) suggested that such a log-normal probability relationship is universal enough that a complete sieving curve can be obtained for UF membranes from experimental measurements with only two macromolecules of different Einstein-Stokes radius.

Figure 3.14 shows dextran MWCO profiles of several commercial membranes. Although not exactly the same as previous profiles done with proteins, it is good enough for a first cut when selecting membranes for a particular application, and it does provide an easy tool for in-house characterization and quality control.

3.D.
FACTORS AFFECTING RETENTIVITY OF MEMBRANES

3.D.1.
SIZE OF THE MOLECULE

This is the most critical factor with UF and MF membranes. Because of the manner in which they are characterized, it is much easier to select the appropriate MF membrane than the best UF membrane just from manufacturers' specifications. The following guidelines should help with the initial selection:

1. The MWCO rating is still the best guide, even though it has now become even more confusing because of the many markers used (proteins, PVP, PEG, dextrans) and by the lack of information by the manufacturers as to which marker was used and the operating conditions. If the objective is merely gross separation (e.g., proteins from salts), then the membrane of that protein's MW or the next smaller MWCO membrane will suffice.
2. However, if fractionation is the goal (e.g., separating proteins from each other), then the choice is more difficult. It is important to keep in mind that a certain difference in molecular *weight* between two proteins does not mean the same degree of difference in molecular *size*; i.e., proteins that differ in MW by 10× may differ in size by only 3× when in the globular or folded form. UF membranes have very broad pore size/MWCO distributions covering more than one order of magnitude (see Figures 3.5, 3.11–3.14). This means that fine fractionations are impossible on the basis of size alone. For example, when a soy protein hydrolyzate was processed with different membranes, the size distribution of the peptides permeating through 5000, 10,000, and 50,000 MWCO membranes were similar: two fractions averaging 2300 and 1000 daltons in MW (Deeslie and Cheryan 1992). However, the 100,000 MWCO membrane resulted in three fractions of 25,000, 13,000, and 2300 daltons. All permeating fractions were much smaller than the rated MWCO.

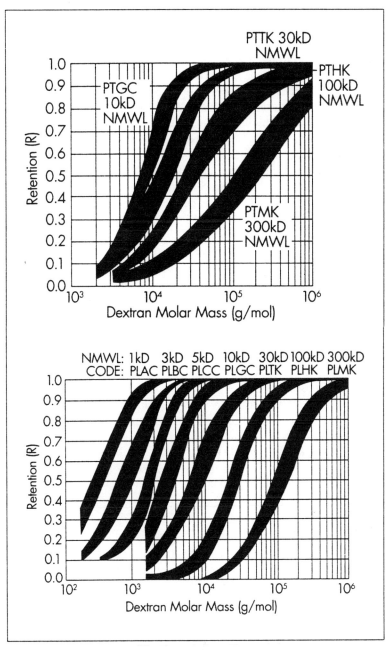

Figure 3.14. Dextran-based MWCO profiles of PT series membranes (polysulfone/polyethersulfone) and PL series regenerated cellulose membranes (Source: Millipore Corporation, with permission).

3. In general, UF is best for enriching or depleting the feedstream of a particular macromolecule in a mixture of macromolecules (whether proteins, dextrans, or polymers), especially if the MW of the target macromolecule is 10× different from the others or the hydrodynamic radius is 3× different. The same rule applies for MF of colloidal suspensions. On the other hand, MF is effective for separating colloidal particles from soluble macromolecules. There are ways to enhance separation of similar sized molecules, as discussed later in this chapter and in Chapters 6 and 7.

4. The final selection of the membrane must be made based on actual tests. For rapid evaluation, it is best to have at least one retention marker and one passage marker; e.g., for a 10,000 MWCO membrane, Cytochrome-C (>95% rejection) or BSA (>99.5%) could be used as retention markers, while vitamin B-12 (~70%) could be used as a passage marker. Similarly, for a 100,000 MWCO membrane, the markers could be immunoglobulin G (>97%) and BSA (~20% rejection).

5. If maximizing flux is the objective rather than separation of components, the best pore size is usually about five to ten times *smaller* than the smallest particle in the mixture. This is discussed in greater detail with specific examples in Section 6.D.1.

3.D.2.
SHAPE OF THE MOLECULE

This is best illustrated schematically in Figure 3.15 and in Table 3.12. The shape and conformation of macromolecules are also affected by ionic strength, temperature, and interactions with other components. Differences in shape, especially under conditions of high shear prevalent at the membrane surface, could also be a reason why the MWCO profiles are different for proteins, dextrans, and PEGs. The best way to account for—and minimize the effect of—differences in shape is to plot rejection data against a volume-based parameter, such as Stokes radius or hydrodynamic volume, rather than molecular weight.

3.D.3.
MEMBRANE MATERIAL

Different membrane materials with the same nominal MWCO will appear to give different solute rejection. Compare, for example, the rejections of PVP-K30 by the three 100,000 MWCO membranes in Table 3.11. In addition to pore size distribution, the chemical nature of the membrane as it affects solute–membrane interactions (i.e., fouling) is important (see Section 3.D.8. below). Compared to polysulfone membranes, membranes made of cellulose acetate or regenerated cellulose had broader pore size distributions and higher rejections (Kim et al. 1994, Wolber and Dosman 1987) and showed less deviation between observed and true rejections and less effect of transmembrane pressure on rejection. These

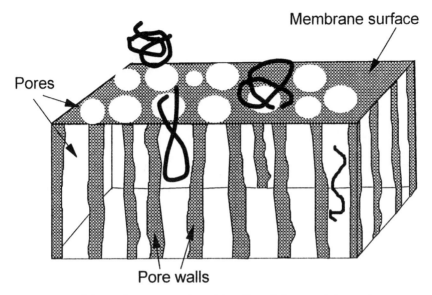

Figure 3.15. *Schematic representation of the effect of shape and size on the passage of solutes through pores. A linear molecule has a greater probability of passing through the pore than a globular molecule of the same molecular weight.*

phenomena are probably related to fouling effects, which in turn are related to hydrophobicity, charge, surface roughness, and the like (Section 6.D.1.). In general, higher fluxes and lower adsorption effects have been observed with hydrophilic materials than hydrophobic membrane, for aqueous-based feeds.

3.D.4.
PRESENCE OF OTHER SOLUTES

In general, low molecular weight solutes (such as sugars and salts), whose molecular size is much smaller than the smallest pore on the membrane, will be freely permeable; i.e., they will have zero rejection, unless they interact with or bind to the impermeable compounds in the feed. As shown in Figure 3.16, the permeability of individual components in a mixture depends on the relative sizes of those components and the pores. If a large pore membrane is used with a feed containing large solutes, which are of the same order of magnitude in size as the pores, then the large solute may be only partially rejected. The smaller solutes (e.g., sugars, salts) will not usually affect the permeability of the large molecules (e.g., proteins), unless they interact with the large molecules and cause molecular changes. However, as noted below, changes in operating

Table 3.12. Effect of size and shape of solutes on their rejection by UF membranes. Molecules above a horizontal membrane line are completely retained by the membrane; below the line, molecules are not or are only partially rejected.

Membrane	Globular Proteins	Branched Polysaccharides	Linear, Flexible Polymers
XM50	Gamma globulin (160,000)* Albumin (69,000)		
PM30	Pepsin (35,000)	Dextran 250 (236,000)	
PM10	Cytochrome C (12,500) Insulin (5700)	Dextran 110 (100,000)	Polyacrylic acid (50,000)
UM10	Bacitracin (1400)	Dextran 40 (40,000) Dextran 10 (10,000)	Polyethylene glycol (20,000)

*Number in parentheses is molecular weight
Source: Adapted from Porter (1979)

conditions such as pressure may force more of the larger solute through the pores, resulting in a decrease in rejection of the large solute.

On the other hand, if the pore size is much smaller than the size of the larger solute (e.g., a tight UF membrane used with a large protein), the large solute will be essentially completely rejected, and its permeability should not change with operating conditions or if other compounds are present. However, the smaller components in that feed may be affected, especially if the size of the smaller solute is of the same order of magnitude or only slightly smaller than the pore. In that case, the large molecule forms a secondary dynamic membrane that inhibits passage of the smaller molecule. As shown in Figure 3.16 on the right, this will be reflected by an increase in rejection of the smaller solute, while the larger solute remains relatively unaffected. With MF membranes, a protein that passes through the membrane at low pressures will be increasingly rejected with the passage of time (due to fouling of the membrane) or if the pressure is increased or if the cross-flow velocity is reduced.

This phenomenon is shown in Figures 3.17–3.20. In a saline environment, the rejection of ovalbumin (MW 43,000), chymotrypsinogen (MW 24,500), and cytochrome-C (MW 12,400) were 75–85, 40–65, and 4–5%, respectively, being greater for the PM30 than the XM50 membrane (Figure 3.17). However, when the solution contained added albumin (MW 67,000), the rejections were only slightly higher for the largest protein (ovalbumin), but much higher for the smallest protein (cytochrome-C): 65% versus 5%. The albumin appeared to have formed a secondary membrane that affected the smaller protein to a greater extent. A similar effect is shown in Figure 3.18: the addition of plasma proteins

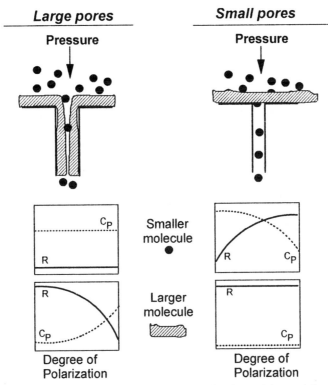

Figure 3.16. Permeability of large and small molecules through large and small pore membranes. Left diagram shows the effect of using a membrane with pores that are relatively large in relation to the larger solutes. Right diagram shows effect of using a membrane with a pore size much smaller than the largest solutes in the feed. Graphs in lower part show effect of the degree of concentration polarization (e.g., higher transmembrane pressure, lower cross-flow velocity, higher feed concentration) on rejection of small molecules and large molecules in the feed. R is rejection and C_p is permeate concentration of the molecule.

created a secondary dynamic membrane on top of the XM50 membrane that caused the membrane to "tighten" up and result in higher rejections.

This phenomenon brings up another question: should single solutes be used to develop the MWCO profile of a UF membrane, or can a mixture of the solutes be used together in one solution? The above discussion clearly suggests the results would be different. In fact, Cherkasov et al. (1995) have shown that successively increasing the number of proteins in a single solution causes a progressive shift in the retention profile to the left. This shift is especially pronounced with intermediate-sized protein markers. Some membrane companies use single-protein solutions, while others use mixed dextrans solutions.

As a general rule, fractionation of polymers can be accomplished to a large degree if there is at least a 10-fold difference in molecular weight. To enhance

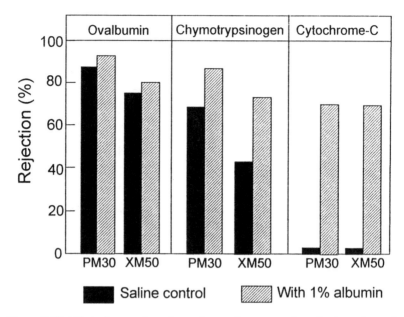

Figure 3.17. Effect of secondary dynamic membranes on the rejection of proteins. Dark bars show rejection with PM30 or XM50 membranes in saline buffer; shaded bars show rejection when 1% albumin was included in the feed sample (adapted from Porter 1979).

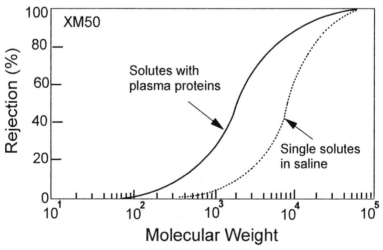

Figure 3.18. Effect of added protein on rejection characteristics of ultrafiltration membranes. The membrane appears to get "tighter" due to the formation of the secondary membrane (adapted from Porter 1979).

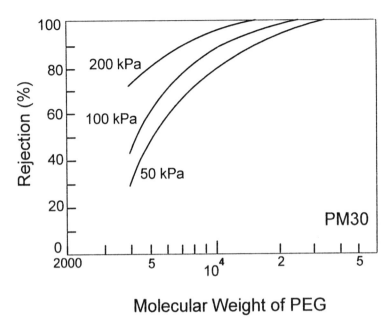

Molecular Weight of PEG

Figure 3.19. Effect of transmembrane pressure on PEG rejection of a PM30 membrane (adapted from Kim et al. 1994).

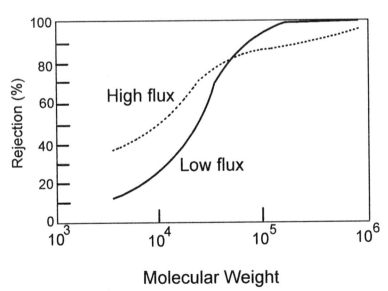

Molecular Weight

Figure 3.20. Change in dextran rejection profile of a cellulosic membrane under different flux conditions (data adapted from Tkacik and Michaels 1991).

separation of similarly sized macromolecules, it might help to dilute the feed to minimize solute–solute interactions and solute–membrane interactions.

3.D.5.
OPERATING PARAMETERS

The biggest problem translating membrane specifications such as MWCO profiles to real situations is duplicating the experimental conditions and operating parameters that were used when the tests were conducted. As mentioned later in Chapter 4, the four major operating variables are transmembrane pressure, turbulence near the membrane surface (provided by stirring in a bench-top test cell or by cross-flow in an industrial module), temperature, and concentration of the solute(s). Additional factors include pH, ionic strength, and other factors that may affect the shape and conformation of the solutes. All these factors affect rejection values. It is most important to distinguish between rejection and adsorption effects and to account for the effect of concentration polarization (defined later in Chapter 4). Concentration polarization is more pronounced at higher pressures, lower velocities, and any other conditions which bring solute to the membrane surface very rapidly (e.g., a very porous MF membrane). For example, as mentioned earlier, if the concentration of a rejected species is high enough, the secondary membrane formed on the membrane may impede the passage of lower molecular solutes. In addition, higher concentrations lead to a decrease in the apparent MWCO (Cherkasov et al. 1995).

Similarly, high pressures may aggravate polarization effects, which will increase the rejection, as shown in Figure 3.19. A similar increase in rejection at higher pressures was observed with dextran markers with other membranes (Kim et al. 1994). This additional layer formed next to the membrane surface will cause the local concentration of the solute at the membrane surface (C_M or C_{wall}) to be higher than in the bulk solution where samples are taken. Thus, the true rejection (R_t) will differ from the apparent rejection (R_a), where

$$R_a = 1 - \frac{C_P}{C_R} \tag{3.5}$$

$$R_t = 1 - \frac{C_P}{C_M} \tag{3.6}$$

where C refers to concentration and the subscripts P, R, and M refer to permeate, retentate, and membrane surface, respectively (the term C_M has the same meaning as C_G in Chapter 4). Since $C_M > C_R$, R_t will be greater than R_a. It is easy to measure C_P and C_R, and thus most published rejection data is actually R_a. These R_a values would only be valid under similar operating and polarization conditions, which in many cases is not known or difficult to duplicate on large industrial systems.

One way to account for polarization and mass transfer effects in rejection data is to use the film theory (described in Chapter 4). The true rejection can then be calculated from the apparent (measured) rejection by the following relationship:

$$\ln\left[(1 - R_a)/R_a\right] = \ln\left[(1 - R_t)/R_t\right] + J/k \qquad (3.7)$$

where J is the flux and k is the mass transfer coefficient (see Chapter 4 for explanations of these terms). Thus, whenever rejection data are reported, the mass transfer characteristics for the apparatus used should be known. This will allow the value of k to be known, along with the volume flux J. Equation (3.7) could then be used to determine the true rejection R_t, which should be the data reported. Potential users could then use these R_t values, together with the k and J values expected for their equipment, to determine the R_a values they could expect under actual operating conditions.

Figure 3.20 shows the effect of polarization in terms of the flux through the membrane. This data was obtained with a cellulosic membrane using dextrans. Under high flux conditions, which results in more polarization, rejection of the larger dextrans decreases. However, there is also an increase in rejection of the smaller dextrans at high flux, in keeping with the concepts shown in Figure 3.16.

3.D.6.
LOT-TO-LOT VARIABILITY

Although membrane manufacturers make every attempt to be consistent in their manufacturing methods, some variation in the quality and specifications of different batches or lots of membranes is to be expected. In the case of MF membranes that must meet HIMA or other standards, it is fairly easy to guarantee filter ratings because the validation and integrity tests are well known and quite easy to do routinely. In addition, it is only necessary to obtain an indication of the largest pore. However, if it is necessary to do a complete pore size or MWCO profile or to obtain realistic flux data with small pieces of a membrane cut from a larger sheet, then a major problem is obtaining a truly representative sample with which to do the testing. Even if hundreds of pieces of a particular membrane are tested, it is unlikely that identical or even close pore profiles or water fluxes will be obtained, even if the pieces come from the same manufacturing batch. However, actual process flux may not have as much variation since the rate-controlling step would not be the membrane, but instead it would be the polarized, dynamic, or fouling layer of rejected molecules (see Chapter 4). Also, as the test scale increases, i.e., if large industrial size modules are tested instead of small 4- to 6-cm diameter pieces, the data becomes more reproducible from module to module and from batch to batch.

3.D.7.
MEMBRANE CONFIGURATION

On the same lines as above, rejections and flux for the same membranes could be different in different module designs. For example, compare the flat-sheet rejections with hollow fiber data (Tables 3.10 and 3.11).

3.D.8.
FOULING AND ADSORPTION EFFECTS

Solute–membrane interactions that result in a physical adsorption of the solute by the membrane, whether on the surface or in the pores, will obviously cause fairly serious losses and decrease the yield of the solute. As will be seen later in Chapter 6, the binding/adsorption of solutes depends on the nature of the solutes and the type of membrane. While low molecular weight sugars appear to have no effect on flux and separation properties of MF or UF membranes, proteins have extreme effects, which depend on the micro-environment (pH, ionic strength) and operating parameters. For example, proteins tend to foul membranes most around their isoelectric pHs, a condition where proteins themselves are unstable, resulting in low fluxes.

As shown earlier in Figure 3.16, the relative sizes of the adsorbing solute and the pore size will determine whether an increase or decrease in rejection is observed. This adsorption phenomenon can also affect apparent rejection values, since the solute will not appear in the permeate in its full concentration until the adsorption sites on the membrane are completely saturated. Table 3.13 shows the effect of different membrane materials on the adsorption of cytochrome-C, a particularly "sticky" protein. Cellulosic membranes, in general, and regenerated cellulose, in particular, such as the Amicon YM series and Millipore PL series, have especially low protein adsorption properties (Table 3.14). On the other hand, commercial pleated MF capsules made of PVDF and cellulose acetate (CA) membranes adsorbed much less bovine serum albumin (BSA) and

Table 3.13. Solute adsorption by various UF membranes (all have a nominal MWCO of 10,000). The solute is cytochrome C, a protein of ~12,500 molecular weight.

Membrane	Material	Adsorption Loss (%)	Rejection (%)
Nuclepore C-10	Regenerated cellulose	0.8	97
Amicon YM-10	Regenerated cellulose	2.3	97
Amicon UM-10	Polyelectrolyte complex	4.3	26
Millipore PTGC	Polysulfone	11.3	99
Nuclepore A-10	Polysulfone	12.4	85
Nuclepore F-10	Polysulfone	22.5	99
Amicon PM-10	Polysulfone	24.6	53

Source: Data from Nuclepore Corporation

Table 3.14. Binding of proteins to UF and MF membranes. Data given as micrograms of protein bound per square centimeter of membrane surface area. Top half of table shows 10,000 MWCO membranes. Bottom half shows commercial 0.2-μ MF capsules exposed to 0.25 g/L protein solutions.

Membrane Material	Membrane	Bovine Serum Albumin	Immunoglobulin G
Regenerated	Millipore PLGC	43.2	0.09
cellulose	Amicon YM-10	56.5	0.21
Cellulose acetate	Sartorius	50.4	0.37
Polysulfone	Millipore PTGC	203.2	1.05
PVDF	Millipak 100	2	3
Cellulose acetate	Sartobran PH	5	5
Nylon	Pall Ultipor N66	31	126

Source: Top half, data from Millipore Corporation (1990); bottom half, data from Brose and Waibel (1996)

immunoglobulin G (IgG) than certain nylon modules, even though unmodified PVDF is supposedly more hydrophobic. PVDF and polyethersulfone (PES) membranes can be modified to reduce their protein-binding tendencies, e.g., the Omega PES series from Filtron.

Membrane–protein interactions could cause changes in the structure of the adsorbed molecule; e.g., BSA adsorbed on the surface of a hydrophilic regenerated cellulose membrane had a globular structure very much like it would have in free solution. However, on the surface of a polysulfone membrane, the protein appeared long and filamentous, more "open" and denatured (Sheldon et al. 1991). Similarly, the activity of carbohydrase enzymes, such as α-galactosidase, β-galactosidase, and glucose isomerase was seriously affected by polysulfone membranes (see Section 8.N.4.b.). Bouhallab and Henry (1995) explained the unexpected high rejection of casein peptides as caused by adsorption by the carbon support of Carbosep membranes, a phenomenon that did not occur with alumina membranes.

3.D.9.
THE MICROENVIRONMENT

Figure 3.21 shows how the permeability of a solute can be affected by its micro-environmental conditions, such as the pH and ionic strength of the solution. This should not be unexpected since pH and ionic strength will affect the conformation and shape of the solute molecule, which, as discussed in

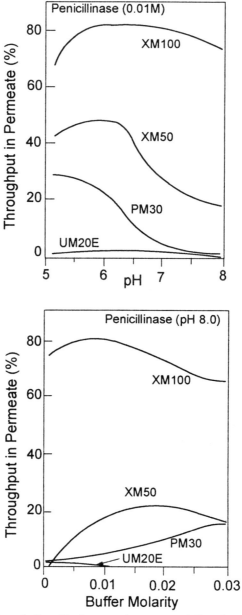

Figure 3.21. Effect of pH and ionic strength on permeability of a solution of penicil-linase: top: effect of pH at 0.01M ionic strength; bottom: effect of ionic strength at pH 8.0. Variable in the graphs is the type and MWCO of membranes (adapted from Melling and Westmacott 1972).

Section 3.D.2., will also affect rejection of the solutes. This phenomenon can be exploited to enhance separation of one protein from another. For example, if the pH of the solution is adjusted so that one protein is at its isoelectric point and the other is the same charge as the membrane, the latter protein's retention would be high while the former protein would be less retained by the membrane. This could explain the permeability data in Figure 3.21. Greater passage of the protein was observed with larger MWCO membranes, as expected. Less passage was observed as the pH was increased, probably because the protein became increasingly negatively charged and was repulsed by the negatively charged membranes.

Similarly, separation of the smaller cheese whey proteins (α-lactalbumin and β-lactoglobulin) from the larger proteins (BSA, lactoferrin and immunoglobulins) was much better with a 100 K MWCO membrane when the whey was adjusted from its native pH 5 to pH 8 (Mehra and Donnelly 1993). Similarly, the glycomacropetide from κ-casein (MW of 7000) could be isolated from whey by a double UF with different MWCO membranes at different pH (Kawasaki et al. 1993). UF with a 50 K MWCO membrane under acidic conditions allowed the peptide to pass through into the permeate. Subsequent UF with a 20 K membrane at pH 7 concentrated it (see Section 8.B.5. and Figures 8.20 and 8.21).

High ionic strength can shield ionic interactions; this is also demonstrated in Figure 3.22, which shows the UF of adenosine derivatives with YM membranes (from Amicon Corporation). Since the membranes were extremely hydrophilic and low-binding, it was expected that the yield of adenosine 5′-triphosphate (ATP, with a formula weight of 551) would be high, even with the tightest membrane (YM-3, MWCO of 3000). Instead, the rejection was higher than expected and did not reach expected values until a 100,000 MWCO membrane was used. In contrast, the rejection of vitamin B-12 (MW = 1355) was low with all the membranes, as expected. Similarly, when other adenosine derivatives, which are even smaller in MW, were tested in a water environment with the YM-3 membrane, rejections were 20–69%, much higher than expected (Figure 3.22, bottom).

In the presence of 1M NaCl, however, the rejections were much less and the recoveries of the adenosine derivatives much better. It was suspected that base "stacking interactions" were the cause of the high rejections in water, in addition to the number of phosphate groups on the molecule. These ionic interactions were apparently reduced in the presence of high salt concentrations. Similarly, addition of NaCl effectively reduced binding of polyphenols to UF membranes (Saeed and Cheryan 1989). Another example of the effect of the concentration and type of salts used to adjust ionic strength is shown in Figure 8.58.

In the final analysis, many of the above factors can interact and occur simultaneously. Although the guidelines in this chapter can help to narrow the list of membranes to consider and possibly explain some of the phenomenon, there is no substitute for actual testing of the membranes for a particular application.

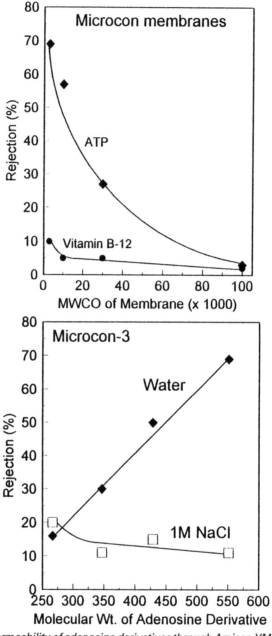

Figure 3.22. Permeability of adenosine derivatives through Amicon YM membranes in centrifugal cells: top: effect of MWCO of the membrane on rejection of adenosine 5'-triphosphate (ATP, MW = 551) and on vitamin B-12 (MW = 1355); bottom: effect of ionic strength on rejection of adenosine (MW = 267), adenosine monophosphate (AMP, MW = 347), adenosine diphosphate (ADP, MW = 429), and ATP. Membrane was the YM-3 in the centrifugal device (data from Amicon Corporation).

REFERENCES

ABATICCHIO, P., BOYYINO, A., CAMERA RODA, G., CAPANNELLI, G. and MUNARI, S. 1990. *Desalination.* 78: 235.

ADHAM, S., JACANGELO, J. G. and LAINE, J. M. 1995. *J. AWWA* 87 (3): 62.

ASTM. 1992. *Annual Book of ASTM Standards,* American Society of Testing Materials, Washington, DC.

BOUHALLAB, S. and HENRY, G. 1995. *J. Membrane Sci.* 104: 73.

BROCK, S. D. 1983. *Membrane Filtration. A User's Guide and Reference Manual,* Science-Tech, Inc., Madison, WI.

BROSE, D. J. and WAIBEL, P. 1996. *BioPharm.* 9(1): 36.

CAO, G. Z., MEIJERINK, J., BRINKMAN, H. W. and BURGGRAAF, A. J. 1993. *J. Membrane Sci.* 83: 221.

CHERKASOV, A. N. and POLOTSKY, A. E. 1995. *J. Membrane Sci.* 106: 161.

CHERKASOV, A. N., TSAREVA, S. V. and POLOTSKY, A. E. 1995. *J. Membrane Sci.* 104: 157.

CUPERAS, F. P., BARGEMAN, D. and SMOLDERS, C. A. 1992. *J. Membrane Sci.* 71: 57.

DEESLIE, W. D. and CHERYAN, M. 1992. *J. Food Sci.* 57: 411.

DIETZ, P., HANSMA, P. K., INACHER, O., LEHMANN, H. D. and HERMANN, K.-H. 1992. *J. Membrane Sci.* 65: 101–111.

DUTKA, B. J. 1983. *Membrane Filtration. Applications, Techniques, and Problems,* Marcel Dekker, New York.

FANE, A. G., FELL, C. J. D. and WATERS, A. G. 1981. *J. Membrane Sci.* 9: 245.

FERRY, J. D. 1936. *Chemical Rev.* 18 (12): 373.

FRITZSCHE, A. K., AREVALO, A. R., MOORE, M. D. and O'HARA, C. 1992. *J. Membrane Sci.,* 81: 109.

GEKAS, V. and ZHANG, W. 1989. *Process Biochem.* 24: 159.

KAWASAKI, Y., KAWAKAMI, H., TANIMOTO, M., DOSAKO, S., TOMIZAWA, A., KOTAKE, M. and MAKAJIMA, I. 1993. *Milchwiss.,* 48: 191.

KEATING, P., LEVY, R., PAYNE, M., PROUIX, S., ROWE, P. and PEARL, S. 1992. *BioPharm.* 5(1): 36.

KIM, K. J., FANE, A. G., BEN AIM, R., LIU, M. G., JONSSON, G., TESSARO, I. C., BROEK, A. P. and BARGEMAN, D. 1994. *J. Membrane Sci.* 87: 35.

KIM, K. J., FANE, A., FELL, C. J. D., SUZUKI, T. and DICKSON, M. R. 1990. *J. Membrane Sci.,* 54: 89.

MCGUIRE, K. S., LAWSON, K. W. and LLOYD, D. R. 1995. *J. Membrane Sci.* 99: 127.

MEEKER, J. T., HICKEY, E. W., MARTIN, J. M. and HOWARD, G. 1992. *BioPharm.* 5(2): 41.

MEHRA, R. K. and DONNELLY, W. J. 1993. *J. Dairy Res.,* 60: 89.

MEIRELES, M. BESSIERES, A., ROGISSART, I., AIMAR, P. and SANCHEZ, V. 1995. *J. Membrane Sci.* 103: 105.

MELLING, J. and WESTMACOTT, D. 1972. *J. Appl. Chem. Biotechnol.* 22: 951.

MELTZER, T. H. (ed.). 1987. *Filtration in the Pharmaceutical Industry.* Marcel Dekker, New York.

MERIN, U. and CHERYAN, M. 1980. *J. Applied Polymer Sci.* 25: 239.

MICHAELS, A. S. 1980. *Sep. Sci. Technol.,* 15(6): 1305.

Millipore Corporation. 1990. *Tangents* (3).

MULDUR, M. 1991. *Basic Principles of Membrane Technology*. Kluwer Academic Publishers, Dordrecht, The Netherlands.

NAKAO, S. 1994. *J. Membrane Sci.* 96: 131–165.

PERSSON, K. M., GEKAS, V. and TRAGRADH, G. 1995. *J. Membrane Sci.* 93: 105.

PORTER, M. C. 1979. In *Handbook of Separation Techniques for Chemical Engineers*, P. A. Schweitzer (ed.), McGraw-Hill, New York.

ROCHE, K. L. and LEVY, R. V. 1992. *BioPharm.* 5(3): 22.

ROCHE, K. L., MEIER, P. M. and LEFVY, R. V. 1993. *Amer. Soc. Brewing Chemists J.*, 51 (1): 4.

SAEED, M. and CHERYAN, M. 1989. *J. Agric. Food Chem.*, 37: 1270.

SHELDON, J. M., REED, I. M. and HAWES, C. R. 1991. *J. Membrane Sci.* 62: 87.

SOURIRAJAN, S. and MATSUURA, T. 1985. *Reverse Osmosis/Ultrafiltration Principles*. National Research Council Canada, Ottawa, Canada.

TKACIK, G. and MICHAELS, S. 1991. *Bio/Technology.* 9: 941.

VIVIER, H., PONS, M. N. and PORTALA, J. F. 1989. *J. Membrane Sci.* 46: 81.

WATERHOUSE, S. and HALL, G. M. 1995. *J. Membrane Sci.* 104: 1–9.

WU, Q. and WU, B. 1995. *J. Membrane Sci.* 105: 113.

WOLBER, P. and DOSMAN, M. 1987. *Pharmaceutical Technol.* 11(9): 26.

ZEMAN, L. 1992. *J. Membrane Sci.* 71: 233–246.

CHAPTER 4

Performance and Engineering Models

Several mathematical models are available in the literature that attempt to describe the mechanism of transport through membranes. Although the operating techniques of microfiltration (MF), ultrafiltration (UF), nanofiltration (NF), and reverse osmosis (RO) are similar, the latter two are almost certainly not separation merely by size alone. Microfiltration and ultrafiltration, on the other hand, due to their relatively large pores, have most frequently been visualized as sieve filtration. The approach taken in this chapter is to briefly describe typical models and possible mechanisms of flow through the membrane and to illustrate how several important operating variables can affect flux. Only *cross-flow* operations are considered: dead-end filtration, especially when used with large (>1 μ) particles and/or with MF membranes, can adequately be described by well-known cake-filtration models and are not discussed here.

4.A.
THE VELOCITY BOUNDARY LAYER

It will be useful to briefly discuss some elementary aspects of fluid flow that are important in understanding the development and the use of the mathematical models. "Boundary layers" develop whenever any fluid flows past a solid surface (whether the surface is porous or not). Friction between this surface, or "wall," and the liquid nearest the wall will cause a retardation of flow near the wall, while causing a speed-up of the fluid layers in the middle of the fluid channel. The line that separates the region of lower velocity from the region of uniform bulk velocity is known as the boundary layer. The development of the momentum boundary layer is shown schematically in Figure 4.1. The shape of the fully developed velocity profile for laminar flow in cylindrical geometry can be modeled using the Hagen-Poiseuille law (see Section 4.C. later). The velocity momentum boundary layer grows in thickness as the fluid flows downstream, reaching a constant asymptotic value at the far downstream end. The channel length L_v required to attain this constant value, i.e., to attain

Longitudinal velocity profile

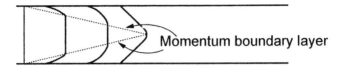

Concentration profile

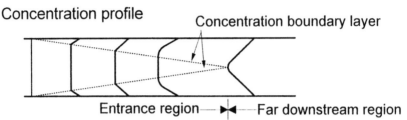

Figure 4.1. *Schematic of velocity and concentration profiles in a developing boundary layer (adapted from Hwang and Kammermeyer 1975).*

fully developed laminar flow, is given by

$$L_v = Bd_h\text{Re} \tag{4.1}$$

where d_h is the hydraulic diameter, Re is the Reynolds number (defined later), and B is a constant that ranges from 0.029–0.05.

The shape of the velocity profile under fully turbulent flow conditions and for other channel shapes is more difficult to model. The constant asymptotic thickness is reached much faster and closer to the entrance in turbulent flow. The relationship between the bulk average velocity (V) and the maximum velocity at the leading edge of the profile (V_{max}) is given as

For laminar flow:

$$V_{max} = 2V \tag{4.2}$$

For turbulent flow:

$$V_{max} = \frac{V}{0.662 + 0.336 \log \text{Re}} \tag{4.3}$$

4.B.
THE CONCENTRATION BOUNDARY LAYER

During ultrafiltration and microfiltration, i.e., when the wall is porous, solids in the feed are brought to the membrane surface by convective transport, and a

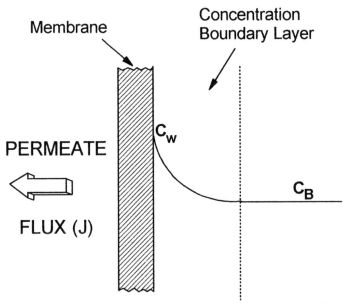

Figure 4.2. *Concentration profile during membrane processing of partially or completely rejected solutes.*

portion of the solvent is removed from the fluid. This results in a higher local concentration of the rejected solute at the membrane surface as compared to the bulk, regardless of whether the solutes are partially or completely rejected by the membrane. This is shown schematically in Figure 4.2. This solute buildup is known as "concentration polarization" and is chiefly responsible for the marked deviation in flux compared to pure water flux (see later, Section 4.D.). Analogous to the velocity boundary layer, there will also be a concentration boundary layer that separates the region of higher concentration near the wall (i.e., near the membrane surface) from the lower, more uniform concentration in the bulk of the liquid (Figure 4.1). This concentration boundary layer will be thinner than the corresponding velocity boundary layer since mass transfer by molecular diffusion is generally a much slower process than momentum transfer. In addition, the length over which the concentration boundary layer develops will be much longer. The length of the concentration profile entrance region (L_c) can be calculated from

$$L_c = \frac{0.1\gamma_w d_h^3}{D} \tag{4.4}$$

where γ_w is the shear rate at the wall and D is the diffusion coefficient. The concentration gradient across the boundary layer is much steeper than the velocity gradient. Thus, it is generally assumed that in the bulk solution outside the boundary layer, the concentration profile is essentially uniform in the direction perpendicular to the membrane surface.

The flow of fluid in the bulk stream influences the back transfer of accumulated solute into the bulk, thus keeping this boundary layer thin. This forms the basis of the film theory. Although the film theory is a good approximation when the boundary layer is thin and uniform (as in turbulent flow), it can also be applied in laminar flow systems provided the physical properties and other quantities can be suitably averaged. The use of the film theory to model flux in mass transfer–controlled UF is discussed in Section 4.E.

4.C.
MODELS FOR PREDICTING FLUX: THE PRESSURE-CONTROLLED REGION

There have been several attempts to model flux as a function of system operating parameters and physical properties. None are wholly satisfactory. The major problem appears to be an inability to precisely model the phenomena occurring near the membrane surface. In an ideal situation, e.g., with uniformly distributed and evenly sized pores in the membrane, with no fouling, negligible concentration polarization, etc., it is generally believed that the best description of fluid flow through microporous membranes is given by the Hagen-Poiseuille law for streamline flow through channels. This model, which relates pressure drop, viscosity, density, and channel dimensions (such as diameter of a tube) to flow rate through the channel, is usually analyzed by means of a momentum balance using cylindrical coordinates (e.g., see derivation by Bird et al. 1960). One form of the model useful in membrane processing is

$$J = \frac{\varepsilon d_p^2 P_T}{32 \Delta x \mu} \tag{4.5}$$

where J is the flow rate through the membrane, i.e., the flux in units of volume per unit area per unit time e.g., liters per square meter per hour (LMH) or gallons per square foot per day (GFD); d_p is the channel diameter (in this case the mean pore diameter: see Section 3.B.); P_T is the applied transmembrane pressure; μ is the viscosity of the fluid permeating the membrane; Δx is the length of the channel (the membrane "skin" thickness); and ε is the surface porosity of the membrane. These parameters are shown schematically in Figure 4.3.

As discussed in Chapter 1, the net driving force for an ideal membrane process [P_T in Equation (4.5)] should actually be ($P_T - \Delta\pi$), as shown in Figure 4.3, where

$$P_T = (P_F - P_P) \tag{4.6}$$

$$\Delta\pi = (\pi_F - \pi_P) \tag{4.7}$$

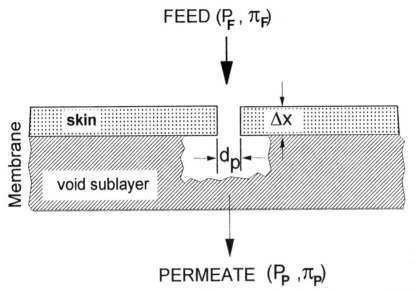

Figure 4.3. *Schematic representation of the cross section of typical asymmetric UF or MF membrane. P_F is applied pressure, P_p is back-pressure on the permeate side (this is zero if the permeate side is open to the atmosphere), and π_F and π_P are osmotic pressures of the feed and permeate solutions.*

where P_P is the pressure on the permeate side of the membrane (the "permeate backpressure"), π is the osmotic pressure, and subscripts F and P refer to feed and permeate, respectively. In practice, for most UF and MF applications, osmotic pressures of the retained solutes are negligible due to the high molecular weights, and using P_T alone in Equation (4.5) is quite adequate, except under certain circumstances (see Section 4.G. later).

There are two ways to operate filtration equipment: the dead-end or the cross-flow mode (Figure 4.4). Some MF equipment such as the pleated cartridges are operated in the dead-end mode, in which the feed is pumped directly towards the filter. There is one stream entering the filter module and only one stream (the permeate or filtrate) leaving the filter. However, most UF and MF modules are operated in the cross-flow mode, in which the feed is pumped across or tangentially to the membrane surface. In this mode of operation, there is one stream entering the module and two streams leaving the module—the retentate and the permeate. If the feed contains relatively high solids and/or if the solids need to be recovered easily, cross-flow is advantageous in that it limits the buildup of the solids on the membrane surface (shown in Figure 4.4 as the "cake"). The solids are kept in suspension in the flowing feedstream, resulting in less cake buildup and less cake resistance on the membrane, thus resulting in a higher average flux during operation. The cross-flow gives rise to a pressure

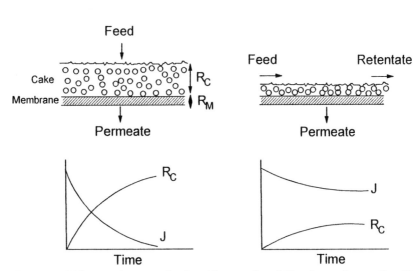

Figure 4.4. *Difference between dead-end (conventional) filtration and cross-flow filtration. R_C is the resistance of the cake formed on the membrane by the impermeable solutes, R_M is the resistance of the membrane, and J is the flux.*

drop from the inlet to the outlet of the module as shown in Figure 1.5 earlier. The feed-side pressure P_F is expressed as

$$P_F = \frac{P_i + P_o}{2} \qquad (4.8)$$

where P_i is inlet pressure of the membrane module and P_o is outlet pressure.

Several assumptions have been made in deriving the model shown in Equation (4.5):

1. The flow through the pores is laminar; i.e., Reynolds number is less than about 2100. This is a valid assumption, considering the pore statistics and experimental data for various UF membranes (Chapter 3).
2. The density is constant; i.e., the liquid is incompressible.
3. Flow is independent of time (steady-state conditions).
4. The fluid is Newtonian.
5. End-effects are negligible.

According to the model, flux is directly proportional to the applied pressure and inversely proportional to the viscosity. Viscosity is primarily controlled by two factors: solids concentration (or feed composition) and temperature

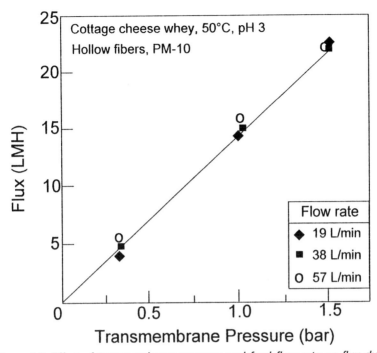

Figure 4.5. *Effect of transmembrane pressure and feed flow rate on flux during ultrafiltration of acidified cottage cheese whey in a hollow fiber module (HF15-43-PM10, Romicon). The whey was pasteurized, pH adjusted to 3, prefiltered in a 25-μ depth filter, and recycled in the module for 60 min before taking flux readings (adapted from Cheryan and Kuo, 1984).*

(for non-Newtonian liquids, it is also affected by shear rate and velocity: see Section 4.I.2.). Thus, increasing the temperature or increasing the pressure should increase the flux. This is largely true, as shown in Figures 4.5–4.8, which are typical examples from the literature. However, this is true only under certain conditions such as (a) low pressure, (b) low feed concentration, and (c) high feed velocity. When the process deviates from any of these conditions, flux becomes independent of pressure, sometimes even at quite low pressure. This is shown in Figures 4.9–4.12. The asymptotic pressure–flux relationship is due to the effects of concentration polarization. Under these conditions, the Hagen-Poiseuille model no longer adequately describes the membrane process, and mass transfer limited models must be used.

The general effect of pressure on flux is shown in Figure 4.13. At low pressures, low feed concentrations, and high feed velocities, i.e., under conditions where concentration polarization effects are minimal, flux will be affected by the transmembrane pressure. Deviations from the linear flux–pressure relationship of Equation (4.5) will be observed at higher pressures (regardless of

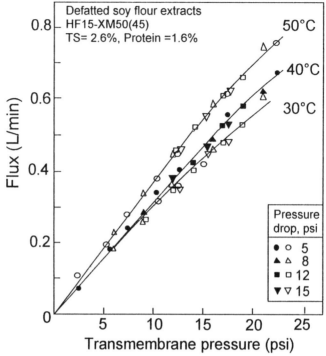

Figure 4.6. *Effect of transmembrane pressure, temperature, and flow rate (expressed in terms of pressure drop) on flux of water extracts of defatted soybeans. UF module was the Romicon HF15-45-XM50 (adapted from Nichols and Cheryan, 1981).*

other operating conditions) because of consolidation of the polarized layer of solute as depicted later in Section 4.D. Pressure independence occurs at lower pressure when the flow rate is lower or when the feed concentration is high. It is important when comparing the data in Figures 4.5–4.12 that the pressure scales be kept in proper perspective. For example, in Figures 4.5–4.7, the hollow fibers appear to operate principally in the pressure-controlled region for the feeds shown only because its maximum pressure limit is about 1.7 bar. At such low pressures, even the spiral-wound module will appear to operate in the pressure-controlled region (Figures 4.8 and 4.11). It is only when the pressure is increased to the limits of the spiral-wound unit that the mass transfer–controlled region becomes apparent.

4.D.
CONCENTRATION POLARIZATION

Concentration polarization is an additional complication that arises when hydrocolloids, macromolecules (such as proteins) and other relatively large

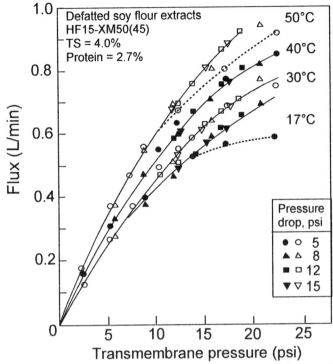

Figure 4.7. *Effect of transmembrane pressure, temperature, and flow rate (in terms of pressure drop) on flux. Feed was defatted soy flour extracts at a higher concentration than in Figure 4.6. Greater deviations from linearity can be observed compared to Figure 4.6 (adapted from Nichols and Cheryan, 1981).*

solutes or particles are filtered. These compounds, being largely rejected by the membrane, tend to form a layer on the surface of the membrane. Depending on the type of solid, this layer could be fairly viscous and gelatinous. Thus, a further resistance to the flow of permeate is encountered, in addition to those of the membrane and the boundary layer. This is shown schematically in Figure 4.14. This additional layer of solute is known by various terms such as the "gel layer," "CP layer," "cake," or "polarization layer." This layer is not to be confused with the *fouling* layer that occurs because of membrane–solute interactions, a phenomenon discussed in Chapter 6.

As shown in Figures 4.8–4.12, concentration polarization can have a major impact on ultrafiltration performance. For example, ultrafiltration of skim milk in a dead-end cell with no agitation results in a drop in flux to 5% of its initial value in 25 seconds (Bruin et al. 1980). Reduction in flux is thought to occur by one of two mechanisms: in one view, the increased solute concentration on the membrane surface results in a significantly higher osmotic pressure, causing a decrease in the driving force ($P_T - \Delta\pi$) and flux. While this may

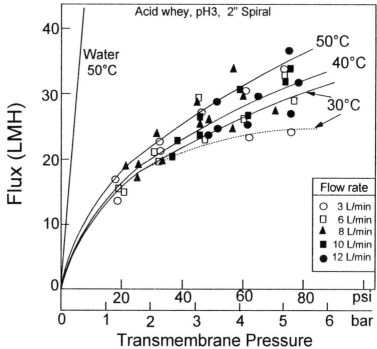

Figure 4.8. *Ultrafiltration of cottage cheese whey in a spiral-wound module (Osmonics 192-SEPA 20K). Feed was the same as that in Figure 4.5 (adapted from Cheryan and Kuo, 1984).*

be valid for reverse osmosis of small molecules in solution, osmotic pressures during MF and UF are generally small because of the large molecular sizes of the rejected solutes (Section 1.C.3.). However, osmotic pressure within the polarized layer could become important if the local solute concentration is high enough, because of the importance of the second and third virial coefficients in the osmotic pressure. This is discussed later in Section 4.G.

The alternate view is that the lower flux in polarization-limited systems is due to the hydrodynamic resistance of the boundary layer. Initially, as a result of convective transport of solute to the membrane, solute buildup will cause a steep concentration gradient within the boundary layer. This causes a back-transport of the solute into the bulk because of diffusion. Eventually, a steady state is reached where the two phenomena balance each other. Solute concentration in the gel layer reaches a maximum; i.e., no more solute molecules can be accommodated because of the "close-packed" arrangement and restricted mobility of solute molecules. This concentration, denoted as C_G in Figure 4.14, could—if high enough—cause the solute to precipitate out and foul the membrane.

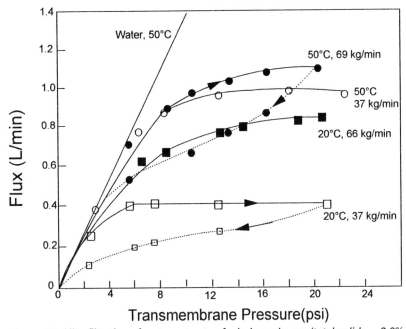

Figure 4.9. *Ultrafiltration of water extracts of whole soybeans (total solids = 3.6%, fat = 0.9%, protein = 1.8%) in hollow fibers (Romicon HF15-45-XM50). Arrows show direction of changing transmembrane pressure. Broken line indicates experiments when pressures were lowered from highest to lowest (adapted from Cheryan, 1977).*

It is because of this consolidated particle or "gel" layer on the membrane that flux becomes independent of pressure, as shown in Figures 4.8–4.12. Increasing the transmembrane pressure merely results in a thicker or denser solute layer. After a momentary rise, the flux will drop back to the previous value. Evidence for this concept was demonstrated by Altmann and Ripperger (1997), as shown in Figure 4.15. These experiments were conducted with monodisperse spherical silica particles suspended in water. The module had a window in it that allowed in-line observation of particle deposition and layer growth with a laser triangulometer. Initially, at low pressure, shear and lift forces were sufficient to minimize particle deposition, and the flux increased linearly with pressure. As soon as the particles started to deposit on the membrane (at 0.3 bar), the rate of increase in flux decreased. Further increases in pressure increased the thickness of the particle layer without a corresponding increase in flux. When pressure was decreased, the layer had consolidated itself such that the prevailing cross-flow shear forces could not remove the particles from the membrane surface, resulting in the hysteresis effect shown in Figure 4.15. Jaffrin et al. (1997) showed similar evidence of the polarization phenomenon.

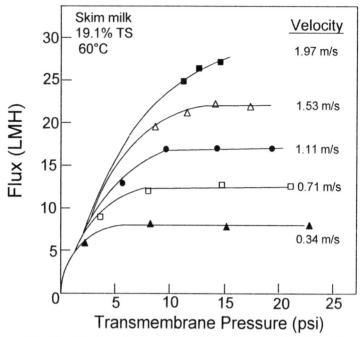

Figure 4.10. Ultrafiltration of concentrated skim milk in a hollow fiber UF module (Romicon HF15-43-PM50). Variable shown in the graph is feed velocity in m/sec (adapted from Cheryan and Chiang, 1984).

The gel-polarized layer is assumed to be dynamic (unlike fouling). Changing the operating conditions, such as lowering pressure or feed concentration, or increasing the feed velocity, should revert the system back to the pressure-controlled operating regime. This may be difficult to achieve in practice, as shown in Figures 4.9 and 4.15: the gel layer may be so well consolidated that lowering the pressure results in hysteresis effects. Of course, the magnitude of the hysteresis will also depend on how long one waits for the flux to rise. If the membrane is fouled, flux will never rise back to the original, except after cleaning. No hysteresis effects were observed in the pressure-controlled systems described in Figures 4.5–4.7. Flux in the pressure-independent region will be controlled by the efficiency of minimizing boundary layer thickness and enhancing the rate of back-transfer of polarized molecules.

4.E.
MASS TRANSFER (FILM THEORY) MODEL

One of the simplest and a widely used theory for modeling flux in pressure-independent, mass transfer–controlled systems is the film theory, as discussed earlier in Section 4.B. This is shown schematically in Figure 4.14. As solution is

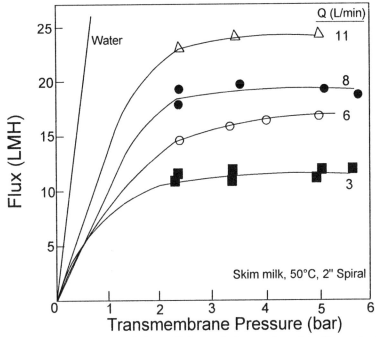

Figure 4.11. *Ultrafiltration of skim milk in a spiral-wound module (Osmonics 192-SEPA 20K). Variable is feed flow rate in L/min (adapted from Cheryan and Chiang, 1984).*

ultrafiltered, solute is brought to the membrane surface by convective transport at a rate, J_s, defined as

$$J_s = J C_B \tag{4.9}$$

where J is the permeate flux in units of volume/time and C_B is the bulk concentration of the rejected solute. The resulting concentration gradient causes the solute to be transported back into the bulk of the solution due to diffusional effects. Neglecting axial concentration gradients, the rate of back-transport of solute will be given by

$$J_s = D \frac{dC}{dx} \tag{4.10}$$

where D is the diffusion coefficient and dC/dx is the concentration gradient over a differential element in the boundary layer. At steady state, the two mechanisms will balance each other, and Equations (4.9) and (4.10) can be equated and integrated over the boundary layer to give

$$J = \frac{D}{\delta} \ln \frac{C_G}{C_B} = k \ln \frac{C_G}{C_B} \tag{4.11}$$

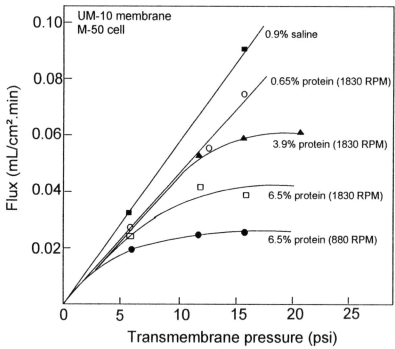

Figure 4.12. Effect of pressure, stirring rate (rpm) and albumin protein concentration on flux in an Amicon stirred cell (adapted from Blatt et al. 1970).

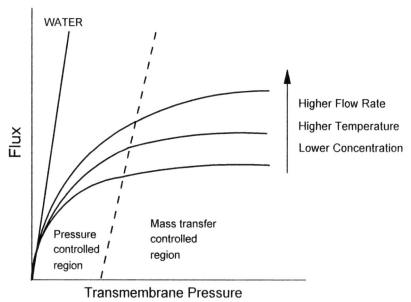

Figure 4.13. Generalized correlation between operating parameters and flux, indicating the areas of pressure control and mass transfer control.

126

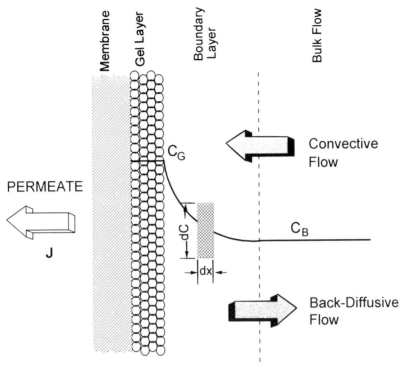

Figure 4.14. *Schematic of concentration polarization during UF of colloidal and macromolecular solutes, showing the buildup of the polarized (gel) layer and associated boundary layer. The same phenomenon occurs with MF membranes, depending on the rejection of the solutes.*

where C_G is the "gel" concentration, i.e., the solute concentration at the membrane surface, and k is the mass transfer coefficient, having the same units as the flux J and is calculated as

$$k = D/\delta \tag{4.12}$$

where δ is the thickness of the boundary layer over which the concentration gradient exists.

Note that, in this model, there is no pressure term. No effect of pressure was assumed or is implied as far as its effect on flux is concerned, and thus, this model will be valid only in the pressure-independent region. In effect, the flux will be controlled by the rate at which solute is transferred *back from* the membrane surface into the bulk fluid. Since, in most operations, the values of C_G and C_B are fixed by physicochemical properties of the feed, flux can only be improved by enhancing k as much as possible, such as by reducing the thickness of the boundary layer. This is why any adjustment that attempts to

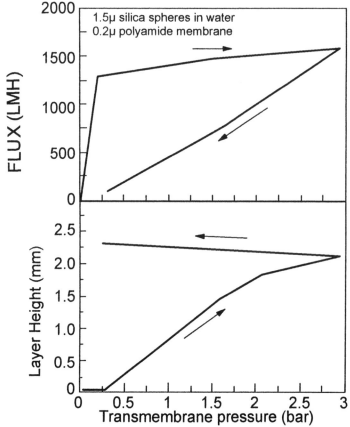

Figure 4.15. Effect of transmembrane pressure on flux (top) and height of the layer of rejected particles on the membrane surface (bottom). The feed was 0.3% silica spheres (1.5-μ size) suspended in water. A polyamide MF membrane with 0.2-μ pores was placed in a 6 cm × 40 cm cell with a channel height of 3 mm. Data was taken at a cross-flow velocity of 2 m/sec. Layer height was measured in situ with a laser triangulometer (adapted from Altmann and Ripperger 1997).

increase flux, such as increasing pressure, without providing a compensating mechanism to increase rate of back-transport, will be self-defeating.

4.E.1.
DETERMINING THE MASS TRANSFER COEFFICIENT

A number of qualitative relationships correlating the mass transfer coefficient to physical properties, flow channel dimensions, and operating parameters exists in the literature. These include, in addition to the film theory, the penetration theory, surface renewal theory, stagnant film theory, three-zone concept, periodic viscous sublayer model, eddy diffusivity model, combination film-surface renewal theory, turbulent boundary layer model, and the surface-stretch theory, among others (Gekas and Hallstrom 1987; Sherwood et al. 1975; Treybal 1981). None are wholly satisfactory; many are continuously being refined or modified to suit particular applications and, thus, are not universally applicable. Under these circumstances, when no good theory exists, dimensional analysis is a powerful tool. Using the π theorem or by analogy with heat transfer, one can obtain a general correlation of the form:

$$\text{Sh} = A(\text{Re})^{\alpha}(\text{Sc})^{\beta} \tag{4.13}$$

where

$$\text{Sh} = \text{Sherwood number} = kd_h/D \tag{4.14}$$

$$\text{Re} = \text{Reynolds number} = d_h V\rho/\mu \tag{4.15}$$

$$\text{Sc} = \text{Schmidt number} = \mu/\rho D \tag{4.16}$$

$$d_h = \text{hydraulic diameter} = 4\frac{\text{Cross section available for flow}}{\text{Wetted perimeter of the channel}} \tag{4.17}$$

For tubes and circular cross sections, d_h = diameter (d). For slits of width a and height b,

$$d_h = 4\frac{ab}{2(a+b)} \tag{4.18}$$

Since $a \gg b$ for slits, $d_h = 2b$.

The dimensionless numbers must be suitably modified for non-Newtonian fluids (see Section 4.I.2.). The Sherwood Number (Sh) is a measure of the ratio of convective mass transfer to molecular mass transfer. It can also be looked at as the ratio of the channel dimensions (in terms of the equivalent hydraulic diameter d_h) to the boundary layer thickness δ. The Reynolds number (Re) is a measure of the ratio of inertia effects to viscous effects and the state of turbulence in a system. In general, Re values less than 1800 are considered to be laminar flow, and Re greater than 4000 is turbulent flow. The Schmidt number (Sc) is a dimensionless measure of the ratio of momentum transfer of mass transfer.

The exponents α and β are constants determined by the state of development of the velocity and concentration profiles along the channel. The Schmidt number dependency (β) is derived from dimensional considerations of the convective diffusion equation under conditions in which the momentum boundary layer is much longer than the diffusion boundary layer, i.e., when Sc $\gg$ 1. For laminar flow systems, if both velocity and concentration profiles are fully developed, both α and β are zero. If velocity profile is fully developed but concentration boundary layer is developing along the entire length of the channel, the Graetz or Leveque solutions can be used with $\alpha = 1/3$ and $\beta = 1/3$. If both velocity and concentration profiles are developing, $\alpha = 1/2$ and $\beta = 1/3$ (Grober et al. 1961). The state of development of these profiles in laminar flow for a particular application can be estimated from Equations (4.1) and (4.4). In almost all cases of interest in UF, the length of the concentration profile entrance region (L_c) will be much larger than the channel lengths used, indicating that a value of 0.33 should always be used for β. For turbulent flow, the Chilton-Colburn or Dittus-Boelter correlation can be used with $\alpha = 0.8$ and $\beta = 0.33$.

The constant A generally reflects physical property variations and other conditions of the system that one cannot explicitly account for from first principles. For example, Kozinski and Lightfoot (1972) express A in terms of parameters that correct for variations in diffusivity, viscosity, and density at the wall, as compared to the bulk [Equation (4.44)]. In situations where the concentration boundary layer is developing down the entire length of the flow channel, the Sherwood number will also be a function of the channel length, L. Thus Equation (4.13) is usually rewritten for laminar flow models as

$$Sh = A'(Re)^{\alpha}(Sc)^{\beta}(d_h/L)^{\omega} \tag{4.19}$$

where A' is 0.664 in the Grober correlation and 1.86 in the Leveque solution. The value of ω is 0.33 in the developing boundary layer (Hwang and Kammermeyer 1975) and 0.5 for fully developed velocity profiles.

In summary, the following models can be used to predict flux:

For turbulent flow, when $Re > 4000$,

$$Sh = 0.023(Re)^{0.8}(Sc)^{0.33} \tag{4.20}$$

For laminar flow, when $Re < 1800$, $L_v < L$ and $L_c > L$,

$$Sh = 1.86(Re)^{0.33}(Sc)^{0.33}(d_h/L)^{0.33} \tag{4.21}$$

For laminar flow, when $L_v > L$ and $L_c > L$,

$$Sh = 0.664(Re)^{0.5}(Sc)^{0.33}(d_h/L)^{0.50} \tag{4.22}$$

An example of the use of these models to predict flux is given below.

4.E.2.
EXAMPLE

Consider the ultrafiltration of milk at 50°C. The important physical properties are: density $(\rho) = 1.03$ g/cm^3, viscosity $(\mu) = 0.8$ cP, diffusivity $(D) = 7 \times 10^{-7}$ cm^2/sec, protein concentration of milk $(C_B) = 3.1\%$ w/v, and

C_G value is 22% w/v. Calculate the flux expected from a hollow fiber unit and a tubular unit with the specifications given below:

	Hollow Fibers	Tubular
Diameter (d), cm	0.11	1.25
Length (L), cm	63.5	240
Number of fibers or tubes (n)	660	18
Cross-flow rate (Q), L/min	38	265
Pressure drop, kg/cm^2	0.9	2

For the hollow fiber module:

$$V = \frac{Q}{(\pi/4)(d^2)n} = \frac{38{,}000/60}{(3.142/4)(0.11)^2 660} = 100 \, \text{cm/sec}$$

$$\text{Re} = \frac{dV\rho}{\mu} = \frac{0.11 \times 100 \times 1.03}{0.008} = 1416$$

$$\text{Sc} = \frac{\mu}{\rho D} = \frac{0.008}{1.03 \times 7 \times 10^{-7}} = 1.11 \times 10^4$$

The Re value indicates this module is operating under laminar flow conditions, as expected, and so the Leveque solution [Equation (4.21)] can be used.

$$\text{Sh} = 1.86(1416)^{0.33}(1.11 \times 10^4)^{0.33}(0.11/63.5)^{0.33} = 54.08$$

From Equation (4.14):

$$k = 54.08 \left(\frac{D}{d_h} \right) = 54.08 \left(\frac{7 \times 10^{-7}}{0.11} \right)$$

$$k = 3.44 \times 10^{-4} \, \text{cm}^2/\text{cm/sec} = 12.39 \, \text{L/m}^2/\text{h}$$

From Equation (4.11):

$$J = 12.39 \, \ln(22/3.1) = 24.3 \, \text{liters/m}^2/\text{hour}$$

(In fact, the actual flux was 35 LMH: this discrepancy is explained later.)
  For the tubular module:

$$V = \frac{265{,}000/60}{(3.142/4)(1.25)^2 18} = 200 \, \text{cm/sec}$$

$$\text{Re} = \frac{dV\rho}{\mu} = \frac{1.25 \times 200 \times 1.03}{0.008} = 32{,}188$$

This is in turbulent flow, and thus Equation (4.20) will be used:

$$\text{Sh} = 0.023(32,188)^{0.8}(1.11 \times 10^4)^{0.33} = 2002$$

$$k = \frac{2002 \times 7 \times 10^{-7}}{1.25} = 1.12 \times 10^3 \text{cm}^2/\text{cm/sec} = 40.37\,\text{LMH}$$

$$J = 40.37\,\ln(22/3.1) = 78.7\,\text{LMH}$$

4.F.
THE RESISTANCE MODEL

Neither of the two models discussed earlier describes the entire pressure–flux behavior observed during typical UF or MF, i.e., pressure-controlled at low pressures, pressure-independent at high pressures. A better approach may be to use the "resistance-in-series" concept that is common in heat transfer. For an ideal membrane and feed solution, Equation (4.5) can be rewritten as

$$J = A\frac{P_T}{\mu} \tag{4.23}$$

where A is a membrane permeability coefficient, which includes the terms characteristic of the membrane itself, and μ is the viscosity of the permeate (not the feed; the Hagen-Poiseuille relationship is for fluids flowing *through* the channel; in the case of the membrane, the "channel" is the pore). For a particular feed solution at a given temperature, viscosity is usually included with the A value and can be written as $1/R_M$ where R_M is the intrinsic membrane resistance determined using pure water as the feed:

$$J_{\text{water}} = \frac{P_T}{R_M} \tag{4.24}$$

In this case, R_M or A values are useful not only for modeling purposes, but also for evaluating the effectiveness of the cleaning procedures (Section 6.A.1.) and for charting the long-term stability of the membrane. R_M should not be used as an indication of a particular membrane's performance under actual operating conditions with a real feed or for comparing the expected performance of several different membrane systems. In actual operation with a real feed, the membrane resistance per se may be only a small part of the total resistance. For example, if significant membrane fouling occurs because of specific membrane–solute interactions, the intrinsic membrane resistance may change. This is accounted for by adding another resistance term because of fouling (R_F) to the model:

$$J = \frac{P_T}{R_M + R_F} \tag{4.25}$$

Since this type of fouling is generally assumed to be because of physicochemical interactions, R_F will be relatively unaffected by operating parameters and is frequently lumped with the intrinsic membrane resistance as R'_M, where $R'_M = R_M + R_F$.

To account for concentration polarization and the boundary layer, an additional resistance term R_G can be added to the flux equation:

$$J = \frac{P_T}{R_M + R_F + R_G} \tag{4.26}$$

where R_G is a function of operating parameters and physical properties. In the context discussed earlier in Section 4.D., R_G will be a function of the permeability of the gel and its thickness, which is a function of applied pressure:

$$R_G = \phi P_T \tag{4.27}$$

The resistance model becomes

$$J = \frac{P_T}{R'_M + \phi P_T} \tag{4.28}$$

This model conceptually fits typical flux–pressure data. At low pressures, the R_G term will be small compared to R'_M, and flux will be a function of pressure. At high pressures, the R_G term becomes relatively large. Flux is less dependent on pressure and approaches the limiting value $1/\phi$. The ϕ term will be a function of the variables affecting mass transfer such as viscosity, shear rate/velocity, and temperature. The optimum pressure to operate such systems is at the point where the resistances R'_M and R_G are equal:

$$(P_T)_{\text{optimum}} = R'_M/\phi \tag{4.29}$$

To illustrate the use of the resistance model, flux–pressure data for several velocities and concentrations of skim milk that were obtained by Chiang and Cheryan (1986) were fitted to Equation (4.28) to determine the values of R_M, R_F, and ϕ. The value of R_M was 0.16 kPa · m² · h/L at 60°C, and the average value of R_F was 0.703 kPa · m² · h/L. R_F was relatively unaffected by operating parameters, and its value was more than four times higher than the intrinsic membrane resistance R_M. However, the analysis also showed that the resistance of the polarized layer (R_G) was much larger than R_M or R_F. The value of ϕ was $0.022 - 0.112$ m² · h/L, being higher at higher concentrations and lower velocities. The R_G value, determined from Equation 4.27, was 3.7 kPa · m² · h/L at the highest flux. This is more than $5R_F$ and $23R_M$.

Table 4.1 lists R values from a variety of applications in the literature. With bovine blood in a hollow fiber plasma separator, $R_F < 0.25R_M$ but R_G was

Table 4.1. Resistance values according to Equation (4.26) for selected applications. Units of R_M, R_F, and R_G are in kPa/LMH (1 kPa/LMH = 3.6 $\times 10^9$ Pa·sec·m^{-1}) .

Feed	Membrane	R_M	R_F	R_G	Reference
Skim milk	Hollow fiber, PM50	0.16	0.7	3.7	Chiang and Cheryan (1986)
Blood, bovine	Hollow fiber, PVA	0.06	0.015	0.16	Ozawa et al. (1987)
Ovalbumin	Plate, PS	0.51	0.91	9.19	Nabetani et al. (1990)
PVP-360					
0.1%	Hollow fiber, P30	0.14	0.53	2.2	Yeh and Wu (1997)
2.0%	Hollow fiber, P30	0.14	6.7	7.9	Yeh and Wu (1997)
Ovalbumin	Plate, PAN	1.2	9.4	7.93	Nabetani et al. (1990)
Ethanol broth	Tubular, ceramic	0.11	2.3	2.0	Saglam (1995)
Wheat starch effluent	Tubular, PS	0.68	4.93	2.7	Harris and Dobbs (1989)
Passion fruit juice	Hollow fiber, PS	0.27	1.39	1.9	Jiraratananon and Chanachai (1996)
Cheese whey	Hollow fiber, PM10	1.0	5.7	0	Cheryan and Kuo (1984)

PAN = polyacrylonitrile, PM/PS = polyethersulfone/polysulfone, PVA = polyvinyl alcohol

$2.6R_M$. The ovalbumin data with the polysulfone (PS) membrane also showed much higher values of R_G compared to other resistances. With polyvinylpyrrolidone (PVP-360), higher feed concentration substantially increased fouling resistance, but polarization and boundary layer effects were still rate-controlling. It is obvious from this analysis that, in these cases, efforts to maximize the flux should focus on reducing polarization effects as much as possible, perhaps by using a different module that allows higher degrees of turbulence to be generated.

In contrast, with the rest of the applications shown in Table 4.1, fouling plays a comparatively larger role. With yeast/ethanol fermentation broth in a Membralox 0.2-μ ceramic membrane, the polarization resistance is comparable to the fouling resistance, caused partly by the extremely high velocities (>8 m/sec) that were used in that case, which kept the R_G value low. The same is true of ovalbumin with the PAN membrane and the wheat starch application. With the passion fruit juice, R_F was slightly affected by pressure, increased with temperature, and was unaffected by flow rate. On the other hand, R_G decreased from 6.4 to 1.9 kPa/LMH when the recirculation flow rate was increased from 600 ml/min to 1200 ml/min. With cheese whey, R_F was about six times larger than R_M. There appeared to be little or no polarization resistance, since the flux was pressure-controlled with little effect of velocity (e.g., see Figure 4.5). Thus, in these five cases, the major focus should be to reduce fouling, since the polarization resistance could be reduced by manipulating the operating conditions.

4.G.
OSMOTIC PRESSURE MODEL FOR LIMITING FLUX

The "gel" model is useful for describing limiting flux in the mass transfer controlled region. However, central to its approach is the assumption of a fairly well-defined gel layer with a concentration of C_G, which is unaffected by operating parameters and the intrinsic membrane resistance, so long as rejection of the solutes remain constant. As discussed in Section 4.H.1. later, experimental C_G values appear to fit this criteria in many cases, but sometimes it varied widely. In some cases, C_G values are much lower than expected; e.g., solutions at that concentration have low viscosities and are still fluid (Jonsson 1984) or are too high, extrapolating to values greater than 100% (typically observed with MF of colloidal suspensions).

In the original expression for the Hagen-Poiseuille model for flux [Equation (4.5)], the driving force term was $(P_T - \Delta\pi)$. For a completely rejected solute, $\Delta\pi$ will be determined only by C_B. In reality, it should be the concentration of rejected solute at the membrane surface (C_M). The $\Delta\pi$ term was dropped from the model because it was assumed that the osmotic pressure for macromolecules is negligible compared to the applied transmembrane pressure. However, in some instances, the osmotic pressure could become significant because of the second and third virial coefficients of the osmotic pressure relationship (Section 1.C.3.). The osmotic pressure model assumes that the deviation from pure water flux occurs solely due to the osmotic pressure at the membrane surface. In this case, the Hagen-Poiseuille equation can be written as

$$J = \frac{P_T - \Delta\pi}{R_M} = \frac{P_T - \pi_M}{R_M} \tag{4.30}$$

where π_M is the osmotic pressure of the solute concentrated at the membrane surface. The osmotic pressure is a function of the concentration and can be expressed in terms of its virial coefficients, as shown earlier in Equation (1.37), as $\pi_M = f(C_M)$.

The simplest way of determining C_M is to use the film theory. From Equation (4.11),

$$C_M = C_B \exp(J/k) \tag{4.31}$$

or

$$J = \frac{P_T - f(C_B \exp(J/k))}{R_M} \tag{4.32}$$

Equation (4.32) also conceptually agrees with observed flux behavior. Increasing transmembrane pressure will at first increase the flux momentarily.

However, due to increased convective transport of solute to the membrane surface, the value of C_M also increases according to Equation (4.31), which in turn increases π_M, thus partly canceling out the increase in driving force $(P_T - \pi_M)$. Because of the nature of the π versus C relationship, there may come a point when a small increase in C_M will lead to a large increase in π_M, canceling out the effect of higher pressure completely or even reducing the flux. (Decreases in flux at very high applied pressures have been observed occasionally and are probably caused by a compaction of the polarized layer. This can lead to increased hydraulic resistance or to higher osmotic pressure at the membrane surface, or both.)

The osmotic pressure model has been tested in a few cases (Jonsson 1984; Kozinski and Lightfoot 1972; Pradanos et al. 1995b; Trettin and Doshi 1981; Vilker et al. 1984; Wijmans et al. 1984). The major problem is lack of osmotic pressure data in the form of Equation (1.37). The mass transfer coefficient (k) can be determined from the appropriate correlations such as Equations (4.20)–(4.22). Thus, those conditions that increase k will decrease the value of C_M and also decrease π_M, which increases the driving force $(P_T - \pi_M)$, thus increasing flux. This is one major deviation from the gel-polarization model. The osmotic pressure model assumes the solute concentration at the membrane surface is a function of all variables, including the applied pressure, while the gel-polarization model assumes C_G is independent of operating conditions. The key factor obviously lies with the nature of the rejected solute and the $\pi = f(C)$ relationship. It could be that with large molecules ($> 100,000$ molecular weight, for example), the thermodynamic limitation will be negligible in the limiting flux region, while it is the controlling factor with smaller molecules. The osmotic pressure model does not apply to colloidal feedstreams.

4.H.
FACTORS AFFECTING FLUX: OPERATING PARAMETERS

There are four major operating parameters that affect the flux: (1) pressure, (2) feed concentration, (3) temperature, and (4) turbulence in the feed channel. The effect of pressure on flux has already been discussed. The other three factors will be discussed below.

4.H.1.
FEED CONCENTRATION

The film theory model [Equation (4.11)] states that the flux (J) will decrease exponentially with increasing feed concentration (C_B). Typical examples are shown in Figures 4.16–4.22. This relationship should hold true regardless of the type of flow or degree of turbulence or the temperature. The film model also implies that, at $C_B = C_G$, $J = 0$; i.e., all data for a particular feed should converge to one point on the concentration scale, which represents the "gel"

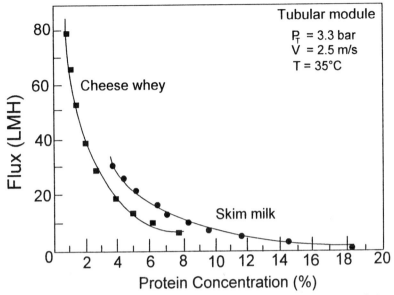

Figure 4.16. *Effect of protein concentration on flux of skim milk and flux of cheese whey (adapted from Kessler 1981).*

concentration C_G. This has experimentally been shown to be true in many cases, as shown in Figures 4.17–4.22. C_G values obtained experimentally for several systems are shown in Table 4.2. The slope of a plot of J versus log of the concentration of the rejected species is the mass transfer coefficient (k) in the same units as the flux. However, it should be remembered that, as the feed concentration changes, the viscosity, density, and diffusivity of the feed solution will change, which will affect all dimensionless groups in Equations (4.20)–(4.22). Not accounting for physical property variations caused by changes in feed concentration is potentially one of the most serious sources of error when using the mass transfer model.

The physical significance of C_G is debatable. According to the assumptions made in deriving the model, C_G denotes the concentration of rejected solute in the so-called gel layer, representing the "close-packed" arrangement of rejected macromolecules on the membrane surface. Assuming hexagonal close-packing, this value will be 74% for solid particles (Zydney and Colton 1986). When the concentration of the feed solution reaches C_G, there will no longer be a concentration gradient in the boundary layer and, thus, no back transport of solute, resulting in zero flux. To obtain accurate C_G values, data should be taken as close as possible to the expected C_G value. In practice, one rarely comes close to the C_G concentration since there will be a huge increase in viscosity of the feed at higher concentrations, which will make pumping or

Table 4.2. Selected values of C_G. Refer to original citations for operating conditions.

Feed	C_G	Reference
Albumin		
Bovine, serum	30%	Porter (1979)
Bovine, serum	20%	Porter and Michaels (1972)
Human, serum	28%, 44%	Porter (1979)
Human, serum	24%	Mitra and Lundblad (1978)
Blood cells		
Porcine	35% protein	Delaney (1977)
Bovine Ht 0%	13% protein	Ozawa et al. (1987)
Bovine Ht 30%	18% protein	Ozawa et al. (1987)
Human Ht 21%	28.7% protein	Isaacson et al. (1980)
Blood plasma		
Porcine	45% protein	Delaney et al. (1975)
Human	60% protein	Porter and Michaels (1972)
Bovine serum albumin (BSA)	30%	Porter (1979)
	20%	Porter and Michaels (1972)
Carbowax 20M	8%	Porter (1979)
Dextran T-500	33.8%	Pradanos et al. (1995b)
Egg white	40% protein	Porter and Michaels (1972)
Gelatin	20–30% protein	Porter and Michaels (1972)
	30% protein	Akred et al. (1980)
Immune serum globulin	19%	Mitra and Lundblad (1978)
Kraft black liquor	30–34% lignin	Woerner (1983)
Microbial cells		
S. cerevisiae	350–450 g/L	Patel et al. (1987), Saglam (1995)
S. cerevisiae	300 g/L	Bell and Davies (1987)
S. cerevisiae	130 g/L	Kavanagh and Brown (1987)
S. cerevisiae	80 volume %	Porter (1979)
A. curvatum	300 g/L	Bell and Davies (1987)
Erwinia carotova	110% w/w	Le et al. (1984b)
E. coli	70% wet cells	Kroner et al. (1984)
B. cereus	40% wet cells	Kroner et al. (1984)
Brevibacterium	60% wet cells	Kroner et al. (1984)
L. bulgaricus	800 g/L	Tejayadi and Cheryan (1988)
A. niger	205 g/L	Sims and Cheryan (1986)
Milk (skim)	22% protein	Kessler (1981)
	20–22% protein	Chiang and Cheryan (1987)
	20% protein	Delaney and Donnelly (1977)
	25% protein	DeBoer and Hiddink (1980)
Milk (whole, 3.5% fat)	9–11% protein	Yan et al. (1979)
Pectin	3.8% w/w	Pritchard et al. (1995)
Polyepichlorohydrin	20–23% w/v	Cooper and Booth (1978)
Polyaminoethylene	12.6% w/v	Cooper and Booth (1978)
Prothrombin complex proteins	7.32% w/v	Mitra and Fillmore (1980)

(continued)

138

Table 4.2. Selected values of C_G. Refer to original citations for operating conditions.

Feed	C_G	Reference
Soybean		
Full-fat water extract	10% protein	Cheryan (1977)
Defatted soy flour extract	20–25% protein	Nichols and Cheryan (1981)
Iso-electric soy protein	35–45% protein	Devereux and Hoare (1986)
Styrene-butadiene latex	70%	Porter (1979)
	42%	Zahka and Mir (1977)
Wheat starch effluent	5–10% solids	Harris and Dobbs (1989)
Whey		
Acid/cottage cheese	30% protein	Cheryan and Kuo (1984)
Sweet/cheddar	20% protein	Donnelly and Delaney (1974)
Sweet/cheddar	28.5% protein	Epstein and Korchin (1981)
Sweet/cheddar	20% protein	Kessler (1981)
Sweet/Gouda	18% protein	Hiddink et al. (1981)
Sweet/Gouda	20% protein	Hanemaaijer (1985)
Sweet/Gruyere	25–35% protein	Foetisch et al. (1987)
Xanthan gum	25% w/w	Pritchard et al. (1995)

agitation of the fluid very difficult. (Sometimes, as discussed later, this increase in viscosity can actually improve the mass transfer coefficient and flux.)

It is important to keep in mind that values of C_G in Table 4.2 were obtained by *extrapolating* the linear portion of J versus log C_B plots. There are several flaws in such published data:

1. In many cases, it appears that no effort was made to keep flow rate or state of turbulence constant during the experiments, especially at higher concentrations.
2. The data were frequently obtained during a continuous concentration experiment, i.e., in one run when the retentate is recycled to the feed tank and the permeate is removed from the system. These data will include effects due to fouling and due to aging of the feed, such as possible denaturation of proteins due to heat or shear. Ideally, separate runs should be made with feeds of different concentration where both retentate and permeate are recycled and with the membrane being cleaned in order to regain the original pure water flux between each run. Alternately, data should be taken in a feed-and-bleed operation.
3. In some cases, flux versus concentration data in both the pressure-controlled and mass transfer–controlled regions are included on the same plot.

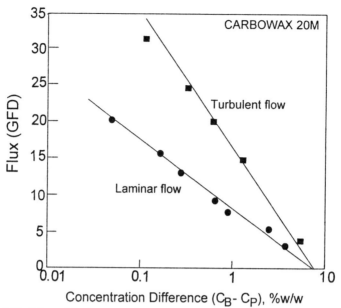

Figure 4.17. Effect of Carbowax 20M concentration on flux in turbulent flow (Abcor tubular module, HFA-300) and in laminar flow (adapted from DeFilipi and Goldsmith 1970).

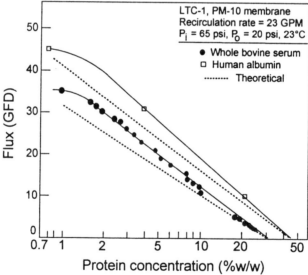

Figure 4.18. Relationship between flux and protein concentration during concentration of human albumin and whole bovine serum. The module was a 15-mil thin-channel laminar flow unit with Amicon PM-10 membrane. Theoretical lines were drawn using the Leveque solution and the film theory, with diffusion coefficients of 4×10^{-7} cm^2/sec for whole bovine serum and 6×10^{-7} cm^2/sec for human albumin (adapted from Porter, 1979).

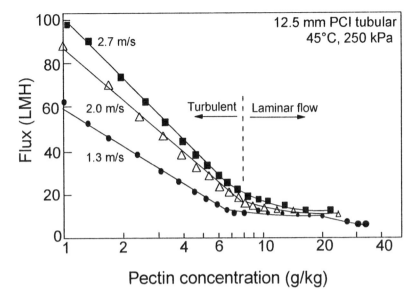

Figure 4.19. *Effect of pectin concentration on flux in a tubular module (PCI, 12.5 mm diameter) with a PES 65,000 MWCO membrane. The Reynolds numbers initially were 22,100–38,900. The change to laminar flow occurred at Re = 2000 for each cross-flow velocity (adapted from Pritchard et al. 1995).*

This makes it difficult to get a true k value, which is valid only for the latter region.

4. Concentration data is sometimes not expressed in terms of concentration of the *rejected* solute, but instead as total solids or Brix, etc.

There are also some instances of biphasic declines in flux. For example, the slope of the line has been known to decrease as zero flux is approached (Cheryan and Kuo 1984), as also shown in Figure 4.19. Flux can also increase with concentration or both increase and decrease (Figure 4.22) or not change, despite the increase in concentration. There are many cases where the flux apparently never really reaches zero, but that could also depend on how long the experiment has been carried out and the accuracy of such low flux measurements.

An increase in slope can be predicted from the mass transfer and osmotic pressure models. From Equation (4.30):

$$J R_M = P_T - \pi_M \tag{4.33}$$

As $J \to 0$,

$$P_T = \pi_M \tag{4.34}$$

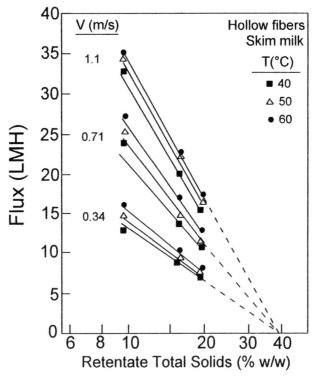

Figure 4.20. *Effect of feed velocity (m/sec), temperature, and total solids of skim milk on flux in a hollow fiber HF15-43-PM50 module (adapted from Chiang and Cheryan 1987).*

Assuming the film theory can accurately represent the environment near the membrane surface, it implies that the so-called gel concentration is nothing more than the osmotic equivalent of the applied pressure. Thus, if higher P_T is used, higher values of C_G should be obtained (it is to be remembered that C_G is an extreme case of C_M). However, there is very little experimental data of J versus C_B as a function of applied pressure. Thus, much of the C_G data in Table 4.2 must be used with caution since pressure conditions were not specified in many cases.

The basic assumption in the film theory is that the mass transfer coefficient remains constant as the bulk concentration increases. In graphical terms, it means the line in the J-log C_B plot is linear. As mentioned earlier, this may not be a realistic expectation, since the physical properties change as concentration increases (see Section 4.I.). The mass transfer models [Equations (4.20) and (4.21)] can be rewritten as

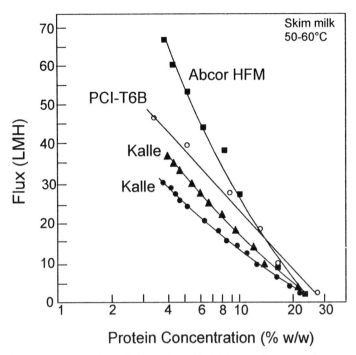

Figure 4.21. Concentration of skim milk at 50–60°C in various membrane modules (adapted from DeBoer and Hiddink 1980).

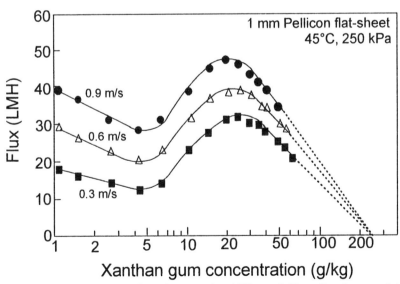

Figure 4.22. Ultrafiltration of xanthan gum in a Millipore Pellicon flat-sheet module with 1-mm channel height. The variable is cross-flow velocity in m/sec (adapted from Pritchard et al. 1995).

For turbulent flow: $k = 0.023 \left(\dfrac{D^{0.67} V^{0.8} \rho^{0.47}}{d_h^{0.2} \mu^{0.47}} \right)$ (4.35)

For laminar flow: $k = 1.86 \left(\dfrac{D^{0.67} V^{0.33}}{d_h^{0.33} L^{0.33}} \right)$ (4.36)

The implication of Equation (4.35) is that, if a turbulent flow system is used (most tubular and spiral modules are turbulent flow), the mass transfer coefficient and the flux will decrease as the concentration increases because of an increase in viscosity, assuming the other parameters are kept constant. However, turbulent flow systems can soon change into laminar flow systems because of an increase in viscosity, as shown in Figure 4.19. In contrast to turbulent flow, laminar flow models show no dependency on viscosity [Equation (4.36)]. If shear rates in the module are not very high, the k values and the flux will not decrease as rapidly with increase in concentration, as shown in Figure 4.19 for the laminar flow region. This phenomenon can also be seen in hollow fiber UF of yeast suspensions (Patel et al. 1987); the flux decreases little compared to the rapid increase in viscosity at higher concentrations.

With high shear rates, another phenomenon may be occurring that could explain data showing more than one slope, such as Figure 4.22. By definition, the shear stress (τ) in a fluid is related to shear rate (γ) as follows:

$$\tau = \mu \gamma \qquad (4.37)$$

where μ is the viscosity. Shear rate at the membrane surface (γ_w) of a tube of diameter d is given in terms of the bulk average velocity (V) by

$$\gamma_w = 8V/d \qquad (4.38)$$

For a slit of height b,

$$\gamma_w = 6V/b \qquad (4.39)$$

For a tube, Equation (4.37) can be rewritten for bulk flow in a channel as

$$\tau_w = \mu_B \left(\frac{8V}{d} \right) \qquad (4.40)$$

According to Equation (4.40), if we try to maintain a constant velocity in the feed channel during a concentration run, the shear stress will increase (in the bulk and at the wall) because of an increase in viscosity. This increase in shear stress will be manifested as an increase in pressure *drop* during the run. The higher shear force at the membrane surface will augment back-diffusion of polarized solute to a greater extent than predicted from the film theory alone. Thus, we may observe the strange phenomenon of *increasing* mass transfer coefficients (and flux) at higher concentrations.

This phenomenon could explain the shape of the data curves in Figure 4.22. Initially, the driving force for mass transfer (ln C_G/C_B) is large, and changes in viscosity and shear stress small. Thus, mass transfer effects will dominate. Between 5 and 25 g/kg, viscosity changes are large enough so that the physical effects associated with increasing shear stress overcome the mass transfer effects, resulting in an increase in flux as explained above. However, as the concentration increases above 25 g/kg, the driving force decreases to a much larger extent, mass transfer effects become more important than shear stress effects, and the flux starts decreasing again. This multiphasic behavior was also reported by Kalyanpur et al. (1985) for UF of *A. chrysogenum* fermentation broth in an open channel plate system using a Millipore 0.2μ GVLP membrane. Sigmodial curves (an initial steep decline, followed by a relatively steady flux, followed by another decline to zero flux) are also common (Kavanagh and Brown 1987; Kroner et al. 1984).

Of course, higher concentrations could also increase diffusivity, which will further enhance flux. Assuming a linear relationship between diffusivity and solute concentration (see Section 4.I.3.):

$$D = D_o(1 + \varphi C) \qquad (4.41)$$

where D_o is the reference diffusivity at a particular concentration C and ϕ is an empirical constant. Substituting Equation (4.41) into the film theory and integrating, and then substituting the dimensionless correlation Equation (4.20):

$$J = A\left(\frac{D_o[1 + \varphi C]}{d_h}\right)(\text{Re})^\alpha(\text{Sc})^\beta\left(\ln\frac{C_G}{C_B} + \varphi[C_G - C_B]\right) \qquad (4.42)$$

For UF of skim milk in hollow fibers, Chiang and Cheryan (1987) fitted Equation (4.42) by nonlinear least-square analysis, assuming $\beta = 0.33$. They obtained $A = 0.087$, $C_G = 20.41\%$, $\alpha = 0.64$, and $\varphi = 0.157$. The positive value of φ was confirmation of enhanced back-diffusion not accounted for by Brownian diffusion. The C_G value was close to what was experimentally observed (see Table 4.2), although the value of α was higher than the Leveque model.

Another unexpected result is the increase in flux with a decrease in *permeable* solids to an extent much greater than could be accounted for by changes in physical properties. Jaffrin (1990) observed that, in the purification of plasma, flux was a function of both solvent (ethanol) and solute (albumin) concentration, increasing with an increase in ethanol concentration, especially when albumin concentration was low. Increase in flux during diafiltration has been observed for perilla extracts (Lin et al. 1989), glutamic acid (Kuo and Chiang 1987), and skim milk (Peri et al. 1973; Rajagopalan and Cheryan 1991). This suggests that the back-diffusion is enhanced by the removal of permeable compounds. In addition, flux changed from being mass transfer–controlled initially to being pressure-controlled at the end of the diafiltration runs even though the concentration of the *rejected* solute had not changed (Rajagopalan and Cheryan 1991).

4.H.2.
TEMPERATURE

In general, higher temperatures will lead to higher flux in both the pressure-controlled region and in the mass transfer–controlled region (Figures 4.6–4.9, 4.20, 4.23, and 4.24). This assumes there are no other unusual effects occurring simultaneously, such as fouling of the membrane due to precipitation of insoluble salts at higher temperatures or denaturation of proteins or gelatinization of starch at the higher temperatures. In the pressure-controlled region, the effect of

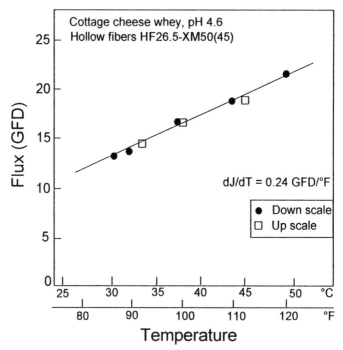

Figure 4.23. Effect of temperature on flux of cottage cheese whey. No hysteresis effects of temperature were observed (adapted from Breslau and Kilcullen 1977).

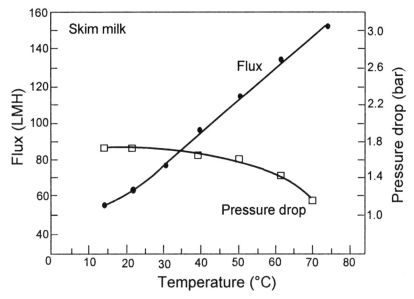

Figure 4.24. *Effect of temperature on UF flux of skim milk and on pressure drop. Module was DDS plate-and-frame system (adapted from Madsen 1977).*

temperature on flux is due to its effect on fluid density and viscosity. Activation energies for both flux and viscosity are similar in the region of 20–50°C, about 3400 kcal/mole. In practical terms, it will take a temperature rise of 30–45°C to double the flux.

As shown in Equations (4.35) and (4.36), k is proportional to $D^{0.67}$ and inversely proportional to $(\mu/\rho)^m$, where $m = 0.47$ for turbulent flow and zero for laminar flow systems. Diffusivity also increases with temperature; e.g., protein diffusivity increases at an average rate of 3–3.5% per °C rise in temperature. Thus, temperature is expected to have a fairly significant effect on flux.

In addition to its beneficial effect on flux, higher temperatures reduce the viscosity of the feed, which lowers pumping energy and horsepower required. This is manifested in a decrease in pressure drop needed to maintain the required velocity (Figure 4.24). High temperatures (>55°C) can also minimize microbial growth problems. For all these reasons, it is best to operate at the highest possible temperature consistent with the limits of the feed solution and the membrane.

4.H.3.
FLOW RATE AND TURBULENCE

Turbulence, whether produced by stirring, pumping the fluid, or moving the membrane, has a large effect on flux in the mass transfer–controlled region.

Agitation and mixing of the fluid near the membrane surface "sweeps" away the accumulated solute, reducing the hydraulic resistance of the "cake" (Figure 4.4) and reducing thickness of the boundary layer. There is also a belief that extremely high shear, such as that obtained with thin-channel and rotary devices, actually reduces the thickness of the "gel" layer. In any case, this is one of the simplest and most effective methods of controlling the effects of concentration polarization.

The effect of turbulence on flux can be seen in Figures 4.5–4.12, 4.19, 4.20, and 4.22. Note that when the system is pressure-controlled, the effect of turbulence appears to be insignificant (Figures 4.5–4.7). This is because the concentration of solute in the feed is not high enough and/or the pressures are too low and/or the velocities are high enough to minimize formation of the boundary or gel layers. When these conditions are not prevailing, the effect of velocity or turbulence will become noticeable (Figures 4.8–4.12). Figures 4.25–4.29 also show the effect of turbulence. In all these examples, turbulence was obtained the classic way: by increasing cross-flow velocity. Turbulence in the feed channel

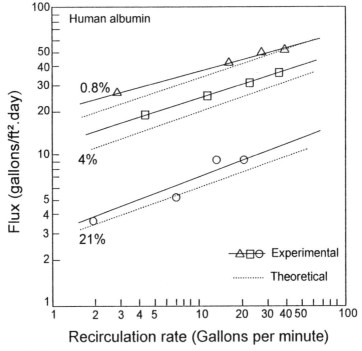

Figure 4.25. Ultrafiltration of human albumin in an Amicon LTC-I, 15-mil channel laminar flow module with XM50 membrane. Variable shown is protein concentration. Broken line indicates theoretical behavior calculated from the Leveque/film model using a diffusion coefficient of 6×10^{-7} cm^2/sec (adapted from Porter 1979).

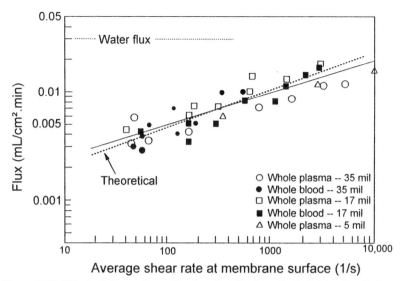

Figure 4.26. *Effect of shear rate on flux for ultrafiltration of whole blood and plasma in an Esmond cell of length L = 13 cm with an Amicon UM-10 membrane. Pressure was 10 psi. Different channel heights, as indicated, were used. The theoretical line is drawn using the Leveque/film model with a diffusion coefficient of 0.6 × 10⁻⁷ cm²/sec (adapted from Blatt et al. 1970).*

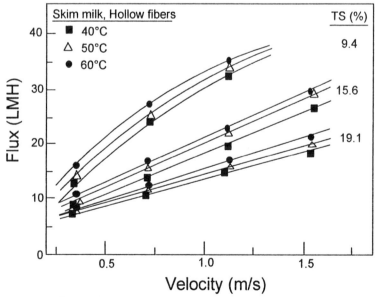

Figure 4.27. *Effect of feed velocity on flux of skim milk in a Romicon hollow fiber HF15-43-PM50 module. Variables are total solids of feed and operating temperature as indicated (adapted from Chiang and Cheryan 1987).*

149

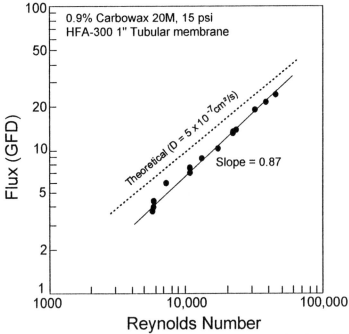

Figure 4.28. *Ultrafiltration flux as a function of Reynolds number in a 1″ diameter Abcor tubular module with HFA-300 membrane. Feed was 0.9% carbowax 20M (adapted from DeFilippi 1977).*

may be expressed as recirculation rate (e.g., gal/min, or gpm), as velocity (cm/min, ft/sec or m/sec), as shear rate (in $\sec^{-1}$), as pressure drop ($P_i - P_o$, psi or bar), or as Reynolds number (dimensionless). All these quantities are interconvertible and are different means of expressing the state of turbulence within a membrane module except, as mentioned later in Section 7.F., the relationship between pressure drop and flow rate through a module is more complex. For this particular discussion, higher pressure drop still means higher flow rates.

The magnitude of the effect of cross-flow rate on k or J is expressed in terms of the exponent on the velocity term, i.e., the value of α in the equation $J = f(V)^\alpha$. This can be obtained from the slope of a plot of log J versus log velocity (or flow rate, Reynolds number, or shear rate, as the case may be). Table 4.3 is a list of α values from the literature. In general, laminar flow units show a wide range of α values, 0.3 to 1.33, with 0.3–0.6 being most frequent. Turbulent flow systems range from 0.47 to 1.23, with 0.8–1.2 being common. As can be seen from Equations (4.35) and (4.36), the theoretical values of α

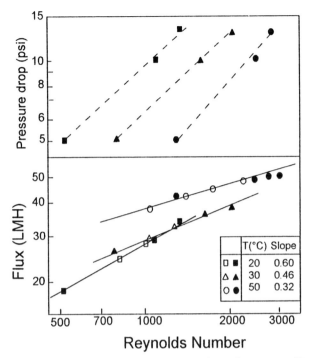

Figure 4.29. *Ultrafiltration of water extracts of whole soybeans: see Figure 4.9 for experimental details. The variable in the graph is operating temperature: top: relationship between pressure drop and Reynolds number; bottom: effect of Reynolds number on flux (adapted from Cheryan 1977).*

should be 0.33 and 0.8, respectively. In practical terms, it means the benefit of increasing cross-flow velocity is greater with turbulent flow than with laminar flow systems, but the price to be paid is in energy consumption, which increases at a faster rate since energy is proportional to V^{2-3} (see Section 7.F.). This is shown in Figure 4.29, where the slopes of the Re-pressure drop lines are steeper than the Re-J lines. This means that by increasing cross-flow velocity, we pay a proportionally greater price for pumping energy than the benefits we get from a higher flux. However, providing higher cross-flow velocities is usually cheaper than the costs associated with greater membrane area at the lower velocities.

The value of α in J versus Re relationships depend on (a) whether the flow is turbulent or laminar; (b) the rheological properties of the fluid, i.e., whether it is Newtonian or non-Newtonian; (c) temperature; and (d) feed concentration. This is because flux is also dependent on other functions such as the Schmidt number, which itself is a function of physical properties and operating parameters. The data will correlate much better, i.e., a more "universal" relationship can be

Table 4.3. Effect of turbulence on flux, expressed in terms of α in the equation $J = f(V)^\alpha$.

Feed	State of Turbulence	α	Reference
Albumin: human, serum	Laminar	0.33–0.4	Porter (1979)
	Laminar	0.61	Mitra and Lundblad (1978)
Blood			
Whole	Laminar	0.60	Porter (1979)
Human	Laminar	0.51	Colton et al. (1975)
Bovine	Laminar	0.47	Ozawa et al. (1987)
Bovine	Laminar	0.74	Jaffrin et al. (1984)
Bovine	Laminar	1.00	Gupta et al. (1986)
Blood plasma, Human	Laminar	0.33	Porter (1979)
	Laminar	0.32	Isaacson et al. (1980)
Casein, 1%	Laminar	0.5	Porter (1979)
Dextran T-500	Laminar	0.33	Pradanos et al. (1995b)
Immune serum globulin	Laminar	0.67	Mitra and Lundblad (1978)
Microbial cells			
L. bulgaricus	Laminar	0.29	Tejayadi and Cheryan (1988)
P. fluorescens	Laminar	0.50	Le et al. (1984a)
Bacteria	Laminar	0.49–0.79	Henry and Allred (1972)
S. cerevisiae	Laminar	0.1	Warren et al. (1991)
S. cerevisiae	Laminar	0.3	Patel et al. (1987)
Milk (skim)	Laminar	0.64	Chiang and Cheryan (1987)
Paint, electrocoat, 15%	Laminar	1.33	Porter (1979)
Prothrombin complex	Laminar	0.44	Mitra and Fillmore (1980)
Soybean: full-fat water extract	Laminar	0.32–0.60	Cheryan (1977)
Styrene-butadiene latex			
1%	Laminar	0.33	Porter (1979)
10%	Laminar	0.53	Porter (1979)
5–50%	Laminar	0.89	Porter (1979)
Albumin, bovine serum (BSA)	Turbulent	0.50	Kozinski and Lightfoot (1971)
Carbowax 20M, 0.9%	Turbulent	0.87	Defilippi (1977)
Microbial cells			
S. cerevisiae	Turbulent	0.50	Patel et al. (1987)
S. cerevisiae	Turbulent	0.57	Riesmeier et al. (1989)
S. cerevisiae	Turbulent	0.24–0.95	Saglam (1995)
S. cerevisiae	Turbulent	0.38	Redkar and Davis (1993)
S. cerevisiae	Turbulent	1.0	Shimuzu et al. (1993)
A. niger	Turbulent	1.0	Sims and Cheryan (1986)
Milk (whole,3.5% fat)	Turbulent	1.66	Yan et al. (1979)
Paint, electrocoat, 7%	Turbulent	1.17	Porter (1979)
Paint, electrocoat, 15%	Turbulent	1.23	Porter (1979)
Styrene-butadiene latex, 1%	Turbulent	1.11	Porter (1979)
Wheat starch effluent, 2–3% TS	Turbulent	0.8-1.25	Harris (1986)

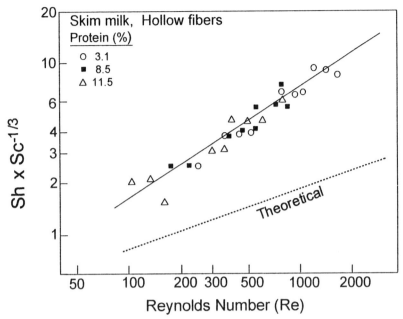

Figure 4.30. *Relationship between Sherwood number, Schmidt number, and Reynolds number for ultrafiltration of skim milk in a laminar flow module (Romicon hollow fiber HF15-43-PM50). The model plotted is Sh · Sc $^{-0.33}$ = A (Re). The broken line is the theoretical relationship calculated from the Leveque solution/film theory (adapted from Chiang and Cheryan 1987).*

obtained, if models of the form of Equation (4.13) are used. Dimensionless quantities (Sh, Re, Sc, and d_h/L) can be calculated from channel dimensions and physical properties. Values of the mass transfer coefficient (k) for use in Sh are obtained from Equation (4.11). Typical plots of Equation (4.13) are shown in Figures 4.30–4.32. Table 4.4 lists values of A, α, and β for several systems. Unfortunately, data in this particular form are rare in the literature since few investigators report sufficient physical property data to enable calculation of the dimensionless groups. An example of the use of these equations in predicting flux was given earlier in Section 4.E.2.

4.H.3.a.
DOES HIGH SHEAR DAMAGE PROTEINS?

There was some concern in the 1970s that the high shear rates common in some membrane modules—up to 14,000 sec^{-1} in hollow fibers and up to 400,000 sec^{-1} in some rotary modules—could denature proteins or otherwise damage fragile molecules. Many studies have shown that high fluid velocities or shear rates per se are not responsible for the observed damage

Table 4.4. Values of constants in the dimensionless mass transfer correlation: $Sh = A\,Re^{\alpha}Sc^{\beta}$.

Feed	State of Turbulence	A	α	β	Reference
Dextran T10,T40,T70	Laminar	1.62	0.33	0.33	Nabetani et al. (1990)
Polyethylene glycol	Laminar	2.0	0.25	0.33	Pradanos et al. (1995a)
Skim milk	Laminar	0.087	0.64	0.33	Chiang and Cheryan (1987)
Soybean extracts	Laminar	0.181	0.47	0.33	Cheryan (1977)
Albumin, bovine	Turbulent	0.15	0.47	0.33	Pace et al. (1976)
serum	Turbulent	0.60	0.50	0.33	Kozinski and Lightfoot (1972)
Polyethylene glycol	Turbulent	1.00	0.96	0.39	Pradanos et al. (1995a)
Polyvinyl alcohol	Turbulent	—	0.88	0.25	Nakao et al. (1979)
Wheat starch effluent	Turbulent	0.02	0.85	0.33	Harris (1986)

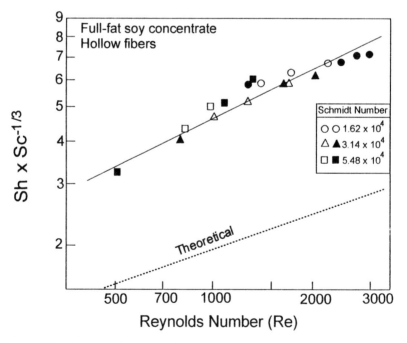

Figure 4.31. *Mass transfer correlation for the ultrafiltration of soybean water extracts. See Figures 4.9 and 4.29 for details. Broken line indicates expected behavior according to the Leveque/film theory model (adapted from Cheryan 1977).*

154

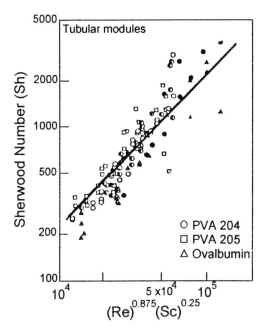

Figure 4.32. *Comparison of experimental data with the Deissler turbulent flow model. Feeds were polyvinyl alcohol PVA 204, PVA 205, and ovalbumin. Tubular modules of 12.5-mm diameter (open points) and 25-mm diameter (closed and half-closed points) were used (adapted from Nakao et al. 1979).*

(Thomas et al. 1979; Virkar et al. 1981). More damage occurs at gas–liquid interfaces, e.g., due to foaming and vortexing in feed tanks (Lee and Choo 1989) and possibly due to improper impeller design in some pumps (Maubois et al. 1986). Thus, if sensitive proteins are being processed, careful attention must be paid to venting all the air and other gases from the system, to minimize vortex formation in the feed tank and to design the piping to minimize cavitation. Also important is adjusting the process volume per unit membrane area so that the number of times the protein recycles through the pump is minimized, as well as the total time of exposure of the protein to possible denaturating conditions.

4.I.
PHYSICAL PROPERTIES OF LIQUID STREAMS

The important physical properties that affect flux are density, viscosity, diffusivity, and osmotic pressure. Physical properties of the fluid will vary as it moves along the transfer path, since the concentration and possibly the state of turbulence is also changing as the feed flows through a module. These properties must be suitably averaged before insertion into a flux prediction model. In most MF and UF applications, however, the fraction of the feed removed during a single pass through a module (termed *recovery* in the RO literature) is very little. For example, with a typical 3″ × 25″ hollow fiber module operating at a velocity of 1 m/sec, the permeate flow may be 1–2 L/min or less while the recirculation rate may be 45–60 L/min. Similarly, an 8″ × 40″ spiral-wound module uses a cross-flow rate of 100–300 gpm, but the permeate flow may be only 0.5–5 gpm. Thus, the change in concentration of feed solids is negligible during passage through a single module, and variations in physical properties of the feed in the axial direction can be neglected. The exception is in certain water treatment applications where the flux is so high in relation to cross-flow rate that the recovery per pass could be as high as 50–70%.

Concentration effects in the radial direction, i.e., perpendicular to the membrane wall, are not negligible. As depicted schematically in Figure 4.14 and from the C_G values shown in Table 4.2, the concentration gradient within the boundary layer is very steep, with polarization moduli (C_G/C_B) frequently in the range of 5–20. Unlike heat transfer, there are few mass transfer correlations that correct for differences in physical properties at the wall. Shen and Probstein (1977) have corrected the film theory model [Equation (4.11)] to include a term for the ratio of diffusivity at the wall (D_G) to bulk diffusivity (D_B):

$$J = A'' \left(\frac{D_G}{D_B} \right)^{2/3} \left(\frac{V D_B^2}{d_h L} \right)^{1/3} \ln \left(\frac{C_G}{C_B} \right) \tag{4.43}$$

Kozinski and Lightfoot (1972) have also modified the Leveque solution as follows:

$$\text{Sh} = \text{Sc}^{1/3} \text{Re}^{1/3} \left(\frac{X_G C_B}{X_B C_G} \right) \left(\frac{D_{av}^2}{\mu_{av}} \right)^{1/3} \left(\frac{3L^2}{2b(d_h/L)} \right)^{1/3} \tag{4.44}$$

where X_i is the mole fraction of the solute.

Ozawa et al. (1987) observed that the mass transfer coefficient for whole blood processed in hollow fibers was a function of shear rate (as expected) and protein concentration, which was not expected (Figure 4.33, open points). This implied that the diffusion coefficient was also a function of concentration (Section 4.I.3.). They applied a viscosity correction factor, as shown in Equation (4.45), and obtained a much better correlation for the mass transfer coefficient (Figure 4.33, closed points).

$$k = 1 \times 10^{-5} \left(\frac{\mu_{\text{water}}}{\mu_B} \right) \gamma_w^{0.47} \tag{4.45}$$

where μ_B is the viscosity of the bulk feed solution, and μ_{water} is the viscosity of water at the same temperature.

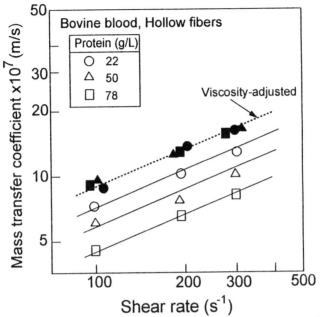

Figure 4.33. *Effect of shear rate on the mass transfer coefficient for the hollow fiber UF of bovine blood. Module was a Plasmacure PVA-SB plasma separator from Kuraray, Japan. Open points indicate k values obtained from slopes of J versus. In C_B plots. Closed points were obtained after correcting for viscosity, as shown in Equation (4.45) (data adapted from Ozawa et al. 1987).*

It is tempting to arbitrarily include a viscosity correction factor in the dimensionless mass transfer correlations analogous to the Sieder-Tate correction factor:

$$Sh = A(Re)^{\alpha}(Sc)^{\beta}(\mu_B/\mu_w)^f \qquad (4.46)$$

A Schmidt number correction factor $(Sc_B/Sc_w)^f$ could be better since it includes both viscosity and diffusivity, both of which vary with concentration (Gekas and Hallstrom 1987). The value of f is typically 0.11–0.14 (by heat transfer analogy) or 0.27 (Pritchard et al. 1995).

Even if good correlations were available, the scarcity of good physical property data, especially at the high concentrations near C_G, is a major limitation to applying many of the mass transfer models.

4.I.1.
DENSITY

Since most UF and MF feedstreams are fairly dilute, their density should be close to that of the solvent (water). In any case, densities of liquids are fairly easy to obtain from standard tables or to experimentally determine using precalibrated pycnometers.

4.I.2.
VISCOSITY

Viscosity (μ) is a parameter that characterizes the flow behavior of liquids. It is defined in terms of shear stress (τ) and shear rate (γ):

$$\tau = \mu \gamma \tag{4.47}$$

Fluids obeying Equation (4.47) are known as Newtonian fluids. There are several types of fluids that do not obey Equation (4.47), in that viscosity is not independent of shear rate or shear stress. Their behavior is commonly modeled according to the power law model:

$$\tau = K \gamma^n \tag{4.48}$$

where K and n are power law constants. To obtain the apparent viscosity (μ_{ap}), combine Equations (4.47) and (4.48):

$$\mu_{ap} = K \gamma^{n-1} \tag{4.49}$$

When $n = 1$, the fluid is Newtonian. All other values of n indicate the fluid is non-Newtonian. Almost all macromolecular solutions and suspensions will display non-Newtonian behavior at sufficiently high concentrations.

One general model for viscosity is the Einstein linear equation for dilute suspensions that has been modified for volume fractions of solids (ε) < 0.40 (Hiemenz 1986):

$$\mu/\mu_o = 1 + 2.5\varepsilon + 10\varepsilon^2 \tag{4.50}$$

where μ_o is solvent (water) viscosity.

In practical terms, the viscosity (a) decreases with increasing temperature; (b) increases with increasing feed concentration; and (c) decreases, increases, or is unaffected by shear rate (i.e., flow rate or stirrer speed), depending on the nature of the liquid. These effects must be taken into account during the design of UF and MF processes. For example, most macromolecular solutes become increasingly pseudoplastic when concentrated. In the initial stages, this may not be a problem since the apparent viscosity will be low at the low solute concentration. However, as the solutes get concentrated in later stages, an undersized pump will not be able to pump the concentrated fluid at a high enough shear rate. This will increase the fluid viscosity, which in turn will increase power consumption, reduce turbulence, and reduce flux to a greater extent than if the pumping system had been properly designed.

To accommodate the apparent viscosity in the quantitative correlation of mass transport phenomena, the relationships are modified as follows for laminar flow:

$$\mathrm{Re} = \frac{d_h^n V^{2-n} \rho}{\frac{K}{8} \left(\frac{6n+2}{n}\right)^n} \tag{4.51}$$

$$\mathrm{Sc} = \frac{d_h^{1-n} V^{n-1} \left(\frac{K}{8}\right) \left(\frac{6n+2}{n}\right)^n}{\rho D} \tag{4.52}$$

4.I.3.
DIFFUSION COEFFICIENTS

Unlike viscosity and density, which are fairly easy to determine experimentally, diffusivity is much more difficult to determine or estimate. Diffusivity data of macromolecules and hydrocolloids are rare in the literature, especially as a function of temperature and concentration. Ionic strength and pH of the solution will affect the state of hydration and conformation of the macromolecule and, thus, will also affect diffusion coefficients. Also unknown is the effect of other solutes in the feed solution and how solute–solute interactions can be taken into account. Table 4.5 lists some diffusion coefficients available in the literature.

Theoretical equations to estimate diffusivity depend on how good the theories are for the structure of liquids. Bretsnajder (1971) has presented a number of semitheoretical and empirical relationships, among them the Wilke-Chang equation:

$$D_{AB} = \frac{(117.3 \times 10^{-18})(\phi M_B)^{0.5} T}{\mu V_A^{0.6}} \tag{4.53}$$

where

D_{AB} = diffusivity of component A in solvent B, m²/sec
M_B = molecular weight of solvent, kg/mole = 18 for water
T = temperature, °K
μ = solution viscosity, kg/m · sec
V_A = solute molal volume at normal boiling point, m³/kmol
ϕ = association constant for solvent = 2.26 for water

The Stokes-Einstein equation for Brownian diffusivity is

$$D = \frac{k_B T}{3\pi \mu d_p} \tag{4.54}$$

Table 4.5. Diffusion coefficients.

Compound	Molecular Weight	Temperature (°C)	Diffusivity × 10^7 (cm²/sec)	Reference
Albumin, human serum	69,000	20	6.43	Walters et al. (1984)
Albumin, bovine serum	66,500	20	6.3	Walters et al. (1984)
	67,000	20	5.9	Smith (1970)
α-Amylase	96,920	20	5.72	Smith (1970)
α-1-Antitrypsin, human	45,000	20	5.2	Walters et al. (1984)
Apoferritin, horse spleen	466,900	20	3.99	Walters et al. (1984)
Casein	24,000	25	1.90	Delaney and Donnelly (1977)
α-Chymotrypsin				
Bovine	21,600	20	10.78	Walters et al. (1984)
Pancreas	21,600	20	10.2	Smith (1970)
Collagen	345,000	20	1.16	Elias (1977)
Concanavalin A, jack bean	96,200	20	5.6	Walters et al. (1984)
Cytochrome C				
Horse heart	13,400	20	11.6	Walters et al. (1984)
Bovine heart	13,370	20	11.4	Smith (1970)
Cytochrome-C1, bovine heart	370,500	20	3.31	Smith (1970)
DNA	6,000,000	20	0.13	Elias (1977)
Ferritin, horse spleen	750,000	20	3.25	Walters et al. (1984)
Fibrinogen, human	339,700	20	2.98	Walters et al. (1984)
Hemoglobin, bovine	68,000	20	7.0	Walters et al. (1984)
Immunoglobulin G, human	153,000	20	4.11	Walters et al. (1984)
Insulin, porcine (Na)	5,800	20	11.68	Walters et al. (1984)
α-Lactalbumin	16,000	25	7.4	Delaney and Donnelly (1977)
β-Lactoglobulin	18,000	25	6.4	Delaney and Donnelly (1977)
Myoglobin, horse muscle	16,900	20	10.27	Walters et al. (1984)
Myosin	493,000	20	1.16	Elias (1977)
Ovalbumin	45,000	25	7.6	Walters et al. (1984)
		25	6.9	Nabetani et al. (1990)
Ribonuclease	13,683	20	11.9	Elias (1977)
Soy protein 7 S	180,000	20	3.85	Koshiyama (1968)
Soy protein 11 S	350,000	20	2.91	Wolf and Briggs (1959)
Soy protein 11 S	350,000	20	3.30	Koshiyama and Fukushima (1976)
Thyroglobulin, porcine	670,000	20	2.65	Walters et al. (1984)
Trypsin, bovine pancreas	15,100	20	10.09	Walters et al. (1984)
Trypsin inhibitor, soybean	22,700	20	9.51	Walters et al. (1984)

where k_B is the Boltzmann constant and T is the absolute temperature. These theories predict very low diffusion coefficients, which means very low flux. In practice, back-diffusion rates in cross-flow MF and UF are much higher, which is usually explained by the shear in the channel augmenting the natural diffusion process, giving rise to a diffusion effect quite different from Brownian diffusion. Eckstein et al. (1977) have suggested the following to account for the effect of shear:

$$D = 0.025\, r_p^2 \gamma_w \qquad (4.55)$$

where r_p is the particle radius. This is valid for solute void fractions greater than 0.2. Bashir et al. (1994) also observed a linear increase in shear-induced diffusivity up to a shear rate of 660 $\sec^{-1}$.

Incorporating Equation (4.55) into the Leveque solution [Equation (4.36)] will give

$$J = A\left(\frac{r_p^{4/3} V}{d_h L^{0.33}}\right) \ln\frac{C_G}{C_B} \qquad (4.56)$$

This predicts a velocity exponent (α) of 1.0, much higher than the one-third value expected from Brownian diffusion considerations and closer to what has been experimentally observed in some cases with thin-channel equipment (Table 4.2; Zydney and Colton 1986).

Young et al. (1980) have developed a correlation for the diffusion coefficient of proteins from the available experimental data:

$$D(\text{cm}^2/\sec) = 7.51 \times 10^{-8} T/(\mu_o v^{0.33}) \qquad (4.57)$$

where μ_o is solvent viscosity, T is absolute temperature, and v is the molal volume. Tyn and Gusek (1990) have correlated diffusivity of proteins with their radius of gyration (R_g, expressed in Å):

$$D(\text{cm}^2/\sec) = 1.69 \times 10^{-5}/R_g \qquad (4.58)$$

Diffusivity of polyethylene glycols (PEG) at 25°C have been correlated by Pradanos et al. (1995a) as

$$D(\text{m}^2/\sec) = 9.82 \times 10^{-9}(\text{MW})^{-0.52} \qquad (4.59)$$

where MW is the molecular weight of PEG.

The diffusivity generally increases with solute concentration. Eckstein et al. (1977) observed a linear increase with concentration up to a particle volume concentration of 0.2. For the following correlation with volume fraction (ε):

$$D(\text{m}^2/\text{sec}) = A\varepsilon^2(1 + Be^{C\varepsilon}) + H \qquad (4.60)$$

Leighton and Avricos (1987) obtained $A = 0.33$, $B = 0.5$, $C = 8.8$, and $H = 0$ for rigid 50-μ spheres with $\varepsilon > 0.3$, while Bashir et al. (1994) obtained $A = 42.3$, $B = 181.8$, $C = 16.2$, and $H = 24.8$ for yeast cells with $0.017 < \varepsilon < 0.052$.

For dextrans in water, Clifton et al. (1984) obtained the following expression:

$$D(\text{m}^2/\text{sec}) = 5.96 \times 10^{-11} + 2.12 \times 10^{-11} \tanh(0.0284C - 1.491) \quad (4.61)$$

where C is the concentration in kg/m^3.

The effect of temperature can be modeled by the Wilke modification of the Stokes-Einstein equation:

$$D_1 = \frac{D_2 \mu_2 T_1}{\mu_1 T_2} \qquad (4.62)$$

where the subscripts refer to two different temperatures. Diffusivity increases at an average rate of 3–3.4% per °C for globular proteins (Keller et al. 1971).

Uncertainties in the diffusivity data is probably the greatest source of error in predicting flux, especially in laminar flow systems since the flux dependency on diffusivity is the highest among all the parameters considered in the dimensionless correlations [Equations (4.35) and (4.36)]. Viscosity and density variations and errors, on the other hand, have a small effect on flux prediction, especially in laminar flow systems, since $J = f(\mu/\rho)^m$, where m has values between zero and 0.16.

4.J.
EXPERIMENT VERSUS THEORY: THE "FLUX PARADOX"

Much of the data shown in the last section indicates that in some cases the mass transfer models appear to underpredict actual experimental flux, sometimes by one to three orders of magnitude. This was shown in Figures 4.30 and 4.31 and also by the α values in Table 4.2 and the A and α values in Table 4.3. Blatt et al. (1970) and Porter (1972, 1979) have presented a number of cases where experimental slopes of plots of J versus turbulence were much higher than expected. Careful examination of the literature in this area over the past 25 years indicates

- The discrepancy between predicted and experimental flux is most noticeable with colloidal feedsteams. Much closer predictions are obtained if the solute being rejected by the membrane is in solution rather than a suspension or emulsion.

- Thin-channel and/or systems operating under high shear conditions show greater deviations than other systems.

The reasons for the discrepancy between theory and experiment are not clear. Madsen (1977) suggested that as long as one operates in the "prefouling" operating region (relatively low pressures and high velocities), the polarized layer is not a true gel and can be easily removed with high shear rates in the feed channel. The assumptions that were made when deriving the original models are also suspect. The film theory is a good approximation only when the boundary layer is thin and uniform. This occurs in turbulent flow situations, but even in the turbulent flow model, as turbulence intensity increases, the exponent on Re in Equation (4.13) should approach unity. As mentioned earlier, the film theory based on Brownian diffusion mechanisms is inadequate, since it does not take into account possible dependency of the diffusion coefficient on shear rate and concentration. In fact, the convective transport of particles to the membrane surface is counterbalanced by three mechanisms: Brownian back-diffusion (which predicts an increase in back-diffusion with decreasing particle diameter, d_p), lateral migration caused by inertial lift (this force is proportional to d_p^3), and surface shear (valid in laminar flow, proportional to d_p^2). The latter two also show a direct dependence on cross-flow velocity. Welsch et al. (1995) explain the anomaly in terms of particle charges or surface potential. They developed a model based on a force balance of particles in the polarized layer and showed a strong dependence of flux on particle charge or surface potential. The flux passed through a minimum as the charge of the particles increased and as the particle size increased.

Another reason for experimental flux being higher than predicted flux is that back-diffusion from the membrane surface to the bulk stream is greater than expected and is controlled by forces other than or in addition to the concentration gradient. Treybal (1981) has discussed situations under high longitudinal velocities, where a separation of the boundary layer from the downstream membrane surface occurs to form a "wake." The average mass transfer coefficient in these cases is obtained by adding the two contributions from the laminar boundary layer and from the wake in the form of two Reynolds numbers terms. Ripples and waves that begin to form at fairly low Reynolds numbers were also not considered in the analysis.

A more elegant explanation is that the enhancement of flux during UF and MF of colloidal suspensions is due to a "tubular pinch" effect (Porter 1972). This phenomenon was first observed by Segre and Silberberg (1961), working with dilute suspensions of rigid spheres. As the particles were flowing through a tube, they were apparently migrating away from the tube wall and the tube axis, reaching equilibrium at some eccentric radial position. The spheres were

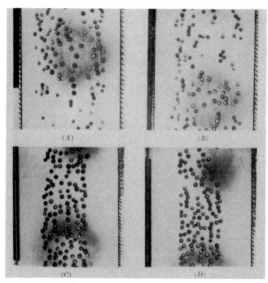

Figure 4.34. *The tubular pinch effect. Particles of 1-mm diameter were flowing in a 15% aqueous glycerin solution. Particle concentration = 5%, channel Re = 900, particle Re = 225 (adapted from Brandt and Bugliarello 1966).*

regularly spaced in chains parallel to the tube axis. Figure 4.34 is spectacular visual evidence of this phenomenon. This phenomenon, whereby colloidal particles migrate away from the tube wall, has been observed experimentally for several other colloidal suspensions, such as carbon black, polystyrene, polyvinyl alcohol (PVA), aluminum-coated nylon rods, alumina particles, silicone oils, and glycerol in various continuous-flowing media (Karnis et al. 1966).

Several mathematical analyses of this phenomenon have appeared and, although based on different assumptions, eventually result in quite similar expressions for the radial migration (i.e., lift) velocity. They take the general form:

$$V_{\text{RM}} = f\left(V, \text{Re}, [r_p/R]^a, r/R\right) \tag{4.63}$$

where

V_{RM} = radial migration velocity
V = average bulk fluid velocity down the channel
Re = Reynolds number
r_p = particle radius
R = channel radius
r = radial position of the particle in the tube
a = constant, whose value lies between 2.84 and 4

The tubular pinch effect can explain why the flux of whole blood is of the same order of magnitude as blood plasma, even though the Leveque solution would predict a much lower flux. In fact, there are patents (U.S. Patent 3,541,006 assigned to Amicon; U.S. Patent 4,751,003 assigned to Henkel) for improving UF flux by deliberately introducing solid particulate materials into the process stream. Equation (4.63) also suggests that smaller channel sizes are better for observing the depolarizing effect with colloidal systems. Secondary flows generated by particle migration could trigger transition from laminar to turbulent flow at Reynolds numbers below 2000. It is difficult to predict exactly the enhanced mass transfer effect, but in addition to the factors listed in Equation (4.63), the shape and extent of deformability of the particles will be important, as well as the ratio of pore size to particle size (Section 6.D.1.).

The back-diffusion of particles away from the membrane surface is supplemented by a lateral migration of particles due to inertial lift, which arises from nonlinear interactions of particles with the surrounding flow field. The flux then becomes

$$J = J_{\text{lateral}} + k \, \ln(C_G/C_B) \qquad (4.64)$$

where J_{lateral} is flux due to the lift velocity, which is a direct function of r_p^{2-3} and γ_w^{1-2} (Belfort 1984; Belfort et al. 1994; Green and Belfort 1984; and several others cited by Zydney and Colton 1986). Many of these models, as well as those incorporating a shear-induced diffusion coefficient such as Equation (4.56), appear to fit experimental data much better than the Leveque model incorporating only Brownian diffusivity.

4.K.
DESIGN FACTORS AFFECTING FLUX

The major emphasis in the design of equipment and its operation is to reduce the effects of concentration polarization. Figure 4.35 shows several such methods. Among the approaches shown in the first tier, reducing solids in the feed may be counterproductive in the long run, since then a greater volume of feed has to be processed. The lowest possible pressure should be used, especially in MF (Section 6.D.3.). To reduce solute concentration at the membrane surface, mixing can be enhanced near the membrane surface by using paddle mixers (commonly done in laboratory-scale apparatus, but impractical for industrial equipment) or by using static mixers (e.g., the kinecs mixer) in large tubular designs, or mesh-spacer turbulence promoters, which are used in spiral-wound and some plate designs (see Chapter 5). "Scouring balls" in large tubular units are also recommended. Substantial improvement in flux can be obtained with these devices (Porter 1979).

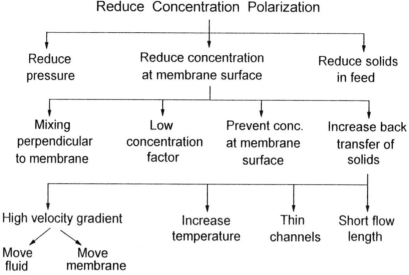

METHODS OF INCREASING FLUX

Figure 4.35. Methods to maximize flux (adapted from Bruin et al. 1980).

The third tier shows some of the more practical and common methods of improving flux. These approaches can be better understood by considering the mass transfer models again [Equations (4.35) and (4.36)]. In both turbulent and laminar flow, flux increases with velocity (V) or flow rate (Q), although the effect of flow rate is more pronounced in turbulent flow due to the higher exponent on V. Both models show a relatively high dependency on diffusion coefficient. The best practical means of enhancing this factor is to increase the temperature.

It is interesting to note how the two models differ. Those systems operating in turbulent flow are affected by kinematic viscosity (μ/ρ), while laminar flow systems are unaffected. Equation (4.36) predicts that shorter channel lengths (L) will result in higher flux. Gupta et al. (1986) confirmed this for laminar flow units, obtaining $J = f(L)^{-0.28 \text{ to} -0.32}$, which is close to the $L^{-1/3}$ predicted by the Leveque/film theory. Romicon hollow fiber modules, which operate in laminar flow, are available in three different lengths; the short cartridges are preferable if the feed is very viscous and where high final product concentrations are required (Chapter 5).

Some equipment designers believe that it is more important to maximize the shear rate rather than Reynolds number or flow rate per se. Also, both turbulent and laminar flow models show that k is inversely proportional to d_h. Thus, the most cost-effective route would be to decrease the channel height to maximize the shear rate. This approach is reflected in the "thin-channel" designs, where the channel height or diameter is about 1 mm or less. In the case of hollow fibers, this has resulted in shear rates as high as $4000–10,000 \text{ sec}^{-1}$. However, energy consumption is proportional to pressure drop $\times$ flow rate, and pressure drop is inversely proportional to the third or fourth power of the channel diameter (Bird et al. 1960). Hence, thin-channel designs are usually restricted to laminar flow conditions to keep the flow rate and the power consumption to a minimum.

In summary, there have been many attempts to develop mathematical models to predict flux. None are completely satisfactory: their biggest failing is the inability to completely describe all regions, from the pressure-controlled to the mass transfer–controlled region (in case of UF) and perhaps to the cake-controlled region (in MF), in a variety of applications. This could be a reflection of the complexity of the feeds that are processed in MF and UF systems. A good model specifically for cross-flow MF is still not available; for dead-end MF, classic cake filtration models seem to work (Porter 1979).

The chief value of any model is that it should give us an appreciation of the phenomena while providing a framework for designing a practical process. It should also show us the relative benefits of manipulating operating variables or changing feed channel dimensions. In that limited role, the models discussed in this chapter are satisfactory. However, there are several other methods of enhancing flux, e.g., by moving the membrane instead of the liquid or subjecting the liquid to different types of perturbations to minimize fouling and polarization. These are discussed in Chapters 5 and 6.

REFERENCES

AKRED, A. R., FANE, A. G. and FRIEND, J. P. 1980. *Polymer Sci. Technol.* 13: 353.

ALTMANN, J. and RIPPERGER, S. 1997. *J. Membrane Sci.* 124: 119.

BASHIR, I., JERABEK, M., MARTIN, T. and REUSS, M. 1994. *J. Membrane Sci.* 95: 229.

BELFORT, G. P. 1984. *Synthetic Membrane Processes,* Academic Press, New York.

BELFORT, G., DAVIS, R. H. and ZYDNEY, A. L. 1994. *J. Membrane Sci.* 96: 1.

BELL, D. J. and DAVIES, R. J. 1987. *Biotechnol. Bioeng.* 29: 1176.

BIRD, R. B., STEWARD, W. E. and LIGHTFOOT, E. N. 1960. *Transport Phenomena,* Wiley, New York.

BLATT, W. F., DAVID, A., MICHAELS, A. S., and NELSON, L. 1970. In *Membrane Science and Technology,* J. E. Flinn (ed.), Plenum Press, New York.

BRANDT, A. and BUGLIARELLO, G. 1966. *Trans. Soc. Rheol.* 10 (1): 229.

BRESLAU, B. R. and KILCULLEN, B. 1977. *J. Dairy Sci.* 60: 1379.

BRETSZNAJDER, S. 1971. *Prediction of Transport and Other Physical Properties of Fluids,* Pergamon Press, New York.

BRUIN, S., KIKKERT, A., WELDRING, J. A. G. and HIDDINK, J. 1980. *Desalination* 35: 223.

CHERYAN, M. 1977. *J. Food Process. Engr.* 1: 269.

CHERYAN, M. and KUO, K. P. 1984. *J. Dairy Sci.* 67: 1406.

CHERYAN, M. and CHIANG, B. H. 1984. In *Engineering and Food, Vol. 1*, B. M. McKenna (ed.), Applied Science Pub., London, U.K. p. 191.

CHIANG, B. H. and CHERYAN, M. 1986. *J. Food Sci.* 51: 340.

CHIANG, B. H. and CHERYAN, M. 1987. *J. Food Engr.* 6: 241.

CLIFTON, M. J., ABIDINE, N. and SANCHEZ, V. 1984. *J. Membrane Sci.* 21: 233.

COLTON, C. K., HENDERSON, L. W., FORD, C. A. and LYSAGHT, M. J. 1975. *J. Lab. Clin. Med.* 85(3): 355.

COOPER, A. R. and BOOTH, R. G. 1978. *Sep. Sci. Technol.* 13: 735.

DEBOER, R. and HIDDINK, J. 1980. *Desalination* 35: 169.

DEFILIPPI, R. P. 1977. In *Filtration. Principles and Practices. Part I*, C. Orr (ed.), Marcel Dekker, New York. p. 475.

DEFILIPPI, R. P. and GOLDSMITH, R. L. 1970. In *Membrane Science and Technology* J. E. Flinn (ed.), Plenum Press, New York. p. 33.

DELANEY, R. A. M. 1977. *J. Food. Technol.* 12: 339.

DELANEY, R. A. M. and DONNELLY, J. K. 1977. In *Reverse Osmosis and Synthetic Membranes. Theory, Technology, Engineering*, S. Sourirajan (ed.), National Research Council, Ottawa, Canada.

DELANEY, R. A. M., DONNELLY, J. K. and BENDER, L. 1975. *Lebensm. Wiss. Technol.* 8: 20.

DEVEREUX, N. and HOARE, M. 1986. In *Food Engineering and Process Applications. Volume 2: Unit Operations*, M. Le Mageur and P. Jelen (eds.), Elsevier, Barking, Essex, U.K. p. 213.

DONNELLY, J. K. and DELANEY, R. A. M. 1974. *Lebens. Wiss. Technol.* 7: 162.

ECKSTEIN, E. C., BAILEY, D. G. and SHAPIRO, A. H. 1977. *J. Fluid Mech.* 79: 191.

ELIAS, H. G. 1977. *Macromolecules, Vol. 1*. Plenum Press, New York.

EPSTEIN, A. C. and KORCHIN, S. R. 1981. Presented at *91st National Meeting, AIChE*, Detroit, MI.

FOETISH, C., VON STOCKAR, U. and BLANC, B. 1987. *Alimenta* 1:14.

GEKAS, V. and HALLSTROM, B. 1987. *J. Membrane Sci.* 30: 153.

GREEN, G. and BELFORT, G. 1984. *Chem. Eng. Sci.* 39: 343.

GROBER, H., ERK, S., and GRIGULL, U. 1961. *Fundamentals of Heat Transfer*, McGraw-Hill, New York.

GUPTA, B. B., JAFFRIN, M. Y., DING, L. H. and DOHI, T. 1986. *Artificial Organs* 10 (1): 45.

HANEMAAIJER, J. H. 1985. *Desalination* 53: 143.

HARRIS, J. L. 1986. *J. Membrane Sci.* 29: 97.

HARRIS, J. L. and DOBBS, M. 1989. *J. Membrane Sci.* 41: 87.

HENRY, J. D. and ALLRED, R. C. 1972. *Dev. Ind. Microbiol.* 13: 177.

HIDDINK, J., DEBOER, R. and NOOY, P. F. C. 1981. *Milchwiss.* 36: 11.

HIEMENZ, P. C. 1986. *Principles of Colloid and Surface Chemistry*. Marcel Dekker, New York.

HWANG, S. and KAMMERMEYER, K. 1975. *Membranes in Separations*, John Wiley, New York.

ISAACSON, K., DUENAS, P., FORD, C. and LYSAGHT, M. 1980. *Polymer Sci. Technol.* 13: 507.

JAFFRIN, M. Y. 1990. *4th Symposium on Protein Purification Technologies*, Clermont-Ferrand, France.

JAFFRIN, M. Y., DING, L. H., COUVREUR, C., KHARI, P. 1997. *J. Membrane Sci.* 124: 233.

JAFFRIN, M. Y., GUPTA, B. B., DING, L. H. and GARREAU, M. 1984. *Trans. Am. Soc. Artif. Intern. Organs* 30: 401.

JIRARATANANON, R. and CHANACHAI, A. 1996. *J. Membrane Sci.* 111: 39.

JONSSON, G. 1984. *Desalination* 51: 61.

KALYANPUR, M., SKEA, W. and SIWAK, M. 1985. *Dev. Industrial Microbiology* 26: 455.

KARNIS, A., GOLDSMITH, A. L. and MASON, S. G. 1966. *Can. J. Chem. Eng.* 44: 181.

KAVANAGH, P. R. and BROWN, D. E. 1987. *J. Chem. Tech. Biotechnol.* 38: 187.

KELLER, K. H., CANALES, E. R., and YUM, S. I. 1971. *J. Phys.Chem.* 75:379.

KESSLER, H. G. 1981. *Food Engineering and Dairy Technology*, Verlag, A. Kessler, W. Germany.

KOSHIYAMA, I. 1968. *Cereal Chem.* 45: 394.

KOSHIYAMA, I. and FUKUSHIMA, D. 1976. *Int. J. Peptide Protein Res.* 8: 283.

KOZINSKI, A. A. and LIGHTFOOT, E. N. 1972. *AIChE J.* 18: 1030.

KRONER, K. H., SCHUTTE, H., HUSTEDT, H. and KULA, M. R. 1984. *Process Biochem.* 19 (4): 67–74.

KUO, W. H. and CHIANG, B. H. 1987. *J. Food Sci.* 52: 1401.

LE, M. S., SPARK, L. B. and WARD, P. S. 1984a. *J. Membrane Sci.* 21: 219.

LE, M. S., SPARK, L. B., WARD, P. S. and LADWA, N. 1984b. *J. Membrane Sci.* 21: 307.

LEE, Y.-K. and CHOO, C.-L. 1989. *Biotechnol. Bioeng.* 33: 183.

LEIGHTON, D. T. and ACRIVOS, A. 1987. *J. Fluid Mech.* 177: 109.

LIN, S. S., CHIANG, B. H. and HWANG, L. S. 1989. *J. Food Engr.* 9: 21.

MADSEN, R. F. 1977. *Hyperfiltration and Ultrafiltration in Plate-and-Frame Systems*. Elsevier, Amsterdam.

MAUBOIS, J. L., PIERRE, A., FAUQUANT, J. and PIOT, M. 1986. *Bulletin of the International Dairy Federation*, Number 212. 24: 154.

MITRA, G. and FILLMORE, K. 1980. *Polymer Sci. Technol.* 13: 480.

MITRA, G. and LUNDBLAD, J. L. 1978. *Sep. Sci. Technol.* 13: 89.

NABETANI, H., NAKAJIMA, M., WATANABE, A., NAKAO, S. and KIMURA, S. 1990. *AIChE J.* 36: 970.

NAKAO, S. I., NOMURA, T. and KIMURA, S. 1979. *AIChE J.* 25: 615.

NICHOLS, D. J. and CHERYAN, M. 1981. *J. Food Process. Preserv.* 5(2): 104.

OZAWA, K., MIMURA, R. and SAKAI, K. 1987. *J. Chem. Engr. Japan* 20: 345.

PACE, G. W., SCHARIN, M. J., ARCHER, M. C. and GOLDSTEIN, D. J. 1976. *Separation Sci.* 11: 65.

PATEL, P. N., MEHAIA, M. A. and CHERYAN, M. 1987. *J. Biotechnol.* 5: 1.

PERI, C., POMPEI, C. and ROSSI, F. 1973. *J. Food Sci.* 38:135.

PORTER, M. C. 1972. *Ind. Eng. Chem. Prod. Res. Dev.* 11:234.

PORTER, M. C. 1979. In *Handbook of Separation Techniques for Chemical Engineers*, P. A. Schweitzer (ed.), McGraw-Hill, New York.

PORTER, M. C. AND MICHAELS, A. S. 1972. *ChemTech.* 2: 56.

PRADANOS, P., ARRIBAS, J. I. and HERNANDEZ, A. 1995a. *J. Membrane Sci.* 99: 1.

PRADANOS, P., DE ABAJO, J., DE LA CAMPA, J. G. and HERNANDEZ, A. 1995b. *J. Membrane Sci.* 108: 129.

PRITCHARD, M., HOWELL, J. A. and FIELD, R. W. 1995. *J. Membrane Sci.* 102: 223.

RAJAGOPALAN, N. and CHERYAN, M. 1991. *Chem. Engr. Comm.* 106: 57.

REDKAR, S. G. and DAVIS, R. H. 1993. *Biotechnol. Progr.* 9: 625.

RIESMEIER, B., KRONER, K. H. and KULA, M. R. 1989. *J. Biotechnol.* 12: 153.

SAGLAM, N. 1995. Ph.D. thesis, University of Illinois, Urbana.

SEGRE, G. and SILBERBERG, A. 1961. *Nature* 189: 209.

SHEN, J. S. S. and PROBSTEIN, R. F. 1977. *Ind. Eng. Chem. Fund.* 16: 459.

SHERWOOD, R. K., PIGFORD, R. L., and WILKE, C. R. 1975. *Mass Transfer.* McGraw-Hill, New York.

SHIMUZU, Y., SHIMODERA, K. I. and WATANABE, A. 1993. *J. Ferment. Bioeng.* 76: 493.

SIMS, K. A. and CHERYAN, M. 1986. *Biotechnol. Bioeng. Symp.* 17: 495.

SMITH, M. H. 1970. In *Handbook of Biochemistry*, 2nd edition, edited by H. A. Sober. CRC Press, Cleveland, OH.

TEJAYADI, S. and CHERYAN, M. 1988. *Appl. Biochem. Biotechnol.* 19: 61.

THOMAS, C. R., NIENOW, A. W. and DUNNILL, P. 1979. *Biotechnol. Bioeng.* 21: 2263.

TRETTIN, D. R. and DOSHI, M. R. 1981. In *Synthetic Membranes: Hyperfiltration and Ultrafiltration Uses*, A. F. Turbak (ed.), American Chemical Society, Washington, DC.

TREYBAL, R. E. 1981. *Mass Transfer Operations.* McGraw-Hill, New York.

TYN, M. T. and GUSEK, T. W. 1990. *Biotechnol. Bioeng.* 35: 327.

VILKER, V. L., COLTON, C. K., SMITH, K. A. and GREEN, D. L. 1984. *J. Membrane Sci.* 20: 63.

VIRKAR, P. D., NARENDRANATH, T. J., HOARE, M. and DUNNILL, P. 1981. *Biotechnol. Bioeng.* 23: 425.

WALTERS, R. R., GRAHAM, J. F., MOORE, R. M. and ANDERSON, D. J. 1984. *Anal. Biochem.* 140: 190.

WARREN, R. K., MacDONALD, D. G. and HILL, G. A. 1991. *Process Biochem.* 26: 337.

WELSCH, K., McDONOGH, R. M., FANE, A. G. and FELL, C. J. D. 1995. *J. Membrane Sci.* 99: 229.

WIJMANS, J. G., NAKAO, S. and SMOLDERS, C. A. 1984. *J. Membrane Sci.* 10: 115.

WOERNER, D. L. 1983. Ph.D. thesis, University of Washington.

WOLF, W. J. and BRIGGS, D. R. 1959. *Arch. Biochem. Biophys.* 85: 186.

YAN, S. H., HILL, C. G. and AMUNDSON, C. H. 1979. *J. Dairy Sci.* 62: 23.

YEH, H. M. and WU, H. H. 1997. *J. Membrane Sci.* 124: 93.

YOUNG, M. E., CARROAD, P. A. and BELL, R. L. 1980. *Biotechnol. Bioeng.* 22: 347.

ZAHKA, J. and MIR, L. 1977. *Chem. Eng. Progr.* 73 (12): 53.

ZYDNEY, A. L. and COLTON, C. C. 1986. *Chem. Eng. Commun.* 47: 1.

CHAPTER 5

Equipment

Industrial users of membrane technology have a choice of six basic designs of equipment: (1) tubular, with inner channel diameters >4 mm; (2) hollow fibers, with inner diameters of 0.2–3 mm; (3) plate units; (5) spiral-wound modules; (5) pleated-sheet cartridges; and (6) rotary modules. In some cases, designers have attempted to use sound hydrodynamic principles and complex mathematical models to optimize the performance of their hardware. Their main constraints have been economics, manufacturing techniques, or sometimes tradition. In this chapter we will describe the various types of equipment available in the market and focus on those aspects that distinguish one type from another. No attempt has been made to exhaustively cover all equipment made by all manufacturers. Any descriptions, photographs, data, and specifications of a particular membrane module are mentioned here only to illustrate features of a particular design. Mention of a brand name should not be construed as an endorsement of that product, nor does it necessarily imply that those not mentioned are unsuitable for a particular membrane application. Equipment development has been done almost exclusively by the manufacturers themselves. Thus, this chapter is based on information provided by equipment manufacturers, as well as the author's own experiences with some of the equipment. A partial listing of manufacturers and their addresses is given in Appendix A.

Pleated sheet cartridges (Figure 5.1) are used mostly for dead-end microfiltration. They work well in their intended applications, and they are relatively uncomplicated in their design and use (Porter 1990; Shucosky 1988; Swiezbin et al. 1996). The focus here is on cross-flow designs, and thus pleated cartridges are not discussed.

5.A.
LABORATORY-SCALE DEVICES

Laboratory (bench-top) devices provide a means of rapidly screening membranes for a particular application or for handling small volumes. The simplest and smallest of the membrane separation units is shown in Figure 5.2. These are essentially dead-end filtration devices (as defined in Figure 4.4) to be used

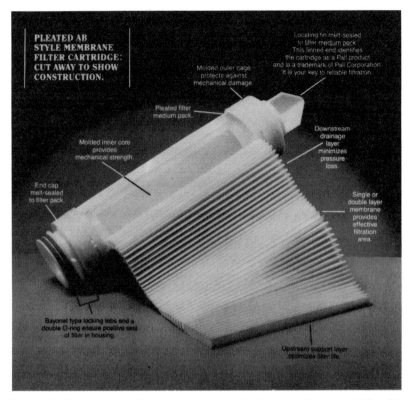

Figure 5.1. Typical direct flow dead-end pleated sheet membrane cartridge filter *(Source: Pall Corporation, with permission).*

in laboratory centrifuges, with the driving force being provided by centrifugal forces. The Centrifree concentrator from Amicon and the Microsep from Pall Filtron are for volumes of 0.15–3.5 mL; similar devices are available for volumes up to 15 mL. The sample to be ultrafiltered is placed in the upper chamber, which contains the membrane, and the whole unit is placed in a fixed-angle rotor centrifuge. Spin times vary from 15–60 min for 10,000 to 100,000 molecular weight cut-off (MWCO) membranes and up to 2 h with 3000 MWCO membranes for desalting biological samples. The Centriflo membrane cone shown in Figure 5.2 is for samples up to 7 mL and is presently available with 25,000 and 50,000 MWCO membranes. The Centrisart devices from Sartorius are similar, with the membrane filter insert made of cellulose acetate or polyethersulfone in a choice of pore sizes. The rigid outer tube may be made of polypropylene. Smaller devices such as the Microcon and Micropure units from Amicon and the Nanosep from Pall Filtron are also available for samples of 50–500 μL. Cost of

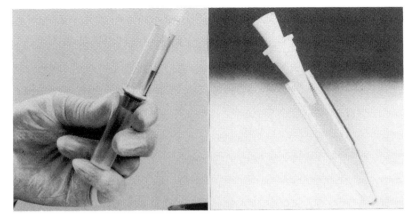

Figure 5.2. *Laboratory-scale UF/MF equipment for extremely small volume samples: left: "Microsep" micropartition device from Pall Filtron for volumes of 0.5–3.5 mL; right: Amicon's "Centriflo" membrane cone shown being inserted into a 50-mL centrifuge tube. The sample to be ultrafiltered is placed in the cone. After centrifugation, the permeate collects in the centrifuge tube and the retentate is retained in the cone.*

these centrifugal devices is $0.90 each for Micropure 0.22-μ units, $1.50 each for the Nanosep, $2.10–$2.30 each for Centricon and Macrosep concentrators, and $9 each for Centriflo cones (all U.S. prices in 1995).

The major use for these small dead-end units is for rapid ultrafiltration (UF) or microfiltration (MF) of small volumes that may be used in clinical and analytical applications, for in vitro diagnostic use and for determining ligand–macromolecule binding parameters as described elsewhere (Cheryan 1986; Cheryan and Saeed 1989). However, being dead-end devices, there is no way to control concentration polarization, and thus they are best used with relatively clean and dilute samples. No flux data can be obtained with these devices.

Larger laboratory devices are shown in Figures 5.3 and 5.4. These are also dead-end cells but have a means for controlling polarization by agitation of the fluid. This is done with a magnetic stirring bar placed as close as possible to the membrane surface. These are commercially available (e.g., from Amicon/Millipore, Pall Filtron) with volumes of 25 mL to 2 L; the common sizes of 100–400 mL cost $400–$800 for the cell assembly alone (shown in Figure 5.3), depending on the size and the manufacturer. Figure 5.4 shows how these cells are used. The cell is placed on a magnetic stirrer, and a pressure source such as a gas cylinder is connected to the cell. If diafiltration or continuous feed is needed, a reservoir is placed between the gas cylinder and the cell via a three-way valve. The cell assembly can be placed in a water bath and then placed on top of the magnetic stirrer if necessary. As will be seen later, agitation results in a vast improvement in polarization control, and flux is much higher.

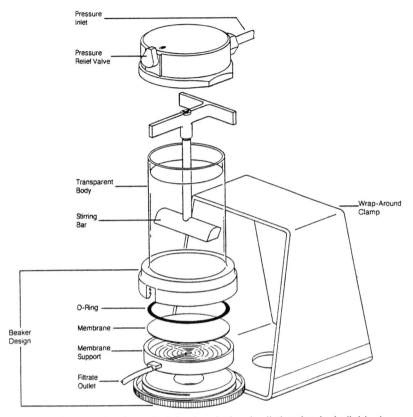

Figure 5.3. Schematic of a typical dead-end stirred cell showing its individual components (Source: Amicon catalog, with permission).

One further step up the polarization control ladder are the thin-channel devices (Figure 5.5), where the retentate is pumped through narrow channels or slits on top of the membrane at high shear rates using a peristaltic pump. The retentate is returned to the feed reservoir. Pressure still has to be provided with an external source such as a gas cylinder. The TCF-10 system shown in Figure 5.5 has a reservoir volume of 600 mL and costs about $2500. It can also be adapted for diafiltration.

For even larger volumes (1–10 L) and more realistic performance data, the devices shown in Figure 5.6 can be used. The basic reservoir–peristaltic pump combination can be connected either to small spiral-wound cartridges (16 cm long, 0.09 m^2) or to hollow fiber cartridges (e.g., H1 series, 20 cm long, 0.03–0.06 m^2), depending on the fittings chosen. The hardware for the units shown in Figure 5.6 costs $3100–3500. The S1 spiral-wound cartridge from Amicon Corporation, available with YM membranes of 1000–100,000 MWCO, is $330

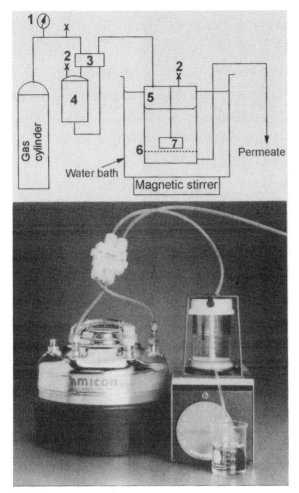

Figure 5.4. Top: Flow diagram for the use of a dead-end stirred cell (1 = pressure gauge; 2 = pressure release valve; 3 = three-way valve; 4 = reservoir for feed; 5 = stirred cell; 6 = membrane; 7 = stirrer bar). Bottom: Amicon stirred cell system showing the reservoir, the three-way valve, and the 200-mL model 8200 stirred cell, which is on top of a magnetic stirrer (adapted from Amicon catalog, with permission).

each. The hollow fiber modules, which are another embodiment of thin-channel devices, are $300 each and made of polyethersulfone. Good scaleup data can be obtained with these devices, since they simulate the cross-flow design of industrial modules.

Figure 5.7 shows some more laboratory-scale modules. The pump capacities needed tend to be higher for these modules, especially the tubular modules.

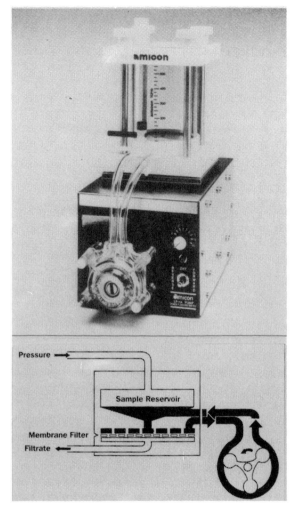

Figure 5.5. *Amicon thin-channel cell for laboratory applications. The cell assembly is shown on top of a peristaltic pump, which recycles the process fluid through shallow spiral channels in a plate placed just above the membrane surface to minimize concentration polarization effects (adapted from Amicon catalog, with permission).*

However, these modules can provide even more realistic data than the stirred cell devices, if operated in the appropriate manner.

Figure 5.8 shows a comparison of the performance of dead-end unstirred cells, stirred cells, and thin-channel devices for the MF of turbid bovine serum. Thin-channel devices (labeled "TCF" in the graph) performed better than the other devices, especially when operated in the recirculation mode as shown

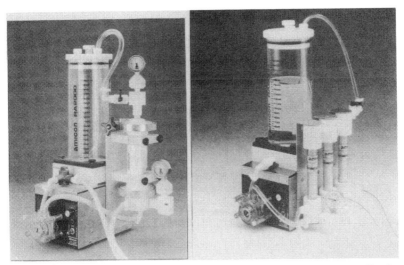

Figure 5.6. Cross-flow membrane equipment for bench-top laboratory applications (model CH2): left: spiral-wound cartridge (Amicon's S1 type) coupled to peristaltic pump and 2-L feed reservoir; right: Amicon's H1 type hollow fiber cartridges in a similar CH2 system.

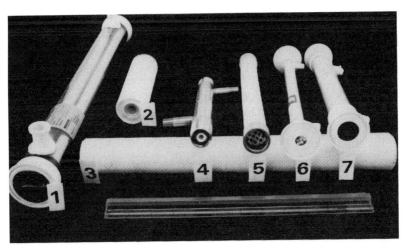

Figure 5.7. Laboratory-scale cross-flow membrane modules. 1 = Koch tubular (HFM-180-TOS, 0.05 m²); 2 = Hoechst spiral (5SFXABU, 0.1 m²); 3 = Koch spiral (S2-HFK131-FYV, 0.18 m²); 4 = USFilter single tube ceramic (1T1-70-125-LI, 0.027 m²); 5 = CeraMem LMC with 4-mm channels (0.06 m²); 6 = A/G Technology TurboTube (CFP-5-K-4XT2C, 0.046 m²); 7 = A/G Technology hollow fiber module (UFP-5-C5, 0.28 m²).

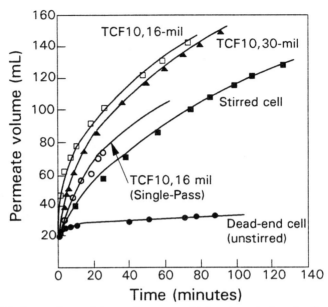

Figure 5.8. *Comparison of the performance of laboratory devices for the ultrafiltration of turbid bovine serum. Pressure was 25 psig (1.72 bar) (adapted from Porter 1979).*

in Figure 5.5. The TCF10 cell performed better when the channel height was reduced from 30 mils (0.76 mm) to 16 mils (0.4 mm). This is because the shear rate increased (with the same recirculation rate) when the channel height was reduced [see Equations (4.38) and (4.39)], leading to an increase in mass transfer coefficients and flux [Equations (4.35) and (4.36)].

Flux data obtained in stirred cells are only a gross approximation of the flux that can be obtained with larger, industrial-type cross-flow modules. Polarization control is usually better in the larger devices, especially with fouling feedstreams. However, stirred cells are useful in rapidly evaluating the rejection properties of membranes under limited conditions. At the very least it will help narrow the range of membranes that have to be evaluated on a pilot scale. These devices are not recommended for evaluating the engineering parameters of a membrane process.

5.B.
INDUSTRIAL EQUIPMENT

5.B.1.
TUBULAR MODULES

Tubular modules (Figure 5.9) are among the earliest design of industrial-scale membrane equipment. Polymeric membranes are cast on the inside of porous

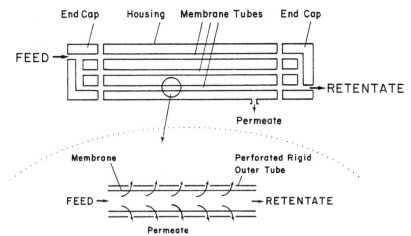

Figure 5.9. Schematic of a tubular membrane designed for ultrafiltration applications.

paper or plastic inserts with internal diameters ranging from 0.5″ (12.5 mm) to 1″ (25 mm) and lengths varying from 2 to 20 ft (0.6–6.4 m). Most ceramic modules also fit the definition of tubular modules but tend to be smaller in diameter, ranging from 2–6 mm, with lengths of individual tubes or multichannel elements up to 1.1 m (Section 2.E.) Tube diameters are essentially a compromise between the optimum size for minimum energy consumption (which may be as low as 1.0 mm diameter or less) and the cost of making membranes and support tubes in such small diameters.

Several of these tubes may be assembled in a housing in a shell-and-tube arrangement. The end-caps will determine whether the feed flows through the tubes in series [typical of reverse osmosis (RO)] or in parallel (typical for UF and MF applications, as shown in Figure 5.9). The permeate collects in the shell side of the housing and is removed through side ports under the pressure gradient from the feed side. This pressure required to remove permeate from the module should be considered in the process design. A significant back-pressure can be created in the permeate line if the permeate piping is not sized properly or if there are too many fittings and elevations in the permeate line. Excessive back-pressure should be avoided in membrane modules, not only because it reduces flux [Equation (4.6)], but it could damage membranes that are not self-supporting. This includes polymeric tubular modules, spirals, and plates. On the other hand, most inorganic modules and hollow fibers can withstand high back-pressures, which is an advantage in MF applications (see Section 6.E.4.).

Figures 5.10 and 5.11 show cross sections of typical tubular modules. Figure 5.12 shows how a 1″ tubular system is assembled. Table 5.1 shows specifications of Koch's tubular membranes.

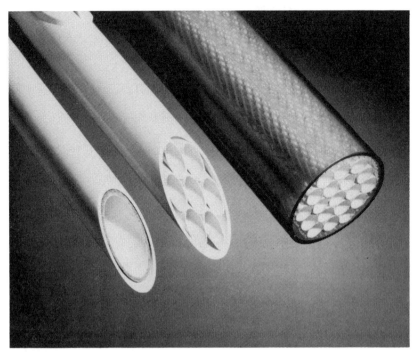

Figure 5.10. Tubular UF/MF module designs available from Koch: left to right: 1" diameter tube, ULTRA-COR module with seven tubes of 0.5" diameter, SUPER-COR module with 19 tubes of 0.5" diameter each.

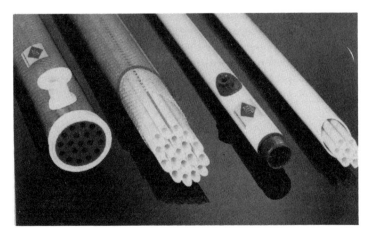

Figure 5.11. Tubular modules from AMT (MAG-19 and MAG-7).

180

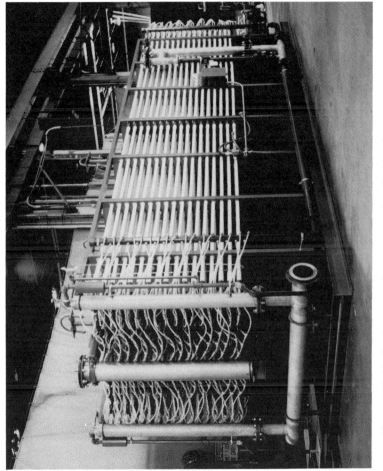

Figure 5.12. Koch tubular membrane system installed in a food waste application.

Table 5.1. Specifications of Koch tubular UF and MF membranes and modules.

Module	Module Diameter (cm/inches)	Length (m/ft)	Membrane Area per Module (m^2/ft^2)
Tubular (1″ diameter single tube)	2.5/1.0	1.5/5	0.10/1.1
Tubular (1″ diameter single tube)	2.5/1.0	3.0/10	0.2/2.2
ULTRA-COR (7 tubes, 0.5″ dia. each)	3.8/1.5	3.0/10	0.68/7.4
SUPER-COR (19 tubes, 0.5″ dia. each)	7.6/3.0	3.0/10	2.2/24.0

Membrane[*]	NMWCO[**]	Maximum Temperature (°C) at pH 6	Operating pH at 25°C	Maximum Pressure (bar/psi)
HFM-100	10,000	90	1–13	10.5/150
HFA-251[a]	15,000	50	2–8	10.5/150
HFM-180	18,000[d]	90	1–13	10.5/150
HFM-183[c]	50,000	90	1–13	10.5/150
HFP-276[b]	35,000	90	1–13	10.5/150
MSD-400[a]	100,000	90	1–13	10.5/150
MSD-181	200,000	90	1–13	10.5/150
MSD-405[a]	250,000	90	1–13	10.5/150
MFK-617	0.3 μ	90	1–13	10.5/150
MMP-406[a]	0.2 μ	90	1–13	10.5/150
MMP-404[a]	0.4 μ	90	1–13	10.5/150
MFK-615	1.2 μ	90	1–13	10.5/150
MMP-516[a]	2 μ	90	1–13	10.5/150
MMP-407[a]	2–3 μ	90	1–13	3.5/50
MMP-600	1–2 μ	90	1–13	3.5/50
MMP-602	2–3 μ	90	1–13	3.5/50

[*]All tubular membranes made of PVDF except MFK-615 and MFK-617, which are PES
[**]Nominal molecular weight cut-off
[a]Hydrophilic membrane; [b]Anionic; [c]Cationic; [d]18,000–180,000, depending on application

The PCI tubular system (Figure 5.13) uses a slightly different approach. The membrane is cast on the inside of a synthetic paper tube while this tube is being continuously manufactured in a helical-spiral design. Membrane formulations are slightly different than those used for manufacture of flat sheets of the same membrane to allow for the stretching that takes place when the tubes are pressurized during operation. These paper membrane inserts fit inside perforated stainless steel support tubes, which are about 14 mm in diameter. This results in an inner (flow channel) diameter of 12.5 mm. In the B1 module of PCI, 18 of these tubes are mounted together in a shell-and-tube arrangement in a housing using end-caps that can link the tubes, either in series or parallel (Figures 5.9 and 5.13). The end-caps are retained by studs screwed into tube plates, which hold the tubes in place. A specially designed, crevice-free rubber seal is used

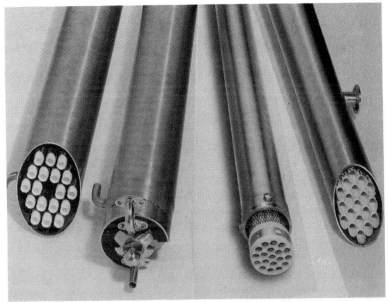

Figure 5.13. *PCI tubular modules: left: The B1 module cross section showing tubular membrane inserts in individual permeate stainless steel tubes; center left: B1 module with end-cap: permeate off-take pipe and feed entry pipe are visible; center right: the A19 module showing the multitube core with retaining bolts; right: A19 cross section; permeate port can be seen.*

to seal the ends of the membrane inserts and the end-caps. Series flow (i.e., the feed enters one of the 18 tubes and flows in series through all 18 tubes) is used when low flow rates can be used, as in RO applications. Parallel flow is more common for UF applications. A combination of series and parallel flow ("twin-entry") is also used, where the flow goes through two tubes in parallel at the same time and then in series through the remaining tubes, i.e., for a total of nine passes through a module.

The A19 design of PCI does not use individual stainless steel support tubes; instead, the 19 tubes are bunched together and cast in epoxy resin at each end. This bundle is then placed in a stainless steel housing (Figure 5.13). The permeate from the individual tubes collects in the shell and is removed through side connections to a central permeate header.

Figure 5.14 shows the arrangement of modules of a typical multistage UF plant (modules are usually horizontal for tubular membranes because of their long lengths). Table 5.2 lists UF and MF tubular membranes available from PCI. Inorganic membranes, which are usually tubular in design, are shown in Figures 5.15 and 5.16. Specifications of some tubular modules are listed in Table 5.3.

Figure 5.14. PCI UF plant at Borregaard Industries, Norway, treating 1400 tonnes per day of spent calcium sulfite liquor.

Table 5.2. Specifications of PCI tubular UF and MF membranes and modules.

Module	Module Diameter (cm/inches)	Length (m/ft)	Membrane Area per Module (m²/ft²)	Maximum Pressure (bar/psi)
B1 (parallel flow and twin-entry)	10/3.9	1.2/4	0.9/10	16/230
	10/3.9	2.4/8	1.7/18	16/230
	10/3.9	3.6/12	2.6/28	16/230
A19 (parallel flow)	8.3/3.25	1.52/5	1.0/11	7/100
	8.3/3.25	3.05/10	2.1/22	7/100
	8.3/3.25	3.66/12	2.5/26.5	7/100

Membrane (and Material)	NMWCO	Maximum Temperature (°C) at pH 6	Operating pH at 25°C	Maximum Pressure (bar/psi)
CA202 (CA)*	2,000	30	2–7	25/360
ES404 (PES)	4,000	80	2–12	30/450
PU608 (PS)	8,000	80	2–12	30/450
ES209 (PES)	9,000	80	2–12	30/450
PU120 (PS)	20,000	80	2–12	15/225
AN620 (PAN)*	25,000	60	2–10	10/150
ES625 (PES)	25,000	80	2–12	15/225
FPA10 (PVDF)	100,000	60	2–10	7/100
FP100 (PVDF)	100,000	80	2–12	10/150
FPA20 (PVDF)	200,000	60	2–10	7/100
FP200 (PVDF)	200,000	80	2–12	10/150
V4000 (PVC)	200,000	50	2–12	10/150
L6000 (FP)	200,000	60	2–12	10/150

* = Hydrophilic membrane
CA = cellulose acetate; PAN = polyacrylonitrile; PES = polyethersulfone; PS = polysulfone; PVC = polyvinyl chloride; PVDF = polyvinylidene flouride; FP = fluoropolymer

5.B.1.a.
CHARACTERISTICS OF TUBULAR MEMBRANES

1. Owing to the relatively large channel diameters, tubular units are capable of handling feed streams and slurries containing fairly large particles. As a general rule of thumb, the largest particle in the feed should be less than one-tenth the channel height. Thus, feedstreams containing particles as large as 1250 μm can be processed in 0.5″ tubular units, while the 1.0″ units can handle particles as large as 2500 μm (particle size is expressed in terms of its largest dimension).

Table 5.3. Specifications of tubular membranes from other manufacturers.

Manufacturer and Module Name	Module Diameter (cm)	Tube Length (m)	Tube Inner Diameter (mm)	Membrane Area per Module (m²)	Membranes Available	Temperature, Pressure and pH (°C/bar)
Diacel/Hoechst						
Molsep MH12	11.4	1.25	14.1	1.0	PAN (10 K, 20 K, 40 K)	45°C/10 bar, pH 2–12
Molsep MH25	11.4	2.5	14.1	2.0		
Membrane Products Kiryat Weizmann	10.1	1.2	12.5	0.9	PS (20 K)	80°C, 15 bar, pH 0.5–14
	10.1	3.6	12.5	2.6	PAN (25 K)	50°C, 15 bar, pH 1–10
Nitto Denko	11.4	3.87	11.5	2.3	PO, PS, PI PP (8 K, 20 K, 100 K)	60°C, 10 bar, pH 2–13

PAN = polyacrylonitrile; PES = polyethersulfone; PI = polyimide; PO = polyolefin; PS = polysulfone; PP = polypropylene; K = thousands MWCO

Figure 5.15. Multiple-stage ceramic membrane system made up of Membralox modules (courtesy of Niro Hudson).

187

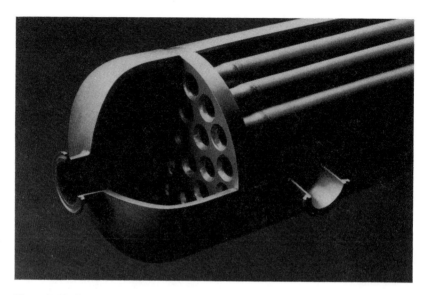

Figure 5.16. *Scepter stainless steel membrane module (courtesy of Graver Separations).*

2. Tubular units with these diameters are operated under turbulent flow conditions with Reynolds numbers usually greater than 10,000. Recommended velocities for UF are 2–6 m/sec, depending on the module (ceramic membranes are operated at the higher velocities). This will result in flow rates of 15–60 L per minute per tube, depending on the tube diameter. For a PCI or Koch tubular module, this means a pump capacity of 16–34 m³/h [70–150 gallons per minute (gpm)] per module. For ceramic membranes, this may require 300 gpm for the CeraMem PMA industrial module and for the USFilter 19P19-40 module.

3. For polymeric 0.5″ tubes of 8–12 ft in length, pressure drop averages 2–3 psi per tube at these velocities. This is also the pressure drop for the module if all tubes are in parallel flow. If the individual tubes are connected in series, typical pressure drops for 0.5–1.0″ polymeric tubes will be about 30–40 psi (2–2.5 bar). For twin-entry modules from PCI, pressure drop may be 15–20 psi per module. For ceramic membranes, pressure drop may be 15–25 psi, depending on the channel diameter and fluid properties. In general, this combination of pressure drop and high flow rates makes tubular modules the highest in energy consumption among all the module types (see Section 7.F. for calculations).

4. The straightforward open tube design and the high Reynolds numbers makes it easy to clean by standard clean-in-place (CIP) techniques. It is also possible to insert scouring balls or rods to help clean the membrane.

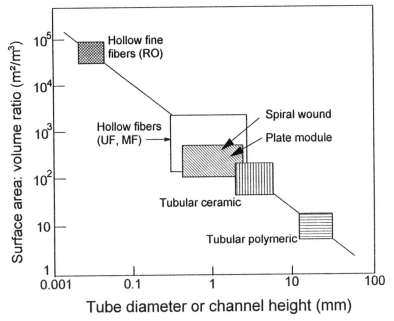

Figure 5.17. Relationship between channel size and surface area:volume ratio of membrane modules.

5. In the PCI B1-type design, the individual membrane can be replaced in the field, resulting in considerable savings in transportation and membrane replacement costs. The newer A19 design allows the entire 19-tube core to be easily removed by loosening retaining screws in the housing for replacement.

6. Tubular units have the lowest surface area-to-volume ratio of all module configurations (Figure 5.17). This results in high floor space requirements. Holdup volume within these units are also high. For the PCI A19 module (8.3 cm diameter × 3.66 m long, 2.5 m² area), holdup is about 8 L in the tube side plus about 6 L in the shroud side. This may limit the degree of concentration that can be achieved during UF or MF when processing relatively small volumes.

7. Tubular designs are relatively expensive. System capital costs for polymeric tubular systems are $1300–1500 per square meter of membrane area. Replacement membrane costs are $8–16 per square foot of membrane area ($85–170/m²) for PCI membranes made of cellulose acetate; $125–170/m² for polysulfone (PS), polyethersulfone (PES), and polyvinylidene fluoride (PVDF); and $280–700/m² for Koch tubes of 1″ diameter. The inorganic membranes have an even higher range of costs. Capital costs are $2200/m² to $6000/m². Membrane replacement costs are $500/m² to about $3000/m².

As inorganic membranes become established and more manufacturers arise, competition will bring down the prices of the more expensive inorganic membranes, but probably never as low as polymeric membranes.

5.B.2.
HOLLOW FIBERS

The hollow fiber membrane configuration is essentially a tubular membrane, but with one important difference: the membrane is in the form of a self-supporting tube with no separate support or backing. For UF and MF applications, the feed is pumped through the inner core of the tube. [In contrast, the DuPont hollow "fine" fibers, which are used for RO applications, the feed is pumped in from the shell (outer) side and permeate removed through the inside of the fiber.] The dense skin layer is on the inside of the tube (Figure 5.18), except for the Microza hollow fibers marketed by Asahi Chemical Industry (Japan) and Pall, which have skins on both the inside and outside of the fiber (Figure 5.19). This presumably confers additional sturdiness to the membrane. Early hollow fibers (e.g., from Amicon and Romicon) had the classic asymmetric structure with macrovoids in the substructure. On the other hand, fibers from A/G Technology (Figure 5.18) and Microgon do not possess macrovoids but are more dense and uniform. The Dow SelectFlo and the X-flow capillary membrane show a fairly smooth surface and consistent pore size and spacing and appears to have a high surface and bulk porosity.

Hollow fibers are available in a wide range of diameters from 8–120 mil (0.2–3 mm). The fibers have a cross-sectional thickness of about 100–400 μ. Bundles containing 50–3000 individual fibers (depending on the diameters of the fibers and the cartridge shell) are sealed into hydraulically symmetrical housings in a shell-and-tube arrangement and bonded on each end with epoxy. The cartridge housings/shells may be constructed of clear see-through PS, translucent polyvinyl chloride (PVC), or stainless steel (Figures 5.20–5.23). Each cartridge is provided with a process (feed) inlet, process outlet, and a pair of permeate outlets on either end of the cartridge. These fittings mate with the appropriate adapters and manifolds for connection to process piping.

Tables 5.4 and 5.5 list specifications of various hollow fiber cartridges available commercially. Many industrial cartridges are 3″ (7.6 cm) in diameter and are available in several lengths, from 18 cm to 1.2 m; some are shown in Figures 5.20–5.23. The longest modules are used primarily for dilute, non-fouling streams that operate in the pressure-dependent, flow rate–independent regime. The shorter cartridges are used for viscous fluids and where very high final concentrations are required. These applications may require high cross-flow rates and corresponding high pressure drops, which may not be possible given the limited pressure ratings of most hollow fibers. In addition, the laminar

Figure 5.18. *Ultrastructure of the cross section of hollow fibers: top: asymmetric structure with macrovoids typical of Romicon and Amicon hollow fibers; bottom: dense microporous structure of A/G Technology hollow fibers (courtesy A/G Technology).*

flow mass transfer model [Equation (4.36)] suggests that shorter lengths should lead to higher flux. Figure 5.24 is a typical industrial installation containing several hollow fiber cartridges.

Hollow fibers used for MF and UF should not be confused with those used for RO, such as the DuPont B-9 and B-10 permeators (Figure 5.25). The RO hollow fibers have a much smaller diameter (40 μ, as thin as a human hair), they can withstand much higher pressures of 600–900 psi, and the feed is pumped

Table 5.4. Specifications of Romicon hollow fiber modules.

Module diameter, inches (cm)	1 (2.5)		2 (5.1)	3 (7.6)	5 (12.7)
Maximum inlet pressure, psi (bar)	75 (5.1)		75 (5.1)	100 (7)	100 (7)
Maximum TMP, psi (bar)	30 (2)		30 (2)	40 (2.8)	40 (2.8)
Cartridge length, inches (cm)	7 (18)	18 (45.7)	25 (63.5)	43 (109)	43 (109)
Membrane areas* (ft²)	0.2–0.6	0.6–2	3.6–10	16–53	40–132
Recirculation flow* (gpm)	1.3–7	1.3–7	5–21	10–50	24–120

Romicon Hollow Fiber Membranes Available

Membranes	NMWCO	Fiber Diameters (mm)
PM1	1,000	1.1
PM2	2,000	1.1
PM5	5,000	1.1
PM10	10,000	0.5, 1.1, 1.5
PM30	30,000	1.1
PM50	50,000	0.5, 1.1, 1.5
PM100	100,000	0.8, 1.1
PM500	500,000	1.1, 1.9, 2.7
PMF0.1	$0.1\ \mu$	1.1

*Depending on fiber diameter

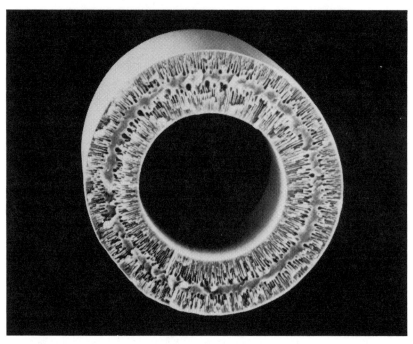

Figure 5.19. Ultrastructure of a Microza hollow fiber membrane, showing the double skin (courtesy of Pall Corporation and Asahi Chemical Industry Co. Ltd.).

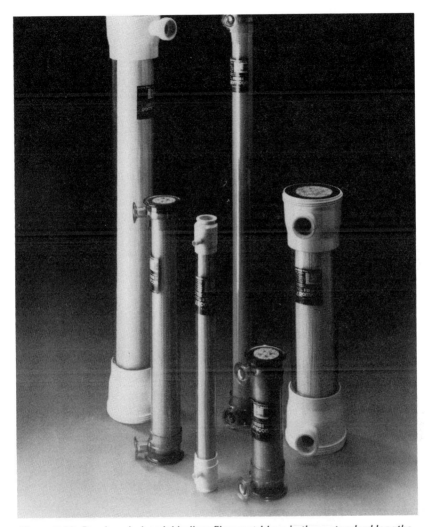

Figure 5.20. *Romicon industrial hollow fiber cartridges in three standard lengths.*

through the shell side (outside of the fibers) with permeate being removed from the inside (tube side).

5.B.2.a.
CHARACTERISTICS OF HOLLOW FIBER MODULES

1. Recommended operating velocity in hollow fibers is 0.5–2.5 m/sec, resulting in Reynolds numbers of 500–3000. Many hollow fibers thus operate in the

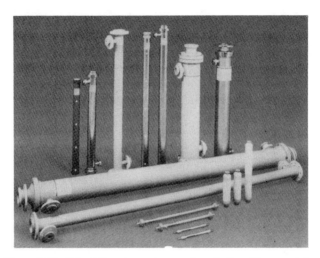

Figure 5.21. *Enka-Microdyn polypropylene hollow fiber modules.*

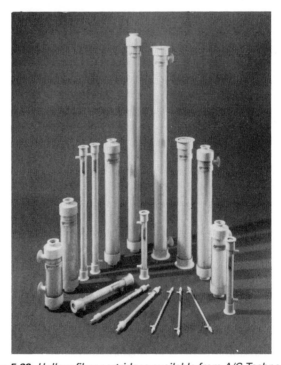

Figure 5.22. *Hollow fiber cartridges available from A/G Technology.*

194

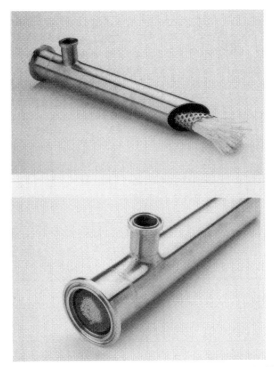

Figure 5.23. Capillary membrane cross-flow microfiltration module (SelectFlo) from Dow.

laminar flow region. Flow rates through 3″ diameter cartridges with 0.2–1.1 mm fibers are usually 10–50 gpm (40–190 L/min). Pressure drops in the various cartridges are typically 5–20 psig (0.3–1.4 bar), depending on flow rates, fiber diameters, and lengths. This combination of pressure drops and flow rates makes hollow fibers one of the more economical in terms of energy consumption (Section 7.F.). However, in some cases (e.g., the RomiPro 5″, 75-mil fiber cartridge and the A/G Technology Maxcell 4.25″ 3-mm fiber module), flow rates required can be up to 120–160 gpm with pressure drops of 13–20 psi. In these latter cases, the flow is probably turbulent. In any case, it is more important to keep the pressure drop within recommended limits rather than try and achieve a certain flow rate that may result in exceeding fiber pressure ratings.

2. Shear rates are very high in hollow fibers because of the combination of thin channels and high velocity. Shear rates at the wall (γ_w) are 2000–16,000 sec^{-1}.

Table 5.5. Hollow fibers from various manufacturers.

Manufacturer	Tradename	Membrane Material	Pore Sizes Available	Fiber Diameters (mm)	Fiber Lengths, cm (and areas, m²)	Temperature and Pressure Ratings	Comments
A/G Technology	Maxcell, MaxiFiber, TurboTube, Xampler, Xpress	Polysulfone	3 K, 5 K, 10 K, 30 K, 100 K, 300 K, 500 K, 0.1 μ, 0.2 μ, 0.45 μ, 0.65 μ	0.25, 0.5, 0.75, 1.0, 2.0, 3.0	18–110 cm (0.002–13 m²)	50–80°C, 1–2 bar (MF) 1.7–3.3 bar (UF)	Some models autoclavable or steam-in-place
Akzo	Microdyn	Polypropylene (ACCUREL)	0.1 μ	0.6	56.0 (2.2)	60°C, 2 bar	Autoclavable 10 times
			0.2 μ	0.6	32.5 (1.0)	60°C, 2 bar	
				0.6	56.0 (2.2)	"	
				1.8	50.0 (1.0)	60°C, 1.7 bar	
				1.8	100 (2.0 or 10)	"	
				5.5	75 (0.036)	60°C, 3 bar	
				5.5	150 (1.0)	"	
				5.5	300 (2 or 8)	"	
		SPES	0.2 μ	1.0	32.5 (1.1)		
				1.0	65 (2.2)		
		RC	10 K	0.2	22 (2.8)		Autoclavable 3 times
				0.2	44 (5.8)		

(continued)

196

(continued)

Table 5.5. Hollow fibers from various manufacturers.

Manufacturer	Tradename	Membrane Material	Pore Sizes Available	Fiber Diameters (mm)	Fiber Lengths, cm (and areas, m²)	Temperature and Pressure Ratings	Comments
Amicon	ProFlux	PES	P3 (3 K), P10 (10 K), P30 (30 K), P100 (100 K)	0.51	20.3 (0.06) 63.8 (0.9 or 2.8) 109.2 (5)	50°C, 1.7 bar	Module diameters 2.3, 4.5, 7.6 cm. H1 cartridges (2.3 × 20.3) are used in CH2 apparatus shown in Figure 5.6.
			P10 (10 K), P30 (30 K), P100 (100 K), MP01 (0.1 μ)	1.1	20.3 (0.03) 63.8 (0.45 or 1.4) 109.2 (2.4)		
Asahi Kasei/Pall	Microza	PAN	6 K, 13 K, 50 K	0.8 0.8 0.8 0.8 1.4 1.4 1.4 1.4	13 (0.017) 34.7 (0.2) 55.2 (1) 112.9 (4.7) 13 (0.012) 34.7 (0.1) 55.2 (0.6) 112.9 (3.1)	50°C, 1 bar 50°C, 5 bar 50°C, 5 bar 50°C, 5 bar 50°C, 1 bar 50°C, 5 bar 50°C, 5 bar 50°C, 5 bar	
			13 K				
		Polysulfone	3 K, 6 K, 10 K	Same as above	Same as above	80°C, same pressures as above	Hot water sanitization

(continued)

(continued from previous page)

Table 5.5. Hollow fibers from various manufacturers.

Manufacturer	Tradename	Membrane Material	Pore Sizes Available	Fiber Diameters (mm)	Fiber Lengths, cm (and areas, m^2)	Temperature and Pressure Ratings	Comments
Dow/FilmTec	SelectFlo	PES-PVP	0.2 μ, 0.5 μ	1.5	104 (1.0 or 9.3)	80°C, 1.4 bar	Similar to X-Flow
	CMF			3.0	104 (0.6 or 5.1)	40°C, 2.7 bar	
Hoechst/Diacel	Molsep	PES	10 K, 30 K, 150 K, 300 K	0.8	36, 41 or 113 (0.26, 1.4 or 5.3)	70 or 98°C 3 bar	Autoclavable under water pressure
			30 K	0.5	36, 41 or 113 (0.53, 2.2 or 8.2)	70 or 98°C 3 bar	
		PAN	30 K	1	36, 41, or 113 (0.16, 1.16 or 4.3)	45°C 3 bar	
			150 K	1.4	36, 41 or 113 (0.11,0.8 or 3.0)	45°C 3 bar	
		CA	150 K	0.8	36, 41 or 113 (0.26, 1.43 or 5.3)	35°C 2 bar	
Microgon	Krosflo	Mixed cellulose ester	0.1 μ, 0.2 μ	0.6	23 (0.5 or 1.0)	82°C, 2 bar	Autoclavable once
	Dynafibre			0.6	46 (2.1)	"	
				0.6	69 (3.3)	"	
			0.2 μ	1.0	23 (0.4 or 0.7)	"	
				1.0	46 (1.5)	"	
				1.0	69 (2.4)		
Nitto-Denko	NTU-3000 NTM-9002	Polysulfone Polypropylene	UF (20 K) MF	0.55, 1.1	107 (3.4–6.3)	95°C, 7 bar	
X-Flow		PES-PVP	50 K–0.8 μ	0.5–4	101.5 (2.6–180)	80°C, 10 bar	

CA = cellulose acetate; PAN = polyacrylonitrile; PES = polyethersulfone; PVP = polyvinylpyrrolidone; RC = regenerated cellulose; SPES = sulfonated polyethersulfone

3. Among the modules, hollow fibers have the highest surface area-to-volume ratio (Figure 5.17). Holdup volume is low, typically a total of 2 L in a 7.6 × 63 cm cartridge with 1–3 m^2 area.

4. Hollow fibers have low pressure ratings: usually the maximum transmembrane pressure is 25 psig (1.8 bar), although there are some that can withstand higher pressures at the appropriate temperatures (Tables 5.4 and 5.5). In general, the shorter modules can withstand higher pressures at lower temperatures. This low pressure rating is not a problem in most MF applications, since low transmembrane pressures generally lead to better long-term performance (Sections 6.D. and 6.E.). However, it may be a disadvantage in some UF applications. There are several process streams that are dilute enough to permit UF operation at pressures much higher than the present 25 psig limiting transmembrane pressure (e.g., see Figures 4.5–4.7). In addition, since the cross-flow rate is proportional to pressure drop, flow rates are sometimes limited because the inlet pressure cannot exceed the fiber rating, which may be as low as 25 psig. Thus, it may be difficult to maintain high flow rates with highly viscous solutions, especially with the longer modules.

5. The small tube diameters make the fibers somewhat susceptible to plugging at the cartridge inlet. To prevent this, the feed should prefiltered to at least one-tenth the fiber inner diameter. This means prefiltration to 100 μm if Romicon 43-mil fibers are being used and down to 20 μm with Akzo's Microdyn 0.2-mm regenerated cellulose (RC) fibers.

6. A big advantage with the hollow fibers is its backflushing capability, which improves its cleanability. This is possible because the fibers are self-supporting. Other methods of minimizing fouling during processing include "lumen-flushing," where the permeate flow is shut off for brief periods during operation, and "permeate pressurization," i.e., maintaining a high permeate back-pressure during operation (Section 6.E.). This self-supporting nature has also enabled the development of continuous plug-flow bioreactors for enzymatic and microbial conversions (see Section 8.N.4.b.).

7. Replacement membrane costs are high. Even if one single fiber out of the 50–3000 in a bundle bursts, the entire cartridge has to be replaced. (Fiber repair kits are available. The leaking fiber is first identified by a modified bubble point test—not an easy task—and the two ends of the fiber are sealed by injecting an epoxy resin with a syringe.)

8. Cost: A large hollow fiber plant (including pumps, controls, fittings, etc.) for UF or MF may cost $1000–1500/m^2 of installed area. Replacement hollow fiber modules are about $600/m^2. Laboratory bench-top modules, such as the H1 series from Amicon (shown in Table 5.5 and Figure 5.6), the A/G Technology CFP-1-K-4X2TC, and UFP-5-C5 cartridges (shown in Figure 5.7), are $300–375 each.

Figure 5.24. *Typical Romicon hollow fiber industrial plant (with 24 cartridges).*

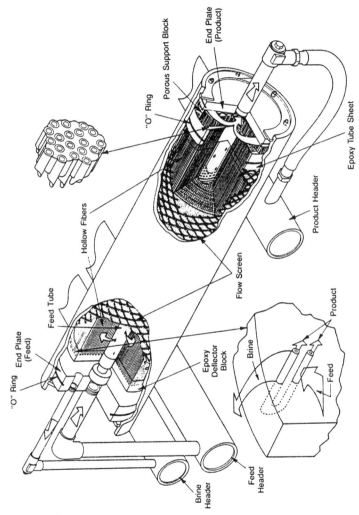

Figure 5.25. *DuPont hollow (fine) fiber module used for reverse osmosis.*

201

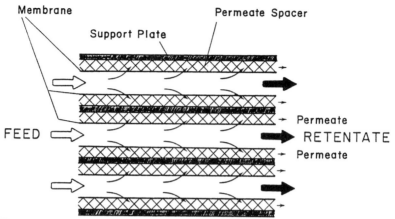

Figure 5.26. Schematic of typical plate type membrane module. The feed channel may also contain a spacer (not shown).

5.B.3.
PLATE UNITS

The plate-type membrane modules were among the earliest configurations in the market. As shown schematically in Figure 5.26, a plate module consists of a rigid, flat plate on which a flat sheet of the membrane is placed. A "scrim," or net-like material, is placed between the membrane sheet and the plate to provide the channel for permeate flow. The sheet is then sealed around the edges, but with a provision for removal of the permeate (usually by a tube). Another membrane sheet and permeate spacer is placed on the other side of the plate. Several of these plates are stacked on top of each other (or arranged next to each other, depending on whether it is a vertical or horizontal stack). Adjacent membrane plates may be separated from each other with another spacer (the feed channel spacer, usually much coarser and thicker than the permeate channel spacer) or by ridges or stubs on the periphery of the plate. Several of these plates are clamped together to form a module or cartridge.

Figure 5.27 shows a plate-and-frame unit manufactured by Tech-Sep (Rhône-Poulenc Group, France), which is of rectangular design. Their PLEIADE UFP 70 and UFP 71 modules are made up by assembling several membrane holding plates between two separating plates to form a subassembly. Several of these subassemblies are fitted together on a frame to result in modules of 0.1–100 m^2 in membrane area. The membrane holding plates have transverse grooves on their faces, onto which the membrane is placed and held by two pairs of sealing rings and a sealing gasket on the periphery of the plate. The grooves serve a dual role: (1) They help to drain the permeate towards the collection channels on their edge of the plate, from which the permeate is removed via two nipples

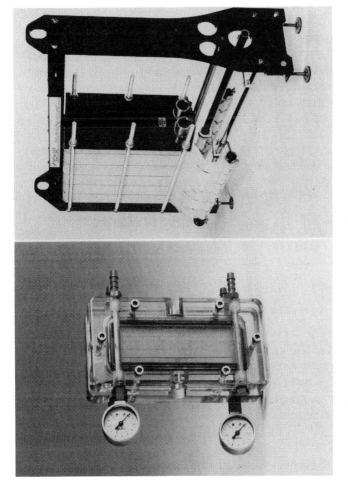

Figure 5.27. Plate systems from Tech-Sep/Rhône-Poulenc: left: Laboratory-scale test unit for flat sheets (Pleiade Rayflow, 0.01 m² with holdup of 44 mL); right: industrial plate-and-frame module (PLEIADE types UFP10-70-71-75-100) (courtesy of Applexion).

203

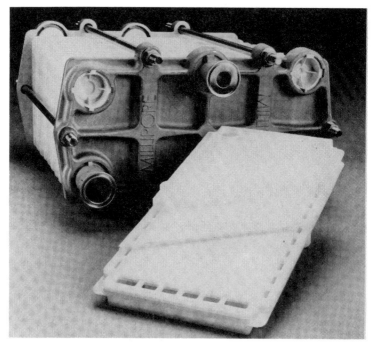

Figure 5.28. Millipore Prostak UF open-channel module.

on each side, and (2) the grooves increase the turbulence in the feed channel. Each plate is 0.11 m^2 in surface area (for the UFP-10 pilot module) or 0.35 m^2 (for the UFP-70, -71, -75, and -100 modules). Feed/retentate channel height is 1.5 mm. The holdup (dead) volume is about 1.3 L/m^2 of membrane area. Maximum pressure for the module is 85 psig (6 atm). Pressure drops are about 0.7 bar per subassembly for a recirculation rate of 50–60 L/min. Table 5.6 shows specifications of several commercial flat-sheet membranes.

The thickness of the feed spacer or the stubs/ridge determine the feed channel height. For example, the Millipore Prostak (Figure 5.28) has a nominal channel height of 21 mils (0.5 mm) and is available with no feed spacer (the Prostak Open-Channel for MF) or with a polypropylene feed spacer (Prostak UF module). The plates are approximately 39 × 19 cm, with an effective membrane area of about 0.09 m^2 each. The stacks are available with 0.17–1.9 m^2 of membrane area, depending on the number of plates per stack. Holdup is about 1–1.5 L/m^2 of membrane area. MF membranes available include PVDF Durapore and PS (0.1–0.65 μ) and UF membranes include PES (10 K–300 K) and RC (1 K–300 K). Operating pressures are 50–60 psi at temperatures of 50°C or less. Prefiltration to at least 150 μ is recommended.

Table 5.6. Specifications of plate systems from various manufacturers.

Manufacturer	Trade Name	Membrane Material	Pore Sizes Available	Areas (m²)	Temperature and Pressure Ratings	Comments
Dow/FilmTec	FilmTec	CA	20 K	Available in sizes to fit any plate modules	50°C, 10 bar	Some membrane designations same as former DDS membranes
		PS	5 K, 10 K, 20 K, 25 K, 50 K, 100 K, 0.1 μ, 0.2 μ, 0.45 μ, 1 μ, 2 μ		75°C, 10 bar (UF), 5 bar (MF)	
		Fluoropolymer	10 K, 20 K, 50 K, 100 K, 0.1 μ, 0.2 μ, 0.45 μ, 1 μ, 2 μ		60°C, 10–15 bar (UF), 5 bar (MF)	
		RC	10 K		60°C, 10 bar	
Hoechst	Nadir	PES	4 K, 20 K, 30 K, 50 K	Available in sizes to fit any plate modules	110°C, 15–40 bar	Available in labscale cassette design (CELTAN)
		PS	100 K		110°C, 10 bar	
		Polyaramide	5 K, 20 K		110°C, 10 bar	
		Cellulose	10 K, 30 K, 100 K		70°C, 10 bar	
		CA	1 K		35°C, 20 bar	
North Carolina SRT, Inc.	Dualport, Quadport	C, CA, PES, PVDF	1 K, 2 K, 4 K, 5 K, 8 K, 10 K, 20 K, 30 K, 40 K, 50 K, 60 K, 70 K, 100 K, 200 K, 250 K, 300 K, 500 K	0.09–93 m², depending on channel height (0.125–1 mm available)		Autoclavable 10 times; Labscale devices available (OPTISEP, CONSEP)
		CA, Nylon, PES, PS, PP, PTFE, PVDF	0.1–0.8 μ			

(continued)

205

(continued from previous page)

Table 5.6. Specifications of plate systems from various manufacturers.

Manufacturer	Trade Name	Membrane Material	Pore Sizes Available	Areas (m²)	Temperature and Pressure Ratings	Comments
Pall Filtron	Minisette, Centramate, Centrasette, Maximate, Maxisette	PES, modified PES, PVDF	1 K, 3 K, 5 K, 10 K, 30 K, 50 K, 70 K, 100 K, 300 K, 500 K, 1000 K, 0.16 μ, 0.3 μ, 0.5 μ	5–10 m²	70–121°C, 4.1–5.2 bar	Some membranes autoclavable
Sartorius	Sartocon II, Sartoflo, Microsart, Ultrasart	CTA, PES, PS	1 K, 5 K, 10 K, 30 K, 50 K, 100 K, 300 K, 0.2 μ	Channel heights of 0.4 mm or	50°C, 2–4 bar	Lab-scale device (MiniUltrasart, EasyFlow, Sartocon-Mini) with 0.005–0.5 m² available
		CA	0.2 μ, 0.45 μ	0.6 mm; 0.3–0.7 m² per module	35°C, 4 bar	
		Polyolefin	0.1 μ, 0.2 μ		121°C, 4 bar	
Membrane Products Kiryat-Weizmann Ltd		PS PAN	20 K 20 K	Available in sizes to fit any plate modules	40°C, 15 bar 50°C, 15 bar	

(continued)

(continued from previous page)

Table 5.6. Specifications of plate systems from various manufacturers.

Manufacturer	Trade Name	Membrane Material	Pore Sizes Available	Areas (m²)	Temperature and Pressure Ratings	Comments
Millipore	Prostak—Open channel	PVDF (PP and PZ series)	0.1, 0.22, 0.45, 0.65 μ, 200 K	0.17–1.9 m² per cassette	50–80°C 2.1–4.2 bar	Autoclavable. 21-mil (0.5 mm) channel, no spacers
	Prostak—UF	PS, PES (PT series, Biomax), RC (PL series)	5 K, 8 K, 10 K, 30 K, 50 K, 100 K, 300 K	0.39–1.9 m² per cassette	55°C, 5.6 bar	21-mil (0.5 mm) feed spacer
	Pellicon	RC, PES	1 K, 3 K, 5 K, 10 K, 30 K, 100 K, 300 K,	0.1–2.3 m² per cassette	50°C, 4 or 7 bar	10-, 20- and 30-mil screens/spacers.
		PVDF	0.1 μ, 0.2 μ, 0.45 μ			100-μ prefiltration
	Pellicon 2	PES (Biomax series), RC	5 K, 8 K, 10 K, 30 K, 50 K, 100 K, 300 K	0.1–2.5 m² per cassette	50°C, 7 bar	
Tech-Sep	IRIS MF	PVDF	0.1, 0.2, 0.4, 0.8, 1.5 μ	0.1–95.2		
	IRIS 30–	SPS	5 K, 10 K, 20 K, 40 K, 100 K			
		PVDF	40 K, 200 K			
		Acrylic	20 K, 40 K, 50 K			
		PES	3 K, 10 K, 30 K, 100 K			

C = cellulose; CA = cellulose acetate; CTA = cellulose triacetate; PAN = polyacrylonitrile; PES = polyethersulfone; PP = polypropylene; PS = polysulfone; PTFE = polytetrafluoro ethylene; PVDF = polyvinylidene fluoride; RC = regenerated cellulose

Figure 5.29. Plate systems from Pall Filtron: top: Maxisette and Centrasette membrane cassettes; bottom left: Centrasette and Centramate modules that can accommodate 0.5–9 m² of membrane area; bottom right: Maxisette AT-50 system.

The Millipore Pellicon 2/Pellicon cassette system works on essentially the same flat-sheet principle, except that there is also a polypropylene mesh spacer (retentate separator screen) in the feed channel. The cassette system can withstand relatively high pressures of 100 psig and is operated under turbulent flow. These cassettes are loaded on "holders" made of stainless steel or acrylic. The cassettes are available with areas of 0.093 to 2.5 m² each, with holdups as low as 500 ml/m². The Pellicon minicassette bench-top system (0.1–0.3 m²) has a very low holdup, about 100 ml/m^2 (low holdup maximizes the recovery of the products).

The Pall Filtron (Figure 5.29) and the North Carolina SRT plate systems are also available with and without separator screens in the feed channel. The ends of the plate are sealed on the permeate side so that the permeate can be directed towards exit ports at the bottom or on the side of the stack.

De Danske Sukkerfabrikker (DDS) plate-and-frame systems were developed at the end of the 1960s and were widely used in the food and dairy industries in the 1970s and 1980s (Cheryan 1986). They are no longer manufactured, but there are still many systems in use. Replacement membranes can be obtained from several flat-sheet membrane manufacturers around the world.

5.B.3.a.
CHARACTERISTICS OF PLATE UNITS

1. Channel heights of most commercial plate units are 0.5–1.0 mm, although some narrower channel units are available. The channel length (the distance between the inlet and outlet ports on a plate) is 10–60 cm. Superficial Reynolds numbers are in the laminar flow region. However, if a screen is present in the feed channel, it is more likely to be in turbulent flow. Screens/spacers can contribute considerably to the degree of turbulence, as shown in Figure 5.30. For example, with the 0.8 mm channel, flux increased from 12 liters per square meter per hour (LMH) to 60 LMH at a superficial velocity of 65 cm/sec (shear rate of 7800 sec^{-1}) when the spacer was inserted. It was observed, however, that pumping energy per unit of permeate flow doubled with the spacer (due to the "parasitic drag" induced by the spacer). But considering the substantial increase in flux, the capital cost and overall operating cost (including membrane replacement and depreciation) would be less with the spacer (Levy and Earle 1994).

2. The permeate from each plate (which contains a pair of membranes, one on either side) can be visually observed in those designs that have plastic tubing coming from each plate (e.g., Figure 5.27). This is convenient for several reasons: for detection of leaks in a particular membrane pair, if samples need to be taken for analysis, or if several different membranes need to be evaluated simultaneously during testing.

3. Replacement of membranes on-site is relatively easy. Care must be taken, however, when closing the stack of plates together again that the previously embedded grooves of the unreplaced plates match exactly as they were previously, or else leakage of feed can occur.

4. Plate systems are intermediate between tubular and spiral units in terms of packing density (Figure 5.17), in energy consumption, and in cost. Flat-sheet membranes themselves are available for $30–40/m^2 for cellulose acetate and $50–100/m^2 for other polymers and composite membranes; however, in a plate system, the plant cost can be $500–$1000/m^2.

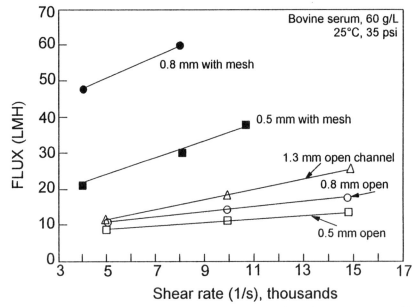

Figure 5.30. Effect of mesh spacers on flux. Amicon Ioplate plates were used with the PM-10 membrane. The open channels were formed with support rails along the outer edges. Spacers were made of polypropylene with a square design of nine squares per inch and strand thickness one-half of the channel height. Shear rates at the wall are based on superficial velocities assuming the flow channel is a slit, neglecting the volume occupied by the spacers (adapted from Levy and Earle 1994).

5.B.4.
SPIRAL-WOUND

This is one of the most compact and inexpensive designs available today. A simplified sketch of the basic concept is shown in Figure 5.31. Like the plate designs discussed earlier, these membrane elements are designed around flat sheets. Two flat sheets are placed together with their active sides facing away from each other. They are separated from each other by a thin, mesh-like spacer material (usually made of Vexar or polypropylene) and are glued together on three sides (Figure 5.32). The fourth open side is fixed around a perforated center tube. Another mesh-like spacer of the required thickness (the feed channel spacer) is placed on one side of this envelope and the whole assembly rolled around the center tube in a spiral or "jelly-roll" configuration. The feed is pumped lengthwise along the unit, while the permeate is forced through the membrane sheets into the permeate channel and spirals towards the perforated center collection tube (Figure 5.33). Multileaf designs are used to

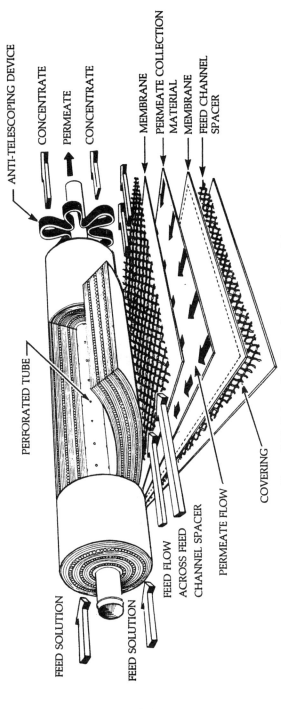

ANTI-TELESCOPING DEVICE

CONCENTRATE

PERMEATE

CONCENTRATE

MEMBRANE

PERMEATE COLLECTION MATERIAL

MEMBRANE

FEED CHANNEL SPACER

PERFORATED TUBE

COVERING

PERMEATE FLOW

FEED FLOW ACROSS FEED CHANNEL SPACER

FEED SOLUTION

FEED SOLUTION

Figure 5.31. Schematic of the spiral-wound module.

Table 5.7. Sizes of spiral-wound membrane modules available.

Diameters of modules (in inches)	1, 2, 2.5, 4 (actually 3.8, 3.9 or 4.3), 6 (5.6, 5.8, 6, 6.3 or 6.4), 8 (7.9, 8, 8.2 or 8.3), 12
Lengths (inches)	6, 12, 16, 22, 25, 33, 35, 38, 39, 40, 60
Permeate tube diameters (inches)	0.675, 0.773, 0.83, 1.1–1.2, 1.625
Feed channel spacers (mil)	22, 24, 26, 28, 30, 31, 38, 43, 45, 46, 48, 50, 60, 65, 71, 80, 90, 100, 120
Membrane area in 3.8″ × 38″ module (Fluid Systems)	26-mil spacer = 81 ft^2 31-mil spacer = 73 ft^2 46-mil spacer = 59 ft^2 60-mil spacer = 45 ft^2 80-mil spacer = 38 ft^2
Membrane area in 6″ × 38″ module (Koch)	28-mil spacer = 210 ft^2 43-mil spacer = 175 ft^2 80-mil spacer = 105 ft^2
Membrane area in 6.3″ × 38″ module (Fluid Systems)	31-mil spacer = 216 ft^2 46-mil spacer = 173 ft^2 80-mil spacer = 116 ft^2
Membrane area in 8.2″ × 40″ module (AMT)	22-mil spacer = 464 ft^2 31-mil spacer = 405 ft^2 48-mil spacer = 319 ft^2 80-mil spacer = 215 ft^2

$1'' = 2.54$ cm; 1 mil $= 1/1000'' = 0.0254$ mm

increase the membrane area without unduly increasing the length of the feed channel or permeate flow path: several layers of the membrane sandwiches are placed on top of each other and rolled together.

Typical finished spiral modules are shown in Figure 5.34. The membrane module is placed inside PVC or stainless steel housings fitted with the appropriate end-caps and manifolds. Each housing (or "pressure vessel") may contain several membrane assemblies in series (Figure 5.35). Figure 5.36 shows a typical spiral-wound installation. Tables 5.7 and 5.8 are dimensions and other specifications of spiral-wound modules currently available.

The space between the periphery of the membrane element and the pressure vessel housing is a potential troublespot. Some spiral elements are "full-fit" designs that expand slightly during use to seal the annular space. However, to facilitate removal and replacement of the membrane element, it is not practical to have a very tight fit of the module in its housing. To minimize feed bypass with these looser designs, one approach is to place a "product seal" around the module at one end. This, however, could lead to cleaning and sanitation problems caused

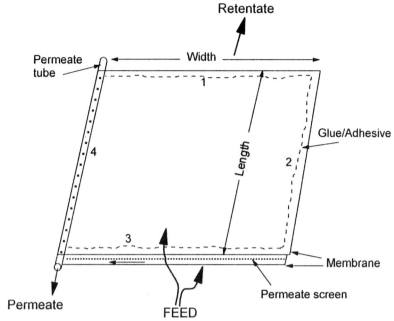

Figure 5.32. *Schematic of an unrolled spiral-wound element. Two flat sheet membranes are placed together with a permeate spacer (screen) between them. Sides 1, 2, and 3 are glued together. Side 4 of the membrane sandwich is glued to the perforated permeate collection tube. A feed channel spacer (not shown) is placed on top of the membrane sandwich before the assembly is rolled into the final spiral form. Feed is pumped across both sides of the membrane sandwich.*

by the dead space behind the product seal. The product seal could be perforated with a few holes to allow a small portion of the feed (usually less than 5%) to continuously flush the annular space to prevent stagnation of the fluid. It is more common to have a mesh-like spacer placed around the module periphery to allow restricted feed bypass through the annular space to keep it clean.

5.B.4.a.
CHARACTERISTICS OF SPIRAL-WOUND MODULES

1. In terms of fluid flow characteristics, spiral-wound modules are basically flat sheets arranged on top of each other to form several narrow slits for fluid flow. The feed channel height is controlled by the thickness of the spacer in the feed channel. Several spacer thicknesses and designs are available, ranging from 22 mil (0.56 mm) to 120 mil (3 mm). Narrow channel spacers are used with clear feedstreams, with little or no suspended matter (e.g., water treatment). The advantage of a narrow channel height is that much more membrane area can be packed into a given pressure vessel. For example, a

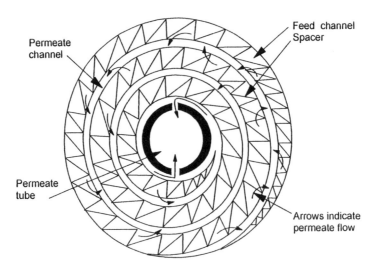

Figure 5.33. Cross section of a spiral-wound element, showing the spiral path taken by the permeate to the central permeate collection tube. Feed is pumped in the feed channel in a direction perpendicular to the plane of the paper. Permeate channel spacer is not shown.

Figure 5.34. Typical spiral-wound modules (2″, 4″, 8″, and 12″). The length of these modules is 40″. The 4″ and 8″ elements are shown with a built-in antitelescoping device (courtesy Osmonics).

214

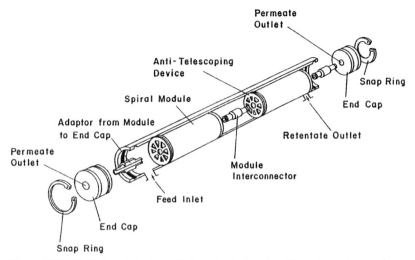

Figure 5.35. *Multiple spiral elements in a single housing. Two elements are shown, but up to four 40" elements could be placed in series in a single housing for MF and UF applications depending on the pressure drop. Six elements in series are used for RO applications. Permeate tubes of each element are connected by module interconnectors.*

3.8″ diameter × 38″ long module will have 75 sq ft (7 m^2) with a 30-mil spacer, but only about 38 sq ft (3.5 m^2) of an 80-mil module (Table 5.7).

2. Spiral modules operate in turbulent flow. The velocity in the feed channel is calculated by dividing the volumetric flow rate by the cross-sectional area. The cross-sectional area is $a \times b$ for slits, where a is the width of the flat sheet (Figure 5.32) minus the glued portion, and b is the channel (spacer) height. On that basis, the velocity in spiral-wound units range from 10 to 60 cm/sec, being higher for the larger mesh spacers. These are "superficial" velocities, however, since the volume occupied by the mesh-like spacer in the feed channel is neglected. These velocities correspond to Reynolds numbers of 100–1300. Technically, this is in the laminar flow region, but the additional turbulence contributed by the spacers, which can be substantial as shown earlier in Figure 5.30, should also be taken into account.

The state of turbulence can be easily determined by the nature of the relationship between pressure drop (ΔP) and flow rate (Q) in the feed channel. The general relationship is

$$\Delta P = Q^n \tag{5.1}$$

For laminar flow, $n = 1$. For turbulent flow, on the other hand, $n = 1.5–1.9$ (Section 7.F.). Figure 5.37 shows such a relationship for hollow fiber and

Figure 5.36. Typical multistage MF plant using spiral-wound modules being assembled. The first two stages (nearest the tank) have six housings each. Each housing contains three 8" × 33" modules in series. The last two stages have four housings each. Each stage has its own recirculation pump (courtesy Koch Membrane Systems).

216

Table 5.8. Specifications of spiral-wound membranes from various manufacturers.

Manufacturer	Trade Name	Membrane Material	Pore Sizes Available	Temperature and Pressure Ratings	Comments
Advanced Membrane Technology	AMT	PS	10 K, 40 K, 200 K	90°C, 10 bar	
		PES	2 K, 5 K, 10 K, 30 K, 60 K, 250 K, $0.1\,\mu$, $1\,\mu$, $3\,\mu$	90°C, 10 bar	
		PVDF	2 K, 5 K, 10 K, 30 K, 70 K, 500 K, $0.1\,\mu$, $3\,\mu$	90°C, 10 bar	
		PVDF (+ charge)	100 K	70°C, 10 bar	
Amicon		RC (YM series)	3 K, 10 K, 30 K, 100 K	55°C; Pressures: S1, S3, S10 = 4.1 bar; S40 = 8.6 bar	All with 28/34 mil spacers; S1 ($2'' \times 6''$; 1 ft²) S3 ($2'' \times 12.6''$; 3 ft²) S10 ($2.7'' \times 12.5''$; 10 ft²) S40 ($5'' \times 12''$; 40 ft²)
		PS (N series)	1 K	55°C; 5 bar	S1 and S10 sizes
Desalination Systems, Inc. (DSI)	Desal-CA	CA	10 K	35°C, 7 bar	
	G-series	Thin-film composite	1 K, 2.5 K, 3.5 K, 8 K, 9 K, 15 K	50°C, 14-28 bar	
	E-series, F-series	PS	10 K, 0.01 μ, 0.04 μ	50–90°C, 7 bar	
	J-series	PVDF	$0.04\,\mu$, $0.40\,\mu$	50–90°C, 7 bar	J- and K-series membranes can be treated to be either hydrophilic or hydrophobic
	K-series	PTFE	$0.02\,\mu$, $0.1\,\mu$, $0.2\,\mu$, $0.45\,\mu$, $1\,\mu$, $3\,\mu$	50–90°C, 1.4 bar	

217

(continued)

(continued from previous page)

Table 5.8. Specifications of spiral-wound membranes from various manufacturers.

Manufacturer	Trade Name	Membrane Material	Pore Sizes Available	Temperature and Pressure Ratings	Comments
Dow Chemical	FilmTec	PVDF	50 K	60°C, 10 bar	30-mil spacer = 80 ft^2 48-mil spacer = 60 ft^2
		PS	20 K	60°C, 10 bar 80°C, 10 bar	30, 48 and 80-mil spacers, 3.8″ and 6.3″ diameters; 33″ and 38″ lengths
		Fluoropolymer	10 K, 20 K, 50 K, 100 K	60°C, 10–15 bar	
Fluid Systems		PES	6–10 K	75°C, 5 bar	10 bar at 25°C
Hoechst	Spira-Cel	All NADIR membranes (see Table 5.6)			2″× 6″; 2.4″ × 40″; 4″ × 26″; 4″ × 40″; 4.2″ × 33″; 6.3″× 38″; 8″ × 40″
Hydranautics	Hydrapaint	Polyolefin		45°C, 6 bar	4″ × 40″ (55 ft^2); 8″ × 40″ (250 ft^2)
Koch	SpiraPro	PES, PVDF	5 K, 10 K, 30 K, 250 K, 0.1 μ, 0.2 μ, 1.2 μ	80°C, 7–9 bar	<0.35 bar permeate pressure, 0.7–1.4 bar pressure drop
Millipore	Helicon UF	RC (PL-series)	1 K, 3 K, 5 K, 10 K, 30 K, 300 K	60°C, 6.8 bar	Module sizes: 2″ × 12″ (3 ft^2) 3′ × 11″ (15 ft^2) 3.9″ × 26″ (40 ft^2)
		PES (PT-series)	10 K, 30 K, 100 K		

(continued)

218

(continued from previous page)

Table 5.8. Specifications of spiral-wound membranes from various manufacturers.

Manufacturer	Trade Name	Membrane Material	Pore Sizes Available	Temperature and Pressure Ratings	Comments
Osmonics	SEPA	S-series	5 K, 10 K, 100 K	60°C, 6.8 bar, pH 2–8	Free chlorine <2 ppm
		J-series	8 K	60°C, 6.8 bar, pH 2–9	Free chlorine <3 ppm
		M-series	2 K, 10 K, 500 K, 0.1 μ	80°C, 6.8 bar, pH 2–12	Free chlorine <0.1 pm
		H-series	5 K, 10 K, 50 K, 120 K	60°C, 13–27 bar, pH 0.5–13	Free chlorine <25 ppm
		A-series	5 K, 20 K	80°C, 13 bar, pH 1–11	Free chlorine <25 ppm
		R-series	10 K, 100 K	100°C, 13–27 bar, pH 0.5–10	Free chlorine <25 ppm
		Y-series	0.02–0.2 μ	100°C, 13 bar, pH 0.5–13	Free chlorine <25 ppm
		W-series	0.2–2 μ	100°C, 13 bar, pH 1–10	Free chlorine <50 ppm

CA = cellulose acetate; PES = polyethersulfone; PS = polysulfone; PTFE = polytetrafluoroethylene; PVDF = polyvinylidene flouride; RC = regenerated cellulose.

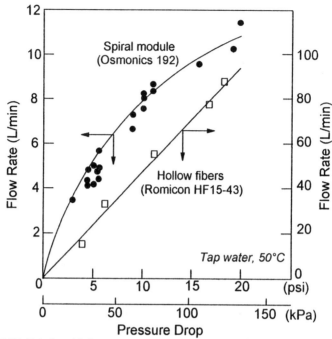

Figure 5.37. *Relationship between pressure drop (ΔP) and flow rate in the feed channel (Q) for hollow fibers (Romicon module, 3″ × 25″, 43-mil fibers) and spiral-wound modules (Osmonics 2″ × 35″, 30-mil spacers). Feed was tapwater at 50° C (adapted from Cheryan and Chiang 1984).*

spiral-wound modules. The former clearly operates in laminar flow, as evidenced from normally prevailing Reynolds numbers (Section 5.B.2.) and the linear ΔP–Q relationship. The spiral module shows a departure from linearity, indicating that the flow is in the turbulent region. Thus, turbulent flow models such as the Chilton-Colburn relationship (suitably modified) will do better predicting flux in spiral modules.

3. Pressure drops in the feed channel are relatively high because of the parasitic drag exerted by the spacer. As shown in Figure 5.37, the pressure *drop* $(P_i - P_o)$ in an Osmonics 192 module is 15–20 psig for a flow rate of 12 L/min (equivalent to a superficial velocity of 25 cm/min). This pressure drop can result in a telescoping effect at high flow rates; i.e., the spiral pushes itself out in the direction of flow. This can damage the membrane. Antitelescoping devices (ATDs) shown in Figures 5.34 and 5.38 are used at the downstream end of the membrane element to prevent this (in practice, it is better to have ATDs at both ends of spiral modules). Even with ATDs, the pressure drop in spirals is limited to 10–20 psi (0.7–1.4 bar) to prevent damage to the module.

In general, smaller feed spacers can tolerate less pressure drop. For example, the Koch 4″ × 33″ module with 28-mil spacers is limited to 15 psi pressure drop (for a corresponding flow rate of 25 gpm with water at 25°C). However, up to 20 psi is allowed with the 43-mil spacer (27 gpm) and the 80-mil spacer (58 gpm). The latter modules in the longer 38″ length will develop up to 24 psi with the same flow rates.

4. One of the frustrating aspects of using spiral-wound modules is the vast number of ATDs and permeate outlet tube sizes and designs available from different manufacturers (Figures 5.38 and 5.39). The feed and retentate ports usually come in at right angles to the housing, as shown in Figure 5.35 (extra length must be provided in the housing for these ports). The permeate exits the housing through one or both end-caps. The connection between the permeate collection tube of the module and the permeate outlet pipe of the housing is where the design differences occur. In those modules where the permeate collection tube extends past the membrane (e.g., industrial designs such as the 2″ and 12″ modules in Figure 5.34), the ATD is placed directly on the extended part of the permeate collection tube, flush against the module. Another pipe fits onto the end of the collection tube, via an O-ring or washer. This latter pipe is usually welded onto the end-cap.

In sanitary "flush-cut" modules, the permeate carrier tube is recessed. In this case, a tube (or a combination tube and ATD) is inserted into the permeate collection tube, the ATD (if separate) is placed on this extension tube flush against the spiral, and the end-cap–pipe combination fits onto the end of the extension tube. It is also possible to have all three (extension tube, ATD, and end-cap) welded together as one fitting, as shown in Figure 5.38(4). Some designs have the ATD separate from the end-cap of the housing and separate from the permeate collection tube. Others (e.g., the Osmonics industrial end-cap shown in Figure 5.39) have a combination feed/retentate port and end-cap and separate ATD.

The Amicon end-caps are opposite in design to the others (Figure 5.39): the feed and retentate ports are aligned axially to the module and the permeate outlet piping makes a right-angle turn after it leaves the housing. All others have these ports coming in at right angles to the module's longitudinal axis, as shown in Figures 5.35, 5.36, and 5.39 (top).

In addition, otherwise identical modules from different manufacturers may have permeate collection tubes of different diameters (Table 5.7). This makes it difficult to substitute one manufacturer's spirals with another's, unless the end-cap/ATD designs are the same (stainless steel end-cap/ATD fittings can cost $200+ each). The sanitary design used in the United States at this time uses flat washers to connect the ATD with the end-cap fitting and the module [Figure 5.38(2)].

5. A potential problem with the mesh spacers in the feed channel is the creation of dead spots directly behind the mesh in the flow path (this is also true of

Figure 5.38. Some of the end-cap assemblies and antitelescoping devices used with spiral modules. (1) From Osmonics: left: end-cap; right: pipe adapter and ATD (they are separate fittings). One end of the pipe adapter with double O-rings fits into the end-cap, and the other end fits into the permeate tube of the module. The ATD (shown) fits on the pipe adapter. Bottom is a damaged ATD from a module that had been subjected to excessive pressure drop. (2) From Niro Hudson: top left is the end cap. Top middle is an ATD plug used to plug the upstream end of the module's permeate tube. Top right is a stainless steel pipe adapter. Bottom row (left) is a plastic ATD from Koch and (right) from Niro Hudson, which are used to connect the end-cap to the module or as a module interconnector. Note the flat washers used in this design for sanitary applications. (3) From Hoechst: left: combined end-cap and permeate tube adapter; right: module interconnector and stainless steel ATD. These fittings use double O-rings instead of washers. (4) One-piece combination end-cap–ATD from Filtration Engineering. The pipe extension with the single O-ring fits directly into the permeate tube of the module. The plastic fittings are from Desalination Systems (two types of ATD using flat washers and a pipe adapter).

plate modules with feed channel spacers or screens). Suspended particles may hang up in the mesh network, which could result in partial blockage of the feed channel (Figure 5.40). In contrast, tubular, hollow fiber, and some plate modules with no feedspacers provide an unobstructed open channel and are preferred for feeds with suspended particles. Spiral modules work best on relatively clean feedstreams with a minimum of suspended matter. The general rule of prefiltering to one-tenth the channel height must be modified for the spiral module because of the presence of the spacer, which

Figure 5.39. Top: Industrial housing adapters from Osmonics; left: feed entry/concentrate port collar (vitaulic style) welded on to each end of housing; right: end-cap with lugs that lock into matching reinforced metal tabs in collar. End-caps could also be retained with snap rings as shown in Figure 5.35. Bottom: Amicon spiral fittings; left: housing; middle: downstream end-cap showing the right-angle pipe that connects directly to the extended permeate tube of Amicon's spiral modules; right: feed entry end-cap and clamp (adapted from Osmonics and Amicon product literature).

reduces the free volume in the channel. Prefiltration of the feed down to 5–25 μm is recommended for the 30-mil spacer, 25–50 μm for the 45-mil spacer channel, and 100 μm for the 120-mil spacer, depending on the type of solids (e.g., sticky solids that tend to agglomerate will need more stringent prefiltration).

There have been some innovations in recent years with corrugated, or "tubular," spacers that alleviate this problem (e.g., Figures 5.41 and 5.42). The larger or special channel spacers are useful with very viscous feeds or feeds containing suspended matter, not only because of less rigorous prefiltration, but also because they will result in lower pressure drops for the

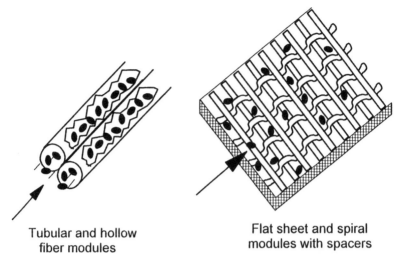

Tubular and hollow   Flat sheet and spiral
fiber modules   modules with spacers

Figure 5.40. *Flow of suspended matter through open tubular channels and through modules with spacer/screens in the feed channel. The latter is more likely to get plugged up.*

same cross-flow rate. In effect, since the pressure drop is limited to 10–20 psi, it means that the cross-flow rate will be less for narrow spacers, which in turn may result in less control of polarization effects, leading to lower mass transfer coefficients, greater fouling, lower flux, and more difficulty with cleaning. Larger spacers will allow us to keep the flow rate at the required level without exceeding the recommended pressure drop, even though it reduces the surface area per module (Table 5.7).

Figure 5.41. *Spiral modules (4" diameter) with spacers of different thickness: left: 38-mil (0.76 mm) conventional spacer. Module was sliced in the middle; middle: 50-mil (1.27 mm) spacer; right: 90-mil (2.29 mm) spacer. The two elements with antitelescoping devices are fitted with tubular spacers from Desalination Systems, Inc.*

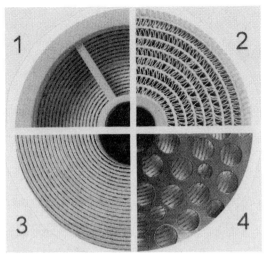

Figure 5.42. *Spacers available for special applications: (1) standard industrial design with rigid outer cover and integral ATD; (2) "tubular" spacer to provide unobstructed flow down the length of the module; (3) full-fit module with no outer-wrap or concentrate seal; (4) high-temperature, full-fit module with stainless steel ATD and stainless steel permeate tube (courtesy Osmonics).*

6. Lengths of individual membrane assemblies vary from 6″ (15 cm) for laboratory-scale devices (Figure 5.6) to 6 ft (72″) for water treatment RO modules. However, 38–40″ (1 m) seems to be the most common length for UF and MF applications. When calculating the surface area of a spiral-wound module, it is convenient to consider it as two flat sheets. For example, the Osmonics 192 module is made with two flat sheets of 37″ square. Thus, the total surface area is $2 \times 37/12 \times 37/12 = 19$ ft^2. However, a 3- to 5-cm band on three sides used for gluing the membrane sandwich and another portion used for fixing the fourth side to the permeate collection tube (Figure 5.32) are not available for permeation. This will reduce the effective membrane area of spiral modules (to 13–14 ft^2 in the case of the Osmonics 192 module). This should be taken into account in all flux and cost calculations. The number of leaves in a module should also be considered. All specifications and designs should be based on effective membrane area.

7. Surface area-to-volume ratio is fairly high, averaging about 200–300 ft^2/ft^3 (Figure 5.17).

8. This combination of low flow rates, pressure drops, and relatively high turbulence makes this module design the lowest in energy consumption (Section 7.F.).

9. Capital costs are lowest among all module designs. Only the membrane element itself needs to be replaced, unlike some other designs where the

membrane and the housing is one integral unit (spirals can also be obtained as one integral unit: this is common with RO water applications). It is possible to recover the permeate tube and the spacers from a spent spiral element. Replacement spiral membranes are priced at $3–4/ft^2 ($30–40/m^2) for cellulose acetate, $4–10/ft^2 ($40–100/m^2) for PVDF and PES membranes, and $50–120/m^2 for thin-film composites. Larger spirals (e.g., 8″) are substantially cheaper per unit membrane area than the smaller size modules. As of this writing, they could be obtained in bulk quantities with MF or UF membranes for about $800–1000 for each module, depending on the membrane.

Note, however, that these prices do not include housings (i.e., pressure vessels). Stainless steel housings, which are preferred over PVC or fiberglass reinforced plastic (FRP) housings, can cost $1350 each for a 1-module (8″ × 50″) industrial 304SS housing with Schedule 5 pipe thickness. The sanitary 316SS Schedule 40 pipe version costs $2800 each. In this case, the housing is more expensive than the module. On the other hand, 4-module housings (8″ × 170″) are $1800 for the industrial and $3300 for the sanitary version, while a 6-module (8″ × 250″) industrial version costs $2200 (these prices include two end-caps and one permeate fitting). Based on the areas available with various spacers (Table 5.7), this will bring the cost to $30–81/m^2 of membrane for 8″ modules placed four to a housing. With 4″ modules, the cost for a 1-module housing + membrane will range from $119–254/m^2 and $80–180/m^2 for 4-long housings + membranes. To obtain an approximate idea of the cost of the whole plant (in the United States in 1996) including pumps, valves, fittings, controls, etc., but without feed or product tanks, multiply the total cost of the membranes + housings by 3.

5.C.
SPECIAL MODULES

Designers of UF and MF modules should have one overriding concern: to minimize the effects of concentration polarization and fouling. Figure 4.35 had shown several methods of minimizing concentration polarization effects. Those concerned with the direct manipulation of operating parameters of the feedstream, such as cross-flow rate, temperature, and/or pressure, were discussed in Chapter 4. Chapter 6 discusses more methods that relate to manipulation of the permeate stream (e.g., backpressurization, backwashing, backpulsing, etc.) or by causing instabilities in the feedstream. However, instead of manipulating the moving fluid, it is also possible to enhance flux by moving the membrane. This has led to module designs that are quite different from conventional equipment discussed so far.

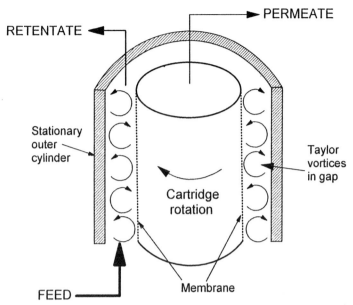

Figure 5.43. *Principle of the rotary (cylinder) module with the membrane on the inner rotating cylinder. Other designs may have the membrane on the outer stationary cylinder or on both cylinders.*

5.C.1.
ROTARY MODULES

Moving the membrane itself, as in a rotary module (Lopez-Leiva, 1979) or centrifugal module (Robertson et al., 1982), has been suggested as being an efficient way to substantially reduce effects of concentration polarization. Rotary modules were commercialized in the mid-1980s. Figure 5.43 shows the principle of the rotary module. The membrane is fitted onto a rotating cylinder (as in the Benchmark unit from Membrex) and placed inside another stationary outer cylinder. Feed is pumped through the annular gap at a low axial velocity and pressure, and permeate is withdrawn from within the cylinder. The inner cylinder rotates at 500–4000 rpm. Above a certain rpm, depending on fluid properties and physical dimensions of the unit, highly ordered fluid flow patterns called Taylor vortices are set up. These are counter-rotating vortices within the annular gap, which can generate tremendous turbulence and transport material away from the membrane surface. Interestingly, shear rates at the wall due to rotation can reach 100,000 sec^{-1}, more than 10 times that attainable in conventional cross-flow devices, yet the average shear rate for the bulk of the fluid is quite low (250–4000 sec^{-1}). This makes such systems well-suited for MF or UF of shear-sensitive materials common in biotechnology applications.

Rotary devices permit independent control of pressures and shear rate and can be operated with no significant pressure drop from feed inlet to outlet. This allows precise control of transmembrane pressure at very low values if needed. This is of special value for MF, where low pressures (<5 psi) may be needed to minimize fouling. Fluxes that are 3–20 times higher than conventional tangential flow systems are claimed, with energy consumption per unit volume permeated that may actually be lower than conventional modules in some cases. Another version of this equipment (Pacesetter Vortex Flow Perfusion unit, also from Membrex) has the membrane fitted to the stationary outer cylinder instead of the rotating cylinder. There is a third cylinder outside of the membrane cylinder to form the annulus through which the permeate flows.

The above two units from Membrex are fairly small: the Benchmark is available with membrane areas of 0.02 m^2 or 0.04 m^2 and is suitable for processing 1–75 L per day. The Pacesetter is available with 0.04, 0.125, or 0.25 m^2, suitable for up to 150–800 L per day. While these may be suitable for pilot-scale feasibility studies and perhaps for small commercial operations in the biotechnology/pharmaceutical industry, they cannot be scaled up for larger industrial-scale applications, primarily due to mechanical instability of large rotating cylinders and the poor surface area:volume ratio, i.e., low membrane packing density.

The rotating flat-plate or disc design is better suited for larger scale operations (though perhaps still not applicable at the scale required in water treatment, chemical, or many food applications). In this design, shown schematically in Figure 5.44, the membrane is cast or fitted onto a hollow polymeric plate or disc with provision for removing the permeate. The membrane disc does not rotate. Rather, solid discs are connected to a central rod and placed a few millimeters away from both sides of the membrane discs. The solid rotors rotate at speeds up to 6000 rpm, resulting in shear rates up to 400,000 sec^{-1}, higher than cylindrical devices or conventional modules. High shear rates at the membrane surface help to keep it free of suspended matter and minimizes concentration polarization. This makes it particularly useful for fine separation and otherwise difficult fractionations, e.g., separation of similarly sized proteins. However, if the membrane is truly fouled by solute–membrane interactions, rotating the membrane, even at a high rpm, will not help much (Jonsson 1993).

Two commercial units working on this principle are shown in Figures 5.45 and 5.46. The Pall Dynamic Membrane Filter (DMF) uses half-moon or quarter-moon membrane discs fitted with the appropriate permeate discharge interconnectors and exit ports. This half-moon shape is convenient since it allows users to change membrane discs without dismantling the rest of the unit. The DMF is available with Pall's Nylon 66 and PVDF microfiltration membranes (0.1–1.2 μ) and can be in situ steam sterilized and integrity tested. A production scale DMF unit would have six membrane discs with a diameter of 16″ (40 cm) and an effective area of 3 m^2. The laboratory-scale unit and the "mini-DMF" use one circular round disc 6″ (15 cm) in diameter. The latter uses a completely

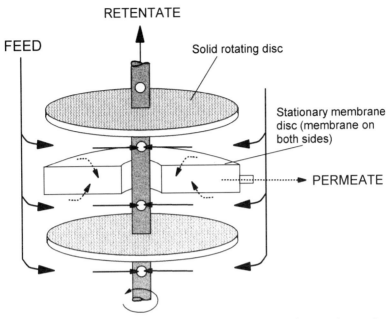

RETENTATE

FEED

Solid rotating disc

Stationary membrane disc (membrane on both sides)

PERMEATE

Figure 5.44. *Principle of the rotating disc module in which the membrane disc is stationary. In this particular design, the retentate is discharged through a perforated hollow central shaft. In other designs, the retentate may be discharged along the periphery of the discs similar to a centrifugal pump.*

sealed disposable unit containing one pair of membranes and a plastic rotating disc and is useful for initial feasibility studies.

Figure 5.46 shows the Discover VFF system from Membrex. This is different from the Pall DMF in that the discs are stacked horizontally. In addition, the Discover system has spiral grooves on the solid discs, which enhance the turbulence in the gap.

As expected, these units are extremely effective in controlling concentration polarization. Figure 5.47 compares the performance of a spiral module with a rotating disc system, each fitted with the same membrane for the same oily wastewater application (described later in Section 8.D.1.). High viscosity, fouling, and high pressure drops limited the spiral module to low concentrations of fats, oils, and greases (FOG) and total petroleum hydrocarbon (TPH) in the retentate. On the other hand, the rotating disc module resulted in flux that was four times higher and that also at much higher solids levels.

The Spin-Tek centrifugal/rotary system is different from the above two systems in that the membrane discs rotate at about 1000–2000 rpm. The membrane "pack" in this system consists of an internal support structure on which the membrane is attached to both sides. The feed enters through the rotor shaft and

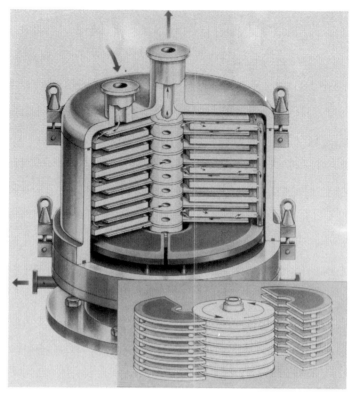

Figure 5.45. *Pall's DMF module with vertically mounted discs. The stationary half-moon membrane discs and the solid rotating discs are shown in the inset.*

moves through holes in the shaft across the spinning discs, and the retentate exits through holes at the end of the pack. Permeate is forced through the membrane into the internal support structure by the applied pressure (up to 150 psi) and exits out the periphery of the unit.

5.C.2.
VIBRATING MODULES

Another variation on the "moving membrane" concept is the vibratory shear enhanced processing (V-SEP) module developed by New Logic International (Figure 5.48) and marketed by Pall as the "Pall-Sep VMF." The module is essentially a series of flat-sheet disc membranes separated from each other by a gasket 1–2 mm thick (prefiltration down to 50 μ is recommended). The units are constructed as a stack of these disc leaf elements, which may be 30–50 cm in diameter. Each leaf element has a membrane placed on a drainage spacer on both sides. The membranes are heat-sealed to the supporting plate. The

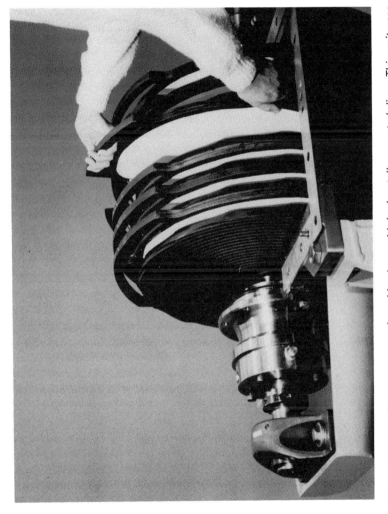

Figure 5.46. *The Discover system from Membrex with horizontally mounted discs. This unit uses stationary half-moon discs with grooved solid rotating plates.*

231

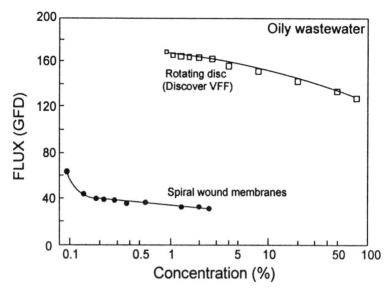

Figure 5.47. *Comparison of the performance of spiral-wound and rotating disc modules. Both modules used the Membrex UltraChem modified PAN membrane. The application was the treatment of 30,000 gallons per day of an oily wastewater (1000 ppm oil and 500 ppm suspended solids) from an engine manufacturing facility. The first stage had spiral membranes that reduced the volume 20×. The second stage was a Discover VFF system, which reduced it further to 150 gallons of sludge that could be incinerated. Permeate FOG and TPH were less than 50 ppm and 10 ppm, respectively (adapted from Rolchigo 1995).*

industrial unit can have up to 150 discs of 18″ diameter (30 m²) and consumes about 10 hp for providing the vibratory forces (additional power is needed for pumping the feed). This is 260 W/m², about the same as cross-flow ceramic modules.

As shown in Figure 5.48, the feed comes in from the top through all the membranes in the stack, flows across the membrane surface, and is removed by a bottom concentrate outlet. The permeate goes through the membrane and underneath the inner mechanical seal and comes out through the center of the stack. The laboratory-scale unit has a membrane area of 0.05 m², with feed/retentate ports coming in from the bottom of the single leaf and permeate exiting from the top. A wide variety of flat-sheet membranes could be used with this module. Operating pressures can be up to 600 psi.

Unlike rotating modules, the V-SEP membrane stack vibrates in a torsional oscillation similar in action to a clothes washing machine. The stack rotates back and forth to a displacement of 1.5″ (3.75 cm) peak to peak at the rim of the stack. The oscillation frequency is 60 times a second: for an 18″ diameter stack, this results in shear rates up to 150,000 sec⁻¹ at the rim. These high shear rates should

FEED

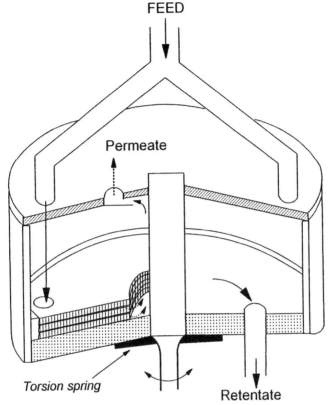

Permeate

Torsion spring

Retentate

Figure 5.48. *Schematic of the V-SEP vibratory module.*

allow concentration of solids that are much higher than conventional modules, as long as the concentrate can be pumped. Since turbulence is generated by vibration and not cross-flow, the unit could be operated in a single-pass mode, with no recirculation of the retentate.

5.C.3.
DEAN VORTICES

If the feed channel is a curved tube, as shown in Figure 5.49, centrifugal forces will push the fluid elements outwards and create a rotary movement of fluid within the channel every few millimeters. Above a critical Reynolds number, this flow is augmented by vortices that twist and spiral in the direction of flow, similar to vortices in rotary modules. These "Dean vortices" could result in enhanced back-migration of solute into the bulk stream, higher wall shear rates, and rapid renewal of the concentration boundary layer, all of which reduce

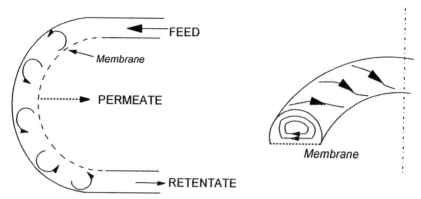

Figure 5.49. *Conceptual diagram of flow in curved channels to take advantage of Dean vortices.*

the chances of particles adhering to the membrane surface. Dean vortices have been shown (in small laboratory-scale devices) to help depolarize the membrane surface and result in high flux (Winzeler and Belfort 1993; Brewster et al. 1993; Chung et al. 1993). In theory, these devices could have the benefits of rotary devices without its disadvantages. As of this writing, this concept has not been commercialized and scaleup issues have yet to be resolved.

5.D.
SUMMARY

Users of membrane equipment are fortunate that there is a wide choice available. There are at least five types of conventional equipment that differ sufficiently in design and operating characteristics that almost any potential application can be covered by one or more types. Nevertheless, since spiral modules tend to be the cheapest and there are more spiral manufacturers than the other types, the trend in recent years has been towards selecting spiral modules over other types. Indeed, the first question to be resolved when selecting a membrane design for a particular application is: why can't a spiral module be used? One reason could be the characteristics of the feed. If it contains suspended particles of 200 μ and larger, it will invariably have to be processed in tubular units. This particular aspect—particle size—does not appear in the models discussed in Chapter 4. It may sometimes be cheaper to prefilter the feed to remove larger particles in order to use spirals rather than the generally more expensive tubular systems. Due to their relatively higher cost, the special-design modules (e.g., rotary) are best used in a hybrid system, e.g., in the back end of a membrane plant handling that portion of the process that would be difficult with conventional modules. The same may be true of inorganic membranes, although in some applications, if the overall economics were considered over a life cycle or over a 5-year or 10-year period, they may become the preferred module. With polymeric systems, a point has now been reached that the cost of the membrane per se is rarely the limiting factor: ancillary equipment, pumps, tankage, controls, piping, and cleaning now assume a greater proportion of the total costs. There are now numerous companies that manufacture membrane equipment. Thus, another important factor to consider is the technical competence of the process engineers and the after-sales service. Regrettably, there is just as wide a range in quality of these two factors as there is in equipment.

The importance of pilot testing cannot be emphasized enough. Scaleup of membrane plants is almost linear: after a certain size is reached, capacity is increased by adding more modules, instead of building larger modules. Thus, one can obtain very good design data with a relatively small pilot plant incorporating just one industrial size module or element, or at least using modules with hydrodynamics similar to the commercial modules that will eventually be used. However, if modules smaller than industrial size must be used for evaluation and testing purposes, there are some guidelines to follow to obtain the most realistic data. With spiral modules, it is more important to use pilot-scale modules of the same *length* as that company's industrial modules, rather than the same diameter modules. With tubular modules, on the other hand, the diameter of the individual tube is more important than the length, since pressure drops are not that high in these modules. With hollow fibers, the length and diameter of the fibers in the test cartridge should be the same as the industrial module, although the test cartridge can have fewer fibers. The pilot system should also be capable of handling the same ratio of process volume to membrane area as expected in the operating cycle of the industrial system.

REFERENCES

BREWSTER, M. E., CHUNG, K. Y. and BELFORT, G. 1993. *J. Membrane Sci.* 81: 127.
CHERYAN, M. 1986. *Ultrafiltration Handbook*, Technomic, Lancaster, PA.
CHERYAN, M. and CHIANG, B. H. 1984. In *Engineering and Food; Vol. 1*, B. M. McKenna (ed.), Applied Science Pub., London U.K. p. 191.
CHERYAN, M. and SAEED, M. 1989. *J. Food Biochem.* 13: 289.
CHUNG, K. Y., EDELSTEIN, W. A. and BELFORT, G. 1993. *J. Membrane Sci.* 81: 151.
JONSSON, A. S. 1993. *J. Membrane Sci.* 79: 93.
LEVY, P. F. and EARLE, R. S. 1994. *J. Membrane Sci.* 91: 135.
LOPEZ-LEVIA, M. 1979. Master of Science Thesis, Lund University, Sweden.
PORTER, M. C. 1979. In *Handbook of Separation Techniques for Chemical Engineers*, P. A. Schweitzer (ed.), McGraw-Hill, New York.
PORTER, M. C. 1990. *Handbook of Industrial Membrane Technology*. Noyes, Park Ridge, NJ.

ROBERTSON, G. H., OLIEMAN, J. J. and FARKAS, D. F. 1982. *AIChE Symp. Ser.* 78(218): 132.

ROLCHIGO, P. M. 1995. Product literature, Membrex Inc. Fairfield, NJ.

SHUCOSKY, A. C. 1988. *Chem. Engr.* 95(1): 72.

SWIEZBIN, J., UBEROI, T. and JANAS, J. J. 1996. *Chem. Engr.* 103(1): 105.

WINZELER, H. B., and BELFORT, G. 1993. *J. Membrane Sci.* 80: 35.

CHAPTER 6

Fouling and Cleaning

A major limiting step in membrane technology is "fouling" of the membrane. Fouling manifests itself as a decline in flux with time of operation. In its strictest sense, the flux decline should occur when all operating parameters, such as pressure, flow rate, temperature, and feed concentration are kept constant. Fouling problems were the single most important reason for the relatively slow acceptance of membrane technology in its early days. Fortunately, substantial progress has been made in the past decade in understanding the mechanism of fouling and alleviating its effects.

6.A.
CHARACTERISTICS OF FOULING

Flux of a real-world feedstream is usually much lower than flux of the pure solvent (e.g., water) for several reasons:

- Changes in membrane properties: These can occur as a result of physical or chemical deterioration of the membrane. Since membrane processing is a pressure-dependent process, it is possible that, under high pressures, the membrane may undergo a "creep" or "compaction" phenomenon, which may change the permeability of the membrane. This usually occurs when pressures are in the hundreds of pounds per square inch and is usually not of concern in microfiltration (MF) and ultrafiltration (UF) where pressures are typically 15–100 psi (1–7 bar). Chemical deterioration, on the other hand, could occur if the pH, temperature, and other environmental factors are incompatible with the particular membrane. Harsh and frequent cleaning regimes, sometimes necessary in food and biological applications, will decrease membrane lifetime significantly.
- Change in feed properties: Solvent transport through porous MF and UF membranes is usually considered a viscous flow phenomenon governed by the Hagen-Poiseuille equation or mass transfer relationships as shown in Chapter 4. Thus, since the feedstream's viscosity and density increase

237

and diffusivity changes as solids levels increase, flux should be lower than that of water, purely from hydrodynamic considerations. This can be easily predicted and taken into account.

- Concentration polarization: Flux-depressing effects due to membrane fouling are frequently confused with flux-lowering phenomena associated with concentration polarization. In theory, concentration polarization effects should be reversible by decreasing the transmembrane pressure, lowering the feed concentration, or increasing cross-flow velocity or turbulence, as discussed in Chapter 4. If this can be done, the cause of lower flux is polarization and not true fouling.

Membrane fouling, on the other hand, is characterized by an "irreversible" decline in flux. Depending on the system, flux may decline in one or more stages, usually rapid in the first few minutes, followed by a more gradual decline in flux (Figures 6.1–6.3). Changes in fluid management techniques may only increase the flux temporarily or mask the decline for a short period. Membrane fouling is due to the deposition and accumulation of feed components—e.g., suspended particles, impermeable dissolved solutes, or even normally permeable solutes—on the membrane surface and/or within the pores of the membrane.

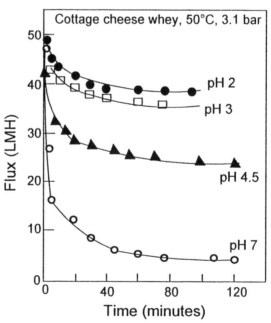

Figure 6.1. *Typical fouling pattern observed in the ultrafiltration of cheese whey (adapted from Kuo and Cheryan 1983).*

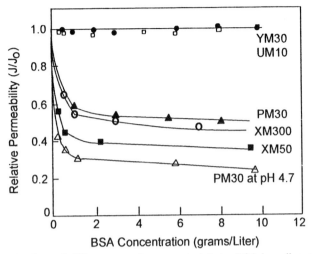

Figure 6.2. *Fouling of different membrane materials by BSA in saline solution at pH 7.4 (the PM-30 was also evaluated at pH 4.7). The UM-10 membrane is a poly-electrolyte complex, YM is regenerated cellulose, PM is polysulfone, XM is Dynel (adapted from Reihanian et al. 1983).*

Almost all feed components will foul the membranes to a certain extent. The nature and extent of membrane fouling is strongly influenced by the physico-chemical nature of the membrane and the solute(s). Surface chemistry, solute–solute and solute–membrane interactions are the key to understanding the fouling phenomenon. In a phenomenological sense, membrane fouling is similar to fouling of heat exchangers, except that the membranes have a more "active" surface; thus, it is a more complicated phenomenon.

6.A.1.
WATER FLUX

The basis of comparison or evaluation of the degree of fouling is the "clean" water flux of a membrane. This implies that the water and the membrane must be physically, chemically, and biologically clean. A reliable value of water flux (J_W) is not as easy to obtain as it may appear (Michaels 1994). From Equation (4.23), the "clean water permeability" (A_W) of a membrane can be expressed as

$$A_W = \frac{J_W}{P_T} \tag{6.1}$$

In practical terms, the units of A_W can be liters per square meter per hour (LMH)/bar, LMH/MPa or gallons per square foot per day (GFD)/psi. Since viscosity is hidden in A_W, it is imperative that temperature be closely controlled

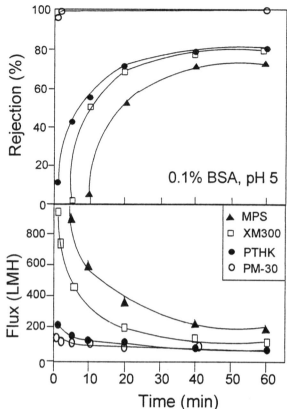

Figure 6.3. Fouling of UF membranes by 0.1% BSA solutions at pH 5, 25°C, 100 kPa. All are polysulfone membranes, except XM300, which is Dynel. MWCO of MPS (Memtec Corporation) and PTHK (Millipore) = 100,000; Amicon's XM300 = 300,000 and PM30 = 30,000 (adapted from Kim et al. 1992).

during water flux measurements. Differences in temperature during measurement can be compensated for by using viscosity–temperature correction factors:

$$J_{25°C} = 0.615 \, J_{50°C} = 1.276 \, J_{15°C} \qquad (6.2)$$

The quality of the water is also important. Tap water and even micron-filtered water may contain organic and inorganic solutes that can interact with and foul the membrane. The best are micron-filtered + deionized water or reverse osmosis (RO) permeate water or others of the quality shown in Table 6.1.

Water flux generally represents the best flux that can be obtained with that membrane. Process flux will be lower for the reasons discussed earlier. Water flux is also used as an indicator of cleaning efficiency: if the cleaning process has been effective and foulants have been removed from the membrane, the

Table 6.1. Suggested water quality for cleaning and water flux measurements.

Index	Concentration
Iron (Fe)	<0.05 mg/L
Manganese (Mn)	<0.05 mg/L
Aluminum (Al)	<0.05 mg/L
Reactive silica (SiO_2)	<2 mg/L
Colloidal silica	None
Calcium (Ca)	<25 mg/L
Magnesium (Mg)	<25 mg/L
Turbidity	<1 JTU
SDI_{15} (fouling index)	<3
Particulate matter	None
Fats, oil, grease (FOG)	None

Source: Data from Michaels (1994)

membrane system will provide the same performance again in the next process cycle. However, the value of A_W for unused membranes may vary by as much as $\pm 25\%$, even if the operating conditions such as temperature and pressure are maintained constant. This may be because of lot-to-lot manufacturing differences, variations in measurement techniques between investigators, and the type of module and device used. The same water flux variability may also occur with used membranes, even if cleaned thoroughly, although it tends to be less variable with usage. Of course, actual process flux will be of a lower magnitude and vary less than water flux, since the contribution of the membrane to the overall resistance is usually much less than other resistances (Section 4.F.).

Different modules or devices can result in different apparent A_W values for the same membrane. Usually, the best (i.e., highest) water flux is obtained with a small test cell such as those shown in Figures 5.3–5.5. The volumes are small, pressure can be adjusted quite well with a gas cylinder, temperature can be accurately controlled by immersing the cell in a water bath, and pressure drops are small due to small fluid path lengths. On the other hand, a membrane in a commercial or pilot-size module will experience a net transmembrane pressure within the module that is lower than that measured with pressure gauges outside the module at the entrance and exit of the housing. This is due to pressure losses arising from fittings and valves outside the housing, changes in feed direction and velocity in the housing and at the fittings, and permeate channel pressure losses that are especially pronounced in a spiral membrane. In addition, whereas a membrane test cell uses a magnetic stirrer for mixing and thus has no pressure drop within the cell, commercial modules use cross-flow, which gives rise to a pressure drop down the length of the module. Even correcting for the pressure drop along the module, as in Equation (4.8), and correcting for permeate back-pressure, as shown in Equation (4.6), will not completely compensate for these pressure losses in a module. It is best to test for water flux at the lowest practical transmembrane pressure and pressure drop.

For these reasons, it is unreasonable to expect a single value of clean water flux or A_W to be provided for a particular membrane in a particular type of module. More likely, a range of the nominal water flux under standard conditions will be provided, e.g., 112–256 LMH/bar for a Millipore PTGC membrane at 25°C (Michaels 1994). When establishing an initial baseline or benchmark water flux with new membranes, it is important to thoroughly flush out all preservatives and moisture retention chemicals such as glycerol that the module may have been shipped with, to wet the membranes fully, and to remove all air in the system (the latter is especially important with ceramic membranes). In subsequent water flux measurements, it is most important to be consistent, i.e., always to take measurements at the same pressures and temperatures and with the same water quality.

MF membranes tend to have extremely high pure water flux compared to UF membranes. This is because of a small number of very large pores through which much of the water permeates initially. With process streams, the high permeation velocity also brings solute to these pores rapidly, causing them to plug up first, resulting in the precipitous drop in process flux that is commonly observed. With high water flux MF membranes, it is generally better to start up the system at low transmembrane pressures to minimize this phenomenon and obtain better long-term flux (see Section 6.E.4.).

6.B.
CONSEQUENCES OF FOULING

An obvious consequence of fouling is higher capital expense caused by the lower average flux over a process cycle. In addition, depending on the nature and extent of fouling, restoring the flux may require powerful cleaning agents, which may reduce the operating life of the membrane. This continues to be a problem with cellulose acetate (CA) membranes, which have a limited pH, temperature, and chlorine tolerance range, although it is less of a problem with other polymeric membranes such as polysulfone (PS), polyethersulfone (PES), polypropylene (PP),and polytetrafluoroethylene (PTFE), and much less with ceramic and other inorganic membranes that can tolerate aggressive cleaning procedures. Some membrane manufacturers will guarantee specific lifetimes of their membranes only if their recommended cleaning procedures are strictly followed by the user. Indeed, membrane lifetime can often be better correlated with the number of cleaning cycles rather than on-line usage time.

Rejection and yields may also be affected. If the buildup of solids on the membrane is significant enough, it may act as a secondary membrane and change the effective sieving and transport properties of the system (Figure 6.3). Bovine serum albumin (BSA) has a molecular weight of 67,000 and a Stoke's radius of 3.6 nm. Thus, it is expected to be completely rejected by the PM30 membrane, as shown in Figure 6.3. The other, more open, membranes initially allowed BSA to pass through but soon showed higher rejections of 70–80% caused by the buildup of protein. Similarly, rejection of polyethylene glycol (PEG 4000) increased substantially in the presence of albumin, with the change in rejection being less at higher temperatures (Busby and Ingham 1980). Membrane-fouled proteins may exhibit molecular and conformational changes (Sheldon et al. 1991). This could affect their subsequent activity and functionality, a phenomenon observed with membrane-immobilized enzyme reactors (Chapter 8). Also, if the volume processed is small in relation to the membrane area, the amount of solids adsorbed on the membrane surface may be significant enough to affect yields (Section 3.D.8.).

6.C.
MATHEMATICAL MODELS OF FOULING

Most models relate the flux to the time or volume permeated and generally take an exponential form, considering the shapes of the fouling curves (Figures 6.1–6.3):

$$J_t = J_0 t^{-b} \tag{6.3a}$$

$$J_t = J_0 e^{-bt} \tag{6.3b}$$

$$J_t = J_0 V^{-b} \tag{6.3c}$$

$$J_t = J_{SS} + k e^{-bt} \tag{6.4}$$

where J_0 is the initial flux, J_t is the flux at any time t, J_{SS} is the limiting or steady-state flux, and V is the volume permeated. The constants k and b characterize the fouling process. Since these models are empirical, they may not help explain the phenomenon itself. Some models predict that flux will be zero at infinite time, which may not occur in practice.

Several models have been developed based on a *pore blocking* mechanism developed for dead-end filtration (Field 1996; Hermia 1982; Jaffrin et al. 1997). In one case, it is assumed that only a fraction of the pores is completely blocked by the particles. This fraction is proportional to the amount of permeate flow through the membrane. This model eventually takes the form

$$J_t = J_0 \exp(-J_0 bt) \tag{6.5}$$

The *cake filtration* version of this model assumes that the entire surface is covered by a layer of foulant and that this layer continues to grow even if back-transport of the solute occurs. Thus, the cake resistance is proportional to the cumulative permeated volume, and the model is

$$J_t = J_0 \left(1 + \frac{2A J_0 b t}{R_M} \right)^{-1/2} \tag{6.6}$$

where A is the membrane area and R_M is the intrinsic membrane resistance. The *internal pore plugging* model assumes that the pores of the membrane get plugged up due to deposition or adsorption of microsolutes. This model takes the final form:

$$J_t = J_0 \left(1 + \frac{J_0 b t}{\varepsilon \lambda} \right)^{-2} \tag{6.7}$$

where ε is the porosity of the membrane (defined in Chapter 4) and λ is the thickness of the membrane. Obviously, more than one mechanism can be occurring with a complex feedstream that contains a variety of particle sizes. For example, the pore blocking model [Equation (6.5)] may describe fouling by a very dilute feedstream, while Equation (6.6) could describe a more concentrated feedstream with solutes that could form a "cake." On the other hand, internal pore plugging could predominate when concentrated solutions of permeable solutes (salts, sugars) are being processed by UF or MF. Model parameters (k or b, the fouling index, and J_0, the initial flux) can be correlated with operating variables, which will allow us to determine a strategy for minimizing fouling. Belfort et al. (1994) reviewed several models for describing fouling of microfiltration membranes. Some models describe only the initial period of flux decline well (Jaffrin et al. 1997).

Another approach is to use the resistance-in-series concept discussed in Section 4.F. From Equation (4.26):

$$J_t = \frac{P_T}{R_M + R_F + R_G} \tag{6.8}$$

Under constant operating conditions of pressure (P_T), flow rate, temperature, and concentration (e.g., in a total recycle mode of operation), R_M and R_G should be constant. Thus, decreases in flux can be correlated to increases in R_F. The application of Equation (6.8) was discussed in Section 4.F. and will also be shown later in this chapter.

6.D.
FACTORS AFFECTING FOULING

Since fouling in its true sense is a result of specific interactions between the membrane and various solutes in the feedstream, and perhaps between the adsorbed solute and other solutes in the feedstream, it is difficult to establish general rules or theories about the nature and extent of fouling that will be universally applicable. Each component of a feedstream will react differently with the membrane: conformation, charge, zeta potential, hydrophobic interactions, and other factors will have a significant bearing on these membrane–solute interactions. Certain process engineering factors, such as cross-flow velocity, pressure, and temperature can also have a bearing on fouling. It should be recognized that all these three general categories—membrane material properties, solute properties, and operating parameters—can interact with each other and give rise to quite different effects in combination than if these factors were studied individually or with model systems. Nevertheless, some general observations can be made that should help understand the mechanisms of fouling.

6.D.1.
MEMBRANE PROPERTIES

The importance of proper selection of the membrane material has been discussed at length in Chapter 2. Some additional factors affecting the fouling behavior of membranes are discussed here:

- Hydrophilicity: With aqueous feedstreams, the ideal membrane should be hydrophilic (water-attracting). If the material is hydrophobic, it will adsorb components that are hydrophobic or amphoteric, resulting in fouling. For example, many proteins have hydrophobic regions within their structure that can interact strongly with hydrophobic materials. Unfortunately, many robust polymeric membranes are relatively hydrophobic (water-repelling, but organic- and oil-attracting). One measure of the relative hydrophilicity of a membrane is the *contact angle*, which is a measure of the wettability of a surface. If a drop of water is placed on a completely hydrophilic material, the water would spread out on the surface, resulting in a zero or low contact angle (Figure 6.4), which can be measured with a goniometer. A hydrophobic material, on the other hand, would repel the water, causing it to have as little contact with the surface as possible, resulting in a high value of the contact angle. Hydrophobic materials would tend to attract oil in an oily wastewater stream, but hydrophilizing the membrane could minimize oil fouling (see Section 6.D.2.).

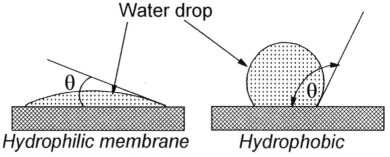

Figure 6.4. Contact angle.

This method of characterizing or predicting hydrophilicity, and presumably the predisposition of a membrane to foul, has attracted considerable interest in the past decade. Table 6.2 lists some contact angles for several membranes. The well-known hydrophobic materials (PTFE, PP) tend to have high contact angles ($>100°$) while cellulosics and ceramics and some specially hydrophilized materials have low values ($<30°$). As shown in Figures 6.2, 6.5, and 6.6 and Tables 3.13 and 3.14, cellulosics tend to foul much less than other, more hydrophobic membranes. The benefits of using cellulosic membranes with protein

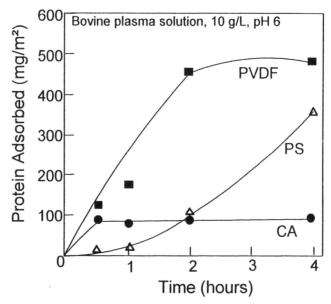

Figure 6.5. Adsorption of proteins from a bovine plasma solution (10 g/L, pH 6). CA is cellulose acetate, PS is polysulfone, and PVDF is polyvinylidene fluoride (adapted from Lockley et al. 1988).

Table 6.2. Contact angles for membrane materials.

Membrane	Contact Angle (°)	Reference
Cellulosic (unidentified)	24	Hodgins and Samuelson (1990)
Nova C-series	12–15	Gekas and Zhang (1989)
Desal CA-UF	45	Oldani and Schock (1989)
Ceramic	30	Rolchigo (1995)
Polyacrylonitrile (unidentified)	46	Hodgins and Samuelson (1990)
Membrex Ultrafilic	4	Hodgins and Samuelson (1990)
Poly(acrylonitrile-*co*-vinyl chloride)		
XM 50, Amicon	40	Jucker and Clark (1994)
XM100A, Amicon	60	Kim et al. (1989)
Polyethersulfone (unidentified)	65	Hodgins and Samuelson (1990)
IRIS UF3028	52	Gourley et al. (1994)
Nova FNS series	72–81	Gekas and Zhang (1989)
(Hydrophilized)	44	Hodgins and Samuelson (1990)
Desal E-100	56	Oldani and Schock (1989)
Polypropylene	108	Rolchigo (1995)
Polysulfone		
PM10, Amicon	38	Oldani and Schock (1989)
PM 30, Amicon	60	Kim et al. (1989)
PM 30, Amicon	42	Jucker and Clark (1994)
PM 30, Amicon	43	Oldani and Schock (1989)
IRIS UF 3026	54	Gourley et al. (1994)
IRIS Rhône-Poulenc	59	Kim et al. (1989)
PTGC, Millipore	65	Kim et al. (1989)
DDS-GR61	44	Oldani and Schock (1989)
DDS-GR81	45	Oldani and Schock (1989)
Desal E-500	81	Oldani and Schock (1989)
Kalle UF PS15	40	Oldani and Schock (1989)
PTFE (unidentified)	112	Rolchigo (1995)
PVDF (unidentified)	66	Rolchigo (1995)
Regenerated cellulose		
YM1, Amicon	96	Jucker and Clark (1994)
YM5, Amicon	48	Kim et al. (1989)
YM10, Amicon	6	Jucker and Clark (1994)
YM30, Amicon	7	Jucker and Clark (1994)
YM30, Amicon	49	Kim et al. (1989)
YM100, Amicon	31	Jucker and Clark (1994)

solutions is well known: e.g., although the water flux of regenerated cellulose (RC) membranes may be less than PES membranes (Table 6.3), the process flux with protein solutions is usually higher (Levy and Sheehan 1991; Marshall et al. 1993; Reed et al. 1987), and RC membranes are also easier to clean. This difference in membrane behavior can be correlated with the relative amount of protein adsorbed by the membranes.

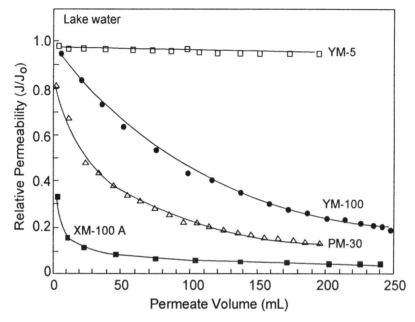

Figure 6.6. *Fouling of UF membranes by lake water. See Figure 6.2 legend for membrane designations. Individual resistances are given in Table 6.3 (adapted from Laine et al. 1989).*

Figure 6.6 shows that the fouling of the YM100 membrane is much greater than the YM5 membrane, even though both are made of the same material. This phenomenon was also observed for BSA solutions (Ko and Pellegrino, 1992). This is partly a reflection of the much higher intrinsic membrane resistance (R_M) of YM5 (Table 6.3), which has much smaller pores, giving rise to a much smaller value of J_0 than the other membranes. Thus, the contribution of the fouling and polarized layer

Table 6.3. *Individual resistances, in units of $(10^{-2} \times \text{LMH/atm})^{-1}$, for ultrafiltration of lake water with Amicon membranes. Fouling patterns are shown in Figure 6.6.*

Membrane	R_M	R_F	R_G	J_0 (LMH/atm)
YM-5	15.00	0.65	0.19	6.7
YM-100	0.74	0.05	2.78	138
PM-30	0.38	0.44	1.85	255
XM-100A	0.35	2.92	5.02	288

Source: Data from Laine et al. (1989)

resistances is masked by the high value of R_M. On the other hand, the larger pores of the YM100 could be partially blocked by the solutes, some of which could even find their way into these pores, resulting in the dramatic drop in flux. On an absolute scale, however, the final flux of the YM100 is higher than the YM5.

Some fouling data may appear to be contradictory. Polyvinylidene fluoride (PVDF) membranes appear to adsorb or foul less with proteins compared to PES, PS, and CA membranes (Table 3.14; Pitt 1987). In other cases (e.g., Figure 6.5), fouling by PVDF membranes is much worse. This could be caused by the varying degrees of hydrophilicity conferred on these membranes by modification, a factor not often considered in published fouling studies. For example, the Millipore GVWP, a surface modified hydrophilic PVDF, adsorbed much less protein (84 mg/m^2) than Millipore GVHP, a hydrophobic PVDF, which adsorbed 1157 mg/m^2 (Persson et al. 1993).

However, when considered as a whole, the relationship between contact angle and fouling tendency is not that clear. This is partly a reflection of the difficulty in obtaining good contact angle data, which is affected by surface roughness, purity of the water, and the technique used by individual investigators. In addition, the nature of the membrane material becomes less important if concentration polarization and total protein deposition are large.

- Surface topography: Another interesting reason why cellulosic membranes foul less compared to other polymeric membranes may be related to the roughness of their surfaces. The surface of CA membranes appears to be smooth and uniform, as shown in Figure 6.7. In contrast, polyamide thin-film composite membranes appear to have protuberances on the surface, which could act as hooks for suspended matter in the feed, thus leading to greater fouling (Figures 6.7 and 6.8). This could explain why polyamide-based membranes tend to foul more than CA membranes, especially with regard to biofouling (Figure 2.15). Similarly, the largest protein deposits occurred with the most heterogeneous membranes (Suki et al. 1984).
- Charge on the membrane: Most membranes have a net negative charge under usual process conditions. The charge on the membrane becomes important if charged particles are being processed. For example, flux of cathodic electrocoat paint improved dramatically when a positively charged membrane was used (Figure 8.3), by taking advantage of the mutual repulsion between the particles and the membrane of the same charge (see Section 6.C.2. below on effect of pH).
- Pore size: The relative size of the pores and the solutes in a feed are very important. There are numerous examples in the literature (e.g., cited by Marshall et al. 1993) where the phenomena shown in Figure 6.9 has

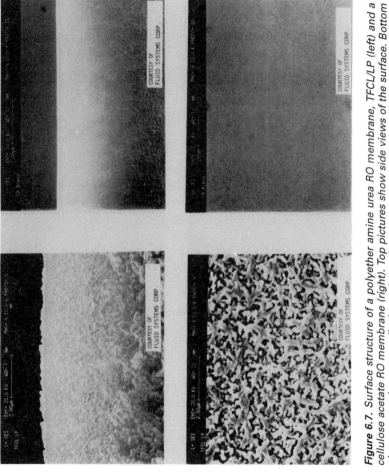

Figure 6.7. Surface structure of a polyether amine urea RO membrane, TFCL/LP (left) and a cellulose acetate RO membrane (right). Top pictures show side views of the surface. Bottom pictures are taken perpendicular to the surface. Note the smooth surface of the CA membrane compared to the thin-film composite membranes (see Figure 6.8 also) (courtesy of William Light, Fluid Systems).

Figure 6.8. Surface structure of polyamide-based, thin-film composite RO membranes: left: CPA2; right: FT30. Note the rough surface compared to the CA membrane shown in Figure 6.7 (courtesy of William Light, Fluid Systems).

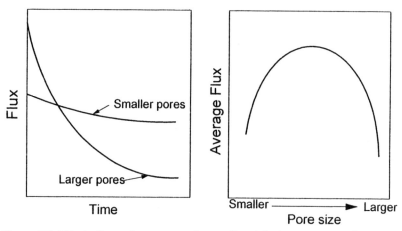

Figure 6.9. *Effect of membrane pore size on flux: left: instantaneous flux versus time, when data is obtained with all operating conditions constant; right: average flux over an operating cycle (e.g., between cleanings) versus pore size, all other factors being equal.*

been observed: larger pore membranes will initially have higher flux than tighter membranes but will eventually (sometimes rapidly) have lower flux. Typical examples with fermentation broths are shown in Figure 6.10. This is especially pronounced with MF applications. One possible reason for this is shown in Figure 6.11. If the size of the particle to be separated is of the same order of magnitude as the range of pore sizes being used, some of the smaller particles in that feed sample could lodge in the pores without necessarily going through them. This physical blockage of the pores will cause a rapid drop in flux in the first few minutes of operation; often, the steady-state process flux is a small fraction of the pure water flux. This is understandable since much of the permeation with pure water—and initially with the process fluid—occurs in the larger pores. For example, if the pore distribution is 0.1-μ pores (10%), 0.2-μ pores (80%), and 0.5-μ pores (10%), 43% of the water permeation will occur through the 0.5-μ pores. As shown previously in Table 3.3, many MF membranes have pores much larger than their ratings, and these are the pores that will get plugged first. There will thus be a steep drop in permeation rate when these large pores are plugged. Higher pressures will aggravate the problem by causing a compression of the adsorbed cake and forcing the partially lodged particles to become even more firmly lodged in the pores, making cleaning more difficult.

In contrast, if the pores are much smaller than the particles to be separated, the particles will not get caught within the pores but will roll off the surface under the shear forces generated by the flow. In fact, if

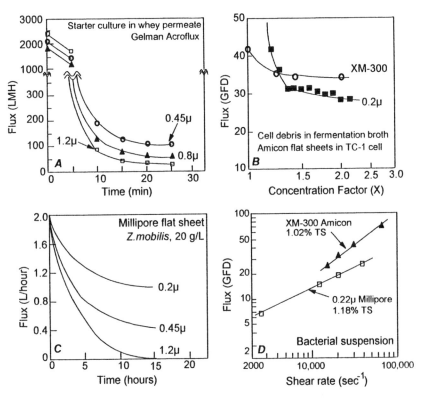

Figure 6.10. Effect of pore size on membrane fouling: (A) starter culture (1–2 g/L) grown in whey permeate (adapted from Merin et al. 1983); (B) fermentation broth in a thin-channel cell with 30-mil channel height at 30 psig (adapted from Porter and Michaels 1971); (C) microbial suspension of Z. mobilis (adapted from Rogers et al. 1980); (D) bacterial suspension (adapted from Henry and Allred 1972).

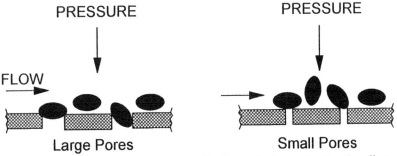

Figure 6.11. Mechanism of membrane fouling by particulates, showing the effect of pore size in relation to particle size.

cross-flow velocity has no effect on flux with a feedstream that initially showed concentration polarization effects, it indicates the above phenomenon has occurred, and the system should be cleaned as soon as possible, since fluid flow management techniques will not be effective.

Thus, although it may seem logical to use membranes with pore sizes that are the same or slightly smaller than the particle size, the short-term advantage of higher initial flux will be outweighed by the long-term problems associated with higher fouling rates. A better rule of thumb is a particle size to pore size ratio of 10; i.e., select a membrane with pores that are one-tenth the particle size (on average), at least as a starting point for testing. However, there is a point of diminishing returns when going to smaller pore sizes (Figure 6.9, right). When pores are small enough to minimize the pore plugging phenomenon, other operating and membrane-related factors, such as those in the flux models discussed in Chapter 4, become important, resulting in the optimum shown in Figure 6.9 (right). One example is shown in Figure 6.12. Similar optima in pore size were observed for harvesting mammalian cells where a $0.6\text{-}\mu$ membrane gave better overall performance than $0.2\text{-}\mu$ or $3\text{-}\mu$ membranes (Radlett 1972). During harvesting of *Erwinia carotovora* cells, the $0.45\text{-}\mu$ membrane gave almost double the flux of a $0.6\text{-}\mu$ membrane, but the $0.2\text{-}\mu$ membrane gave lower flux than the $0.6\text{-}\mu$ membrane (Le et al. 1984). With MF of cheese whey with Nuclepore membranes, the best flux was obtained with a $0.08\text{-}\mu$ membrane rather than smaller pore (0.01, 0.03, 0.05 μ) or larger $0.1\text{-}\mu$ membranes (Piot et al. 1988).

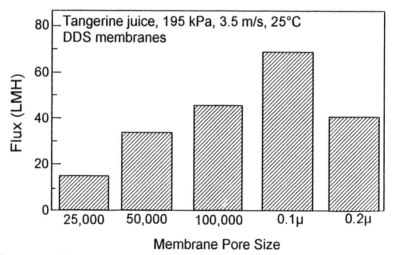

Figure 6.12. *Effect of membrane pore size on flux with tangerine juice (adapted from Chamchong and Noomhorm 1991).*

● Surface modification: A membrane can be made more hydrophilic and, by implication, less prone to fouling by aqueous feedstream components by introducing a large number of hydrophilic functional groups on its surface, i.e., those capable of binding water or forming hydrogen bonds, such as hydroxyl groups (–OH). Methods include surface coating by adsorption, free radical or radiation grafting of hydrophilic or polymerizable hydrophilic monomers, microwave or plasma treatment, chemical conversion of polymer side chains to hydrophilic groups, and coating with oriented monolayers using Langmuir-Blodgett methods (Hildebrandt 1991). One early technique still in use is simply to soak a membrane in a nonionic surfactant or water-soluble polymer and then dry it. This is reportedly particularly effective with PS (Brink and Romjin 1990; Kim et al. 1988) and PVDF membranes, although the coating may not last through several cycles of use. Adding sulfonate groups helps minimize fouling by negatively charged foulants.

Free radical grafting of glycine to the surface of a PVDF membrane under strongly basic conditions can make it hydrophilic. Similarly, grafting a phosphorylcholine derivative substantially improved fouling resistance of PVDF and CA (Akhtar et al. 1995). Blending PVDF with a more hydrophilic polymer such as CA or sulfonated PS also helps.

As noted in Chapter 3, nylon tends to bind large amounts of protein, even though it is inherently wettable and hydrophilic. This fouling tendency can be reduced by radiation grafting hydroxyl group rich polymers (hydroxypropyl acrylate and hydroxypropyl methacrylate) or by coating with hydroxypropyl cellulose and subsequent cross-linking. This is reportedly the method for producing the low protein binding Pall Nylon-66 MF membrane (Hildebrandt 1991).

Chemical conversion of the cyano groups in polyacrylonitrile (PAN) membranes with strong sulfuric acid and formaldehyde to *N*-methylolamides makes it hydrophilic (Hodgins and Samuelson 1990). The applicability of this membrane and its structure is shown later in Figure 6.18. There are several methods of making PS and PES membranes more hydrophilic, most involving blending with hydrophilic polymers [e.g., polyvinylpyrrolidone (PVP)] or chemical modification of the polymer chain (Kai et al. 1989).

Light-activation chemistry is used to covalently bond compounds to membranes to change their hydrophilicity (BSI 1996). These photoreagents are claimed to reduce fouling, improve hemocompatability, modify tissue adhesion as needed (e.g., for artificial organs), and even reduce bacterial fouling.

It should be noted that surface modified versions of membranes may not be as tolerant to aggressive environments as the native materials.

6.D.2.
SOLUTE PROPERTIES

Since fouling in its true sense is a result of specific interactions between the membrane and various solutes in the feedstream, it is important to understand the physicochemical characteristics of individual feed components in relation to membrane performance. Although this has been discussed at length in Chapter 3, some specific examples are discussed here:

- Proteins: It is not surprising that proteins are a major foulant in membrane processing, considering the multiplicity of functional groups, the charge density within protein molecules, the varying degrees of hydrophobicity, and the complex secondary and tertiary structure that allows a protein to interact with other feed components, as well as the membrane itself. Complicating the phenomena is that all these properties—and thus the nature and extent of fouling—are affected by pH, ionic strength, shear, heat treatment, and other environmental factors (Marshall et al. 1993; Nilsson 1990).

  A popular model protein for studying fouling phenomena is *bovine serum albumin* (BSA); some typical fouling data were shown in Figures 6.2 and 6.3. The decrease in flux has been correlated with increasing protein deposition on the membrane (Figure 6.13). Cheese whey has also been widely studied, not only because it was one of the first successful commercial applications of UF, but because membrane fouling by cheese whey is a particularly vexing problem, as discussed in detail in the first

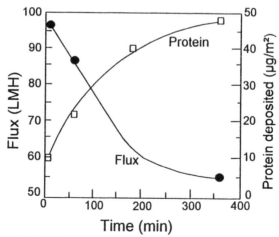

Figure 6.13. *Relationship between protein deposition on the membrane surface and flux decline. PM30 membrane used with 0.1% BSA in a stirred cell at 100 kPa, 25°C (adapted from Kim et al. 1992).*

edition of this book (Cheryan 1986). With acid whey, it appears that fouling is principally caused by a casein fragment that remains in the whey after the cheese curd is removed and that forms a complex with β-lactoglobulin and calcium. On the other hand, with sweet whey, α-lactalbumin had the greatest flux-depressing effect in the short-term and β-lactoglobulin in the long-term (Merin and Cheryan 1980). Scanning electron microscopy showed two types of deposits, one appearing as white clusters on top of another lower layer of denser material (Cheryan and Merin 1980). Transmission electron micrographs showed protein within the voids and substructure of the membrane [Figure 6.14(A)]. This is not unexpected since whey contains proteins/peptides that are small enough to partially permeate the 10,000 MWCO membrane used in these studies. The major whey protein, β-lactoglobulin, showed similar type of deposits in a whey environment [Figure 6.14(B)]. When the salts had been removed by dialysis prior to UF, a different type of deposit was observed [Figure 6.14 (C) and (D)], and flux was higher, showing the importance of salts (primarily calcium in this case) in protein fouling.

Typical adsorption isotherms are obtained, in that an increase in bulk concentration of protein increases the adsorption of protein (Dillman and Miller 1973 for BSA; Nichols and Cheryan 1981 for soy proteins). Protein buildup occurs in multiple layers: 80–200 equivalent monolayers of BSA could be adsorbed on a PM-30 PS membrane after 3–8 h of use (Fane et al. 1982). This will amount to an adsorbed layer thickness of 1–1.5 μm, which is not inconsistent with the electron micrographs shown in Figure 6.14.

It is interesting that skim milk fouls much less than whey. Fouling index parameters (b) for skim milk are typically -0.01 to -0.06 (Cheryan and Chiang 1984), compared to whey, which has values of -0.06 to -0.38 (Kuo and Cheryan 1983). This is curious because milk contains higher solids than whey (8.2% in nonfat milk versus 5.6% in whey), much higher protein (3.3% versus 0.6%) and calcium (30 mM versus 10 mM), all of which are considered to be involved in fouling. One possible reason could be that two-thirds of the protein in milk is in the form of casein micelles that are so large (10–100 nm) that they simply roll off the membrane surface under the high shear forces present in most commercial membrane modules. In addition, most of the calcium in milk is tied up within the micelle, and thus, it cannot form "salt bridges" between the protein and the membrane.

- Salts: Mineral salts have a profound influence on the fouling of membranes. On one hand, they can precipitate on the membrane because of poor solubility or bind to the membrane directly by charge interactions. Figure 6.15 shows data obtained with a simulated solution

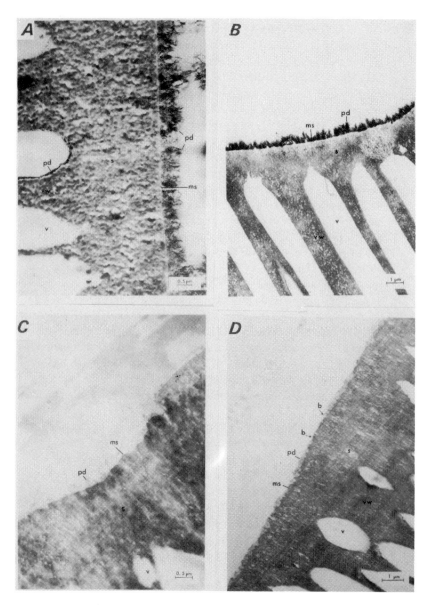

Figure 6.14. Transmission electron micrographs of a UF membrane (Millipore PTGC) fouled by cottage cheese whey (pd = protein deposit; ms = membrane surface; v = voids; s = skin of membrane). (A) acid whey: note protein deposit in the voids; (B) β-lactoglobulin in whey dialysate; (C) acid whey in salt-free environment; (D) β-lactoglobulin in salt-free environment (adapted from Cheryan and Merin 1980).

258

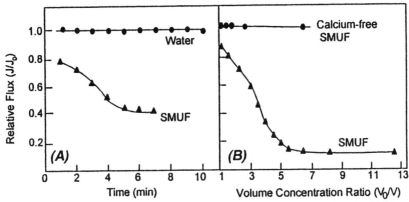

Figure 6.15. *Fouling of UF membranes by milk salts (SMUF = simulated milk ultra-filtrate): (A) PTGC membrane at 40 psig, 50°C (adapted from Merin and Cheryan 1980); (B) AC30 membrane at 50°C (adapted from Hanemaaijer 1988).*

of the salts normally found in bovine milk. The salts alone cause a substantial decline in flux. The higher the pH, the lower is the solubility of salts and thus greater chance of fouling, as shown in Figure 6.1. At high pH, flux is higher if the precipitated salts are removed before UF.

Sodium chloride can change the ionic environment, which in turn affects solute–solute and solute–membrane interactions. At a concentration lower than saturation, it appears to have no effect by itself. In the presence of proteins, it appears to increase protein deposition and loss of flux (Fane et al. 1983). Sodium chloride in the feed solution reduced fouling by polyphenols (Saeed and Cheryan 1989).

Calcium (30 mM in milk) has been identified as a major cause of fouling in dairy streams, not only due to precipitation in the form of tricalcium phosphate, but also because its ionic form could act as a salt bridge between the membrane and proteins, which will lead to faster protein fouling. Flux can be improved by removing the calcium—e.g., by centrifugation, electrodialysis, or ion-exchange—as shown in Figures 6.15(B) and 6.16 or by interacting it with calcium-sequestering agents (e.g., EDTA, citrates). Ca^+ removal will also reduce protein deposition [Figure 6.14(A) and (B) versus (C) and (D)].

Inorganic particulate matter such as aluminum silicate clays (> 10–40 Å in size, negatively charged) are commonly found in natural waters. With these charged particles, maximum flux was obtained at a zeta potential of zero and for very low zeta potentials less than −40 mV (McDonogh et al. 1988).

- pH: In general, flux is lowest at the isoelectric point of the protein and is higher as the pH is moved away from the isoelectric point (Figure 6.16).

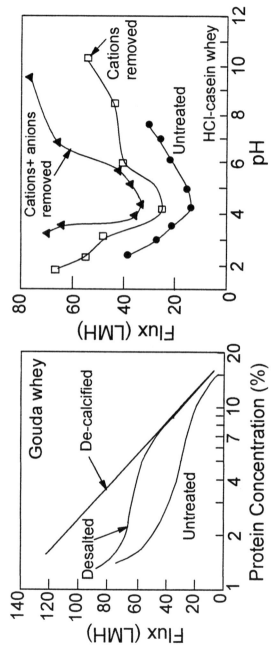

Figure 6.16. Effect of salts on flux of whey: left: UF of preconcentrated Gouda whey (11% total solids) in a tubular module; whey was preheated at 55°C for 30 min, then UF at 55°C, 5 bar, pH 6.6 (adapted from Hiddink et al. 1981); right: UF of HCl-casein whey in Amicon P10 module at 1 bar (adapted from Hayes et al. 1974).

Changes in pH affect solubility and conformation of feed components. The solubility of a protein is generally lowest at the isoelectric point; it increases as the pH is adjusted away from the isoelectric point, and this could explain the flux behavior shown in Figure 6.16. However, if there are salts that decrease in solubility and precipitate on the membrane, then flux could decrease at higher pH (Figure 6.1). Prior removal of insoluble calcium salts from cheese whey at pH 6.4–7 by filtration (Daufin et al. 1992) or centrifugation (Kuo and Cheryan 1983) improved UF flux significantly.

Even the method of adjusting pH prior to UF could be important. With acid whey, which has about twice as much calcium as cheddar (sweet) whey, maximum fouling from calcium salts occurs at pH 5.8. If pH were adjusted upwards rapidly, calcium phosphate would form apatites of a gelatinous nature, which would readily foul membranes. When pH is adjusted slowly and some time is given for equilibrium to be attained, the apatite is of a different physical form and apparently does not foul membranes as easily. However, some membranes, especially the non-celluloics, appear to be more sensitive to this pretreatment than others.

Heat treatment of whey is often combined with pH adjustment to maximize the flux. The heat treatment apparently causes the formation of casein-β-lactoglobulin complexes and/or calcium apatite complexes that are less fouling. Figure 6.17 shows the beneficial effect of specific preheat treatments on UF flux of Gouda cheese whey. Some cheese whey processors incorporate a "warm-hold" operation in their UF plants, where the whey is held for 30–90 min at 50–60°C to stabilize the calcium–protein complex (Section 8.B.4.). Sometimes a more severe heat treatment is given in the plate pasteurizer, e.g., 72–85°C for 15 sec, and then adjustment of pH carefully to 5.6 for acid whey, or the heat treatment alone for sweet whey.

- Lipids, fats, and oils: The evidence seems to indicate that the removal of lipids in whey by centrifugation or microfiltration has a beneficial effect on UF flux. Indirect evidence of the deleterious effect of lipids on flux can be seen by comparing the flux during UF of full-fat soybean water extracts (Figure 4.9) and defatted soybean water extracts (Figures 4.6 and 4.7). In the former case, flux became pressure-independent at relatively low pressures and was much lower, while in the latter case, the flux was pressure-dependent throughout the operating pressure range and was much higher.

  With oil-in-water mixtures, it should be remembered that "like attracts like"; if hydrophobic membranes are used, free oils can coat the membrane, resulting in poor flux (emulsified oil is usually not as much of a problem, unless it is concentrated to such a high level that the emulsion breaks, releasing free oils). As shown in Figure 6.18, oils have a structure

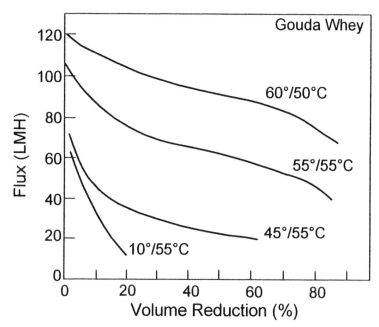

Figure 6.17. Effect of pre heat treatment on flux during UF of Gouda whey (5.5% TS, pH 6.6). A tubular module in a batch mode was used. The whey was preheated for 30 min at the first temperature shown in the figure and then ultrafiltered at the second temperature (adapted from Hiddink et al. 1981).

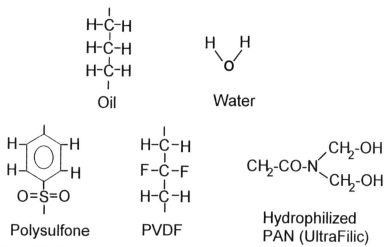

Figure 6.18. Structure of various polymeric membranes in relation to the structures of oil and water (adapted from Rolchigo 1995). PS and PVDF membranes have functional groups similar to oil and, thus, would be attracted to it. The modified PAN membrane would preferentially attract water rather than the oil because of the similarity of their structures.

similar to the functional groups of membranes such as PVDF and PS, thus resulting in considerable fouling. Hydrophilic membranes such as the modified PAN membranes (manufactured by Membrex under the trade name of UltraChem/UltraFilic) have functional groups more like water, which gives it excellent hydrophilicity (Table 6.2). This preferentially attracts water rather than the oil, resulting in much higher flux (Rolchigo, 1995). Oil–water separation data with this membrane was shown in Figure 5.47.

- Antifoams: These are used to suppress foaming in evaporators and fermenters. Many commercial antifoaming agents (e.g., polyoxyethylene polyoxypropylene oleyl ether, polyglycols, silicone oils) severely foul hydrophobic membranes (Cabral et al. 1985; Harris et al. 1988; Khorakiwala et al. 1986; Kloosterman et al. 1988; Kroner et al. 1986; Levy and Sheehan 1991; McGregor et al. 1988; Yamagiwa et al. 1993). Flux was much better when process temperature was lowered below a defoamer's cloud point. Hydrophilic membranes are fouled less by antifoams.

- Humic substances: These are weak acidic electrolytes that are amphiphilic, controlled largely by carboxylic- and phenolic-OH groups. They are micelle-like with molecular weights between 500 and 100,000 and can represent up to 80% of the total organic carbon of natural waters. They become more hydrophobic as pH decreases and, thus, deposit to a greater extent on hydrophobic membranes at lower pH (Jucker and Clark 1994). Fouling of several membranes by lake water was shown in Figure 6.6. Calcium in natural waters enhances adsorption, perhaps by acting as a bridge between the membrane surface and the negatively charged membrane surface and/or between negatively charged carboxyl groups of the humic acid. This appears to be a similar mechanism by which calcium enhances protein fouling in cheese whey.

    Other components that have been reported as fouling agents include microbial slime, polyhydroxy aromatics, and polysaccharides (Defrise and Gekas 1988; Matthiasson and Sivik 1980).

6.D.3.
PROCESS ENGINEERING

In addition to the complicated physicochemical interactions of feed components, process parameters such as temperature, flow rate, pressure, and feed concentration, as well as overall equipment design, have great influence on membrane fouling.

- Temperature: The effect of temperature on fouling is not too clear. According to the Hagen-Poiseuille model on which the pore flow phenomenon is based, increasing temperature should result in higher

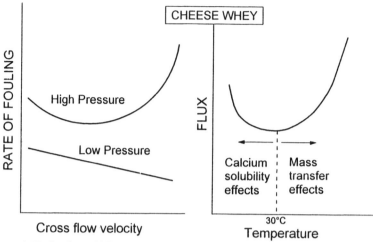

Figure 6.19. *Fouling of UF membranes by cheese whey: left: effect of cross-flow rate and transmembrane pressure on rate of fouling; right: effect of temperature on flux.*

flux, and this is generally borne out in practice. However, it could also result in a decrease in flux for certain feeds such as cheese whey, as shown in Figure 6.19. At temperatures below 30°C, flux decreases with increasing temperature because of a decrease in solubility of calcium phosphate. However, as temperature is increased further, the beneficial effects (lower viscosity, higher diffusivity) will outweigh the detrimental effects and may result in a net increase in flux.

Obviously, for biological systems, too high a temperature will result in protein denaturation and other heat damage, which will result in a lowering of the flux. The adsorption of protein molecules in the temperature range of interest in most UF applications (30–60°C) generally increases with temperature.

- Flow rate and turbulence: High shear rates generated at the membrane surface tend to shear off deposited material and thus reduce the hydraulic resistance of the fouling layer. However, this may not always happen if the transmembrane pressure is high in relation to the permeation velocity (Figure 6.19). As discussed in Section 4.I.3. and 4.J., Brownian diffusion is more important for submicron/dissolved solutes and at low shear rates, since it is independent of shear and inversely proportional to particle size. On the other hand, shear-induced diffusion is more important with larger suspended particles. Depending on the prevailing mechanisms, this shear-induced lift velocity (V_L) is proportional to ($\gamma \cdot d_p^n$), where $n = 1.3$–4. Thus, as the velocity increases, larger particles in the feedstream will experience a greater lateral lift away from the membrane surface. This will cause a stratification of smaller particles

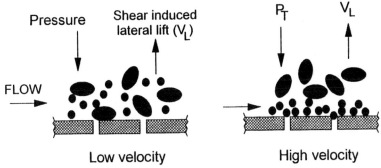

Figure 6.20. *Fouling by particulates of mixed sizes. The forces within the feedstream include shear due to cross-flow, hydraulic pressure (e.g., by a pump), and lateral lift of the particles due to shear. At high velocity, larger particles experience proportionally greater lift velocity, thus providing the smaller particles a greater chance to foul the membrane.*

on the membrane surface, which will foul membranes to a greater extent, perhaps by pore plugging (Figure 6.20).

This phenomenon could explain the greater fouling observed at higher shear rates in some cases (Baker et al. 1985; Kim et al. 1993; Kuo and Cheryan 1983; Mackley and Sherman 1992; Wakeman and Tarleton 1991). There will obviously be an interaction with the applied transmembrane pressure, as shown in Figure 6.19 for cheese whey. At low pressures (i.e., low permeation velocity), shear forces are high enough to minimize deposition of all particles on the membrane surface. At high pressures or high flux, particles move to the membrane surface at a much faster rate than their removal by shear, leading to greater fouling. Superimposed on this is the stratification of particles due to shear-induced lateral forces, resulting in greater rates of fouling at higher velocities.

Tubular modules have flow channels large enough (1.25–2.5 cm diameter) to allow most suspended matter to flow easily. Since flow velocities are usually 1–4 m/sec, this results in fairly high Reynolds numbers in the turbulent region, but pumping cost to maintain the high flow velocity is also quite high, sometimes as much as 100 times more energy per unit permeate volume removed than the spirals or hollow fibers (Chapter 7). At the other extreme are the hollow fiber modules, with inside diameters of 0.5–1.1 mm, which operate in the laminar region but under high shear rates (4000–14,000 sec^{-1}).

- Pressure: When transmembrane pressure is in the pre-gel region, flux increases as pressure increases, though usually not linearly for macromolecular feeds. As pressure increases further, the concentration polarization layer reaches a limiting concentration, and the flux becomes independent of pressure and becomes mass transfer–controlled. Any pressure increase beyond this point will only bring up the flux momentarily, but as soon as the equilibrium is reestablished between the rate of transport of solute to and from the membrane surface, the flux remains essentially unchanged. The situation changes, however, when fouling layers form and they begin to get compressed under the high pressures. Increasing pressure above a critical point may result in a lower flux. This was observed with UF of acid whey (Kuo and Cheryan 1983). Apparently, at high pressures, the fouling layer gets compacted and becomes less permeable. Higher cross-flow rates may aggravate the situation since a higher pressure drop is needed to create the higher flow (Figure 6.19). This can also be explained in terms of the ratios of convective to shear forces (see Section 6.E.4.).

With MF and high flux UF membranes, high pressures may cause severe fouling, as shown in Figures 6.21 and 8.16, probably due to compaction of the fouling layer. Methods to maintain low transmembrane pressures during operation are discussed later in Sections 6.E.3. and 6.E.4. Sometimes it is more important to maintain a low *flux* rather than low pressures to minimize fouling, as discussed in Section 6.F.

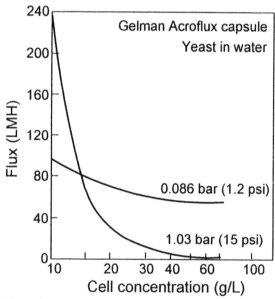

Figure 6.21. *Effect of transmembrane pressure on rate of fouling during concentration of an aqueous yeast suspension. The Gelman Acroflux capsule was a pleated sheet module with 0.09 m² area and pore size of 0.2 μ. Cross-flow rate was 2 L/min, initial volume 8 L, temperature 30°C (adapted from Patel et al. 1987)*

6.E.
FLUX ENHANCEMENT

As shown previously in Figure 4.35, there are several strategies to maximize flux. Most involve moving the fluid (Section 4.H.3.), and some involve moving the membrane (Section 5.C.). Some additional strategies that involve manipulating the net transmembrane pressure and fluid–membrane interactions are discussed in this section.

6.E.1.
TURBULENCE PROMOTERS/INSERTS/BAFFLES

The insertion of rods, wire rings, glass beads, kenics mixers, doughnut-, disc-, and cone-shaped inserts, baffles, or moving balls in the feed channel of membrane modules has been suggested as a way to minimize fouling. In some cases, fluxes that are 50–300% higher have been reported with model systems. Inserts reduce holdup in the feed channel, increase velocities and wall shear rates, and could produce secondary flows or instabilities if the velocity is high enough. Despite intense research in this area, however (see citations in Belfort et al. 1994; Gupta et al. 1995; Winzler and Belfort 1993), placing protuberances or inserts in the feed channel is not yet commercially practiced. One exception is the mesh-like spacer material in spiral-wound and plate modules, as discussed in Section 5.B.4. Although they are used primarily to keep the membrane sheets apart and to form the feed channel, they can also cause considerable turbulence and enhance flux. However, sometimes the spacers can be more harmful for certain feeds containing suspended particles, as shown in Figure 5.40. Of course, insertion of any turbulence enhancers in the flow channel is not without some expenditure in power consumption.

6.E.2.
BACKFLUSHING, -PULSING, -SHOCKING, AND -WASHING

The conventional mode of operating MF modules is to use high velocities to minimize polarization. This requires a high pressure drop (expressed as $Pr_i - Pr_o$ in Figure 6.22, left). Permeate channels are usually operated at low (close to atmospheric) pressures. On the retentate side, the high inlet pressure would cause fouling by compaction at the inlet of the module, while the low pressure at the outlet means that some of the available membrane area towards the outlet was not being utilized effectively. The resulting performance would be as shown in Figure 6.22 (bottom, left): a rapid drop in flux and an increase in rejection of solutes (e.g., Figure 6.3).

Periodic reversal of the filtrate flow back into the feed channel is commonly practiced in filtration. For many years, hollow fiber manufacturers recommended an occasional "lumen flush" during operation, which involved shutting off the permeate flow for a few seconds (Figure 6.22, right). This forces permeate

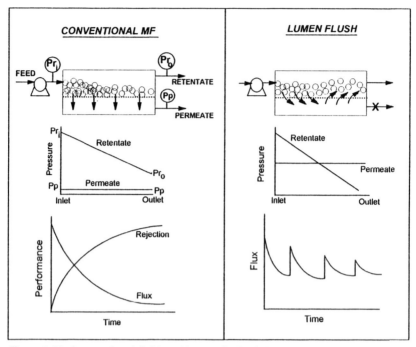

Figure 6.22. Conventional (left) and lumen flush (right) operating modes in cross-flow membrane filtration, showing pressure profiles and expected behavior of flux and rejection. P = pressure, r = retentate, p = permeate, i = inlet, and o = outlet.

back into the feed channel, presumably dislodging accumulated particles from the membrane surface. However, as shown in Figure 6.22, the pressure in the permeate channel would become the average of the feed channel pressures (the feed flow is not shut off during the lumen flush). Thus, only a portion near the outlet of the module would be affected, where the pressure in the permeate channel would be higher than the feed channel pressure. Although effective in some cases (Jonsson 1993), lumen flushing may not be as effective as some of the other pressure manipulation techniques discussed in this section.

Periodic backwash (PBW) is conducted by pumping permeate back into the feed channel to lift deposited material off the membrane surface (Figure 6.23). Its effectiveness also depends on the nature of the deposit (sticky gelatinous deposits and polarization by soluble solutes are less likely to benefit by back-washing). The backwash pressure should be greater than the normal operating inlet pressure to be effective. Most backwashing is done with permeate for 1–5 sec at a frequency of one to ten times per minute at pressures of 1–10 bar. Memtec (Australia) backflushes with pressurized air instead of liquid. When averaged over all operating cycles, PBW can provide higher flux (Figure 6.24), although its effectiveness may diminish with time (Padilla and McLellan 1993),

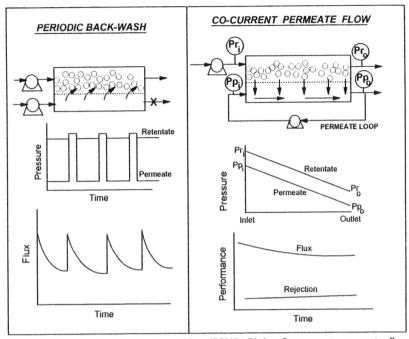

Figure 6.23. Left: Periodic backwashing (PBW). Right: Co-current permeate flow (CPF), also known as uniform transmembrane pressure (UTP).

especially if pore fouling is the main cause. Matsumoto et al. (1987) reported that backwashing for 5 sec every 5 min resulted in 100 LMH with 30 g/L yeast in water and 80 LMH with 32 g/L yeast in fermentation broth. Without backwashing, fluxes were 40 LMH and 20 LMH, respectively, at 10 g/L yeast.

More rapid backwashing—termed "backpulsing," or "backshocking"—is supposed to be even more effective. These pulses are generated by a diaphragm and solenoid valve arrangement placed in line on the permeate fitting of the membrane module. These backpulses are of extremely short duration (0.1 sec or less) and are operated continuously or periodically. Backpulsing may be especially useful with colloidal suspensions and with those streams requiring high protein transmissions through the membrane (Bhave 1995; Milisic and Bersillon 1986; Rodgers and Sparks 1993; Wenten 1995).

6.E.3.
UNIFORM TRANSMEMBRANE PRESSURE/CO-CURRENT PERMEATE FLOW

This is a variation of the PBW mode of operation and is particularly suited for MF of colloidal suspensions. As mentioned earlier, MF membranes quickly

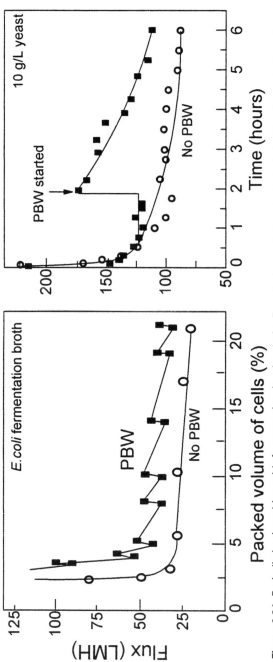

Figure 6.24. *Periodic backwashing with fermentation broths: Left: E. coli fermentation broth in Enka/Membrana hollow fibers (0.3 μ), 1 bar, 2 m/sec; PBW was done at 1.5 bar, 5 seconds every 5 minutes (adapted from Kroner et al. 1984); right: S. cerevisiae cells in water in Membralox ceramic module (0.2 μ), 0.83 bar, 7.3 m/sec, 30°C operated in the total recycle mode. Initial cell concentration 10 g/L. PBW for 5 sec every 5 min (adapted from Saglam 1995).*

lose their separation characteristics and high flux because of the rapid formation of a "dynamic," or secondary, membrane of polarized particles. To minimize the formation of this secondary membrane, the modules have to be operated at extremely high velocities (over 4–6 m/sec with ceramic membranes) and low transmembrane pressure (less than 2 psig/0.2 bar) to minimize compaction of the polarized layer. This appears to be a contradiction since high velocities, especially in narrow-diameter tubes, result in high pressure drops. To overcome this problem of simultaneously obtaining high velocities and low P_T, the module can be operated in the "uniform transmembrane pressure" (UTP) or "co-current permeate flow" (CPF) mode. As shown in Figure 6.23, this requires the simultaneous operation of a retentate pumping loop and a permeate pumping loop, to simulate a pseudo-"backwashing" operation, but in a continuous manner rather than the periodic or intermittent manner. With two parallel flows adjusted so that the pressure *drop* is the same on the permeate and retentate sides of the module, the pressure profile would be more like that shown in Figure 6.23 (right). A schematic of the process is shown in Figure 8.14.

As discussed later in Section 8.B.3., this mode of operation (termed *Bactocatch* by Alfa-Laval for use in the dairy industry and marketed by USFilter for other applications) has produced dramatic results. With low-fat and skim milk, flux of 500–900 LMH can be achieved for several hours at a transmembrane pressure of 0.1–0.8 bar/1.5–12 psi (Figure 8.15). Bacterial retention is 99% with the microbial load usually found in milk; on the other hand, there is no significant change in the concentration of other components, so the permeate is essentially bacteria-free skim milk. In the conventional mode, fluxes are much lower.

Figure 6.25 is another example of UTP/CPF. A Membralox 1P19-40 module was used with yeast cells suspended in water. A velocity of 6 m/sec required a pressure drop of $\sim$24 psi. In the conventional mode, this was obtained with a $Pr_i = 24$ psi and $Pr_o = 0$ psi, for a transmembrane pressure (TMP) of 12 psi (0.83 bar). To obtain a TMP of 12 psi in the UTP mode, the settings shown in the figure were made, so that P_T at the inlet of the module $= Pr_i - Pp_i = 40 - 28 = 12$ psi, and P_T at the module outlet $= Pr_o - Pp_o = 12 - 0 = 12$ psi. This implies that the TMP is uniform throughout the length of the module. This resulted in an average flux of 175 LMH, compared to 100 LMH in the conventional mode. Reducing TMP to 5 psi resulted in even higher average flux, as explained below.

6.E.4.
PERMEATE BACKPRESSURE

With high-flux UF and MF membranes, the lower the TMP, the lower the initial flux, but the long-term flux is better (Figure 6.21). Low TMP or low flux presumably minimize plugging of intermediate-sized pores of the same size range as the particles in the feed (larger pores will get plugged in any

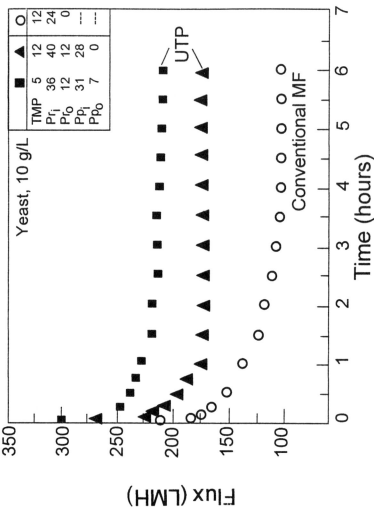

Figure 6.25. Comparsion of conventional and UTP methods of operation with yeast suspended in water (10 g/L). The membrane was Membralox 0.2μ 1P19-40 module, operated at 30°C and 6 m/sec in the total recycle mode (adapted from Saglam 1995).

case). However, it is also important to have shear forces (τ_W) in the axial direction larger than forces drawing the particles in to the pores; i.e., the module should operate at a certain critical value of τ_W/J. This is illustrated in Figure 6.19: fouling is less at low pressures (i.e., low values of J) and high velocity (high τ_W). Fouling is greater when J increases, especially if not accompanied by a proportional increase in τ_W. Even in the UTP mode, reducing the TMP to 5 psi gave better flux than 12 psi (Figure 6.25). With tubular inorganic membranes and dairy feeds, this critical ratio is 0.8–1.2 Pa/LMH (e.g., see Gesan et al. 1995). Needless to say, the manner in which the system is started up is also critical.

Low TMP and/or low flux can be obtained by maintaining very low feed pressures. However, as discussed earlier, high feed pressures may be necessary in some cases, e.g., high viscosity of the feed or to maintain a high cross-flow velocity. A simpler alternative to the backflushing or UTP modes is to maintain a high permeate backpressure. At startup, the permeate valve is completely closed. After the feed pressures have been established, the permeate valve is slowly opened to the optimum backpressure. In some cases, this optimum backpressure may actually be slightly higher than the retentate outlet pressure. An example is shown in Figure 6.26 with stillage, which is a by-product of fermentation ethanol manufacture (Section 8.D.2.). At the same TMP and cross-flow

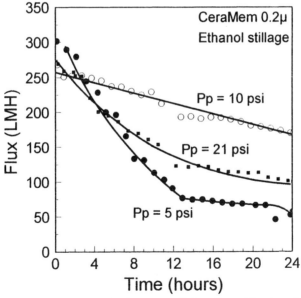

Figure 6.26. *Effect of permeate backpressure (P_p) on flux with ethanol stillage in a CeraMem ceramic module (2-mm channels). The values of Pr_i and Pr_o were adjusted to result in a pressure drop of 15 psi and a TMP of 15 psi at the designated value of P_p. Temperature was 80°C (adapted from Rane and Cheryan 1996).*

velocity, average flux increased substantially when the backpressure (P_P) was increased from 5 to 10 psi, but decreased when P_P was increased to 21 psi, probably because of increased fouling caused by the correspondingly higher feed side pressures.

A note of caution if these pressure manipulation techniques are attempted: The module should be robust and the membrane self-supporting; i.e., it should be able to withstand backpressurization with no damage. This eliminates most polymeric tubular, plate, and spiral modules. These techniques are most effective with rigid modules such as inorganic modules. Hollow fibers have been used but may be less effective with rapid backpulsing techniques because of their lack of rigidity. In addition, backpressure manipulations are more likely to show benefits with high-flux, high-porosity membranes (e.g., MF) than with tight UF membranes and where true fouling by solute–membrane interactions are not controlling.

6.E.5.
INTERMITTENT JETS

Unsteady flows and large vortices can also be generated with an intermittent jet of the feed pumped through a nozzle placed coaxially in a membrane tube. This causes the velocity of the feed to abruptly increase and then decrease, creating a toroidal vortex (Arroyo and Fonade 1993; Geppert and Thielemann 1983). Flux with bentonite suspensions were two to two and a half times higher with intermittent jets compared to conventional flow.

6.E.6.
PULSATILE FLOW

Oscillations and unsteady flows can also be obtained by pulsations in the feed or permeate channels. Pulsatile flow can be obtained with pistons (Gupta et al. 1992; Kennedy et al. 1974); flexible tubing in the feed flow path, either at the inlet or outlet of the module (Bertram et al. 1993; Gupta et al. 1985); a rotating perforated disc placed at the inlet of a tubular module (Spiazzi et al. 1993); and solenoids and valves in the permeate line to generate negative TMP pulsations, similar to backshocking (Rodgers and Sparks 1993; Miller et al. 1993). Increases in flux and protein transmission have been reported with these techniques.

6.E.7.
ELECTRICAL METHODS

Fouling due to charge interactions between charged solutes and the membrane would be relatively unaffected by manipulating velocity and pressures, as discussed above. In "electrofiltration," the electrical properties of charged particles are used to enhance flux. These electrokinetic properties (magnitude and sign of the charges on a molecule) also depend on ionic strength and pH of the solution. An electric field can be applied to the flowing fluid, with one electrode being cast on or placed against the membrane, and the other electrode within the liquid, but away from the membrane surface (e.g., in the center of a membrane tube). Two different phenomena occur: electrophoresis, which causes transport of dispersed solutes away from the membrane, and electroosmosis, which causes electrolytes to pass through the membrane. Charged particles will move away from the membrane surface, depending on the electric field strength applied, thus reducing the extent of concentration polarization and increasing flux.

There are numerous laboratory examples of electric fields enhancing flux in MF and UF (Bowen 1991; Jagannadh and Muralidhara 1996; Vradis and Floros 1995). Henry et al. (1977) identified a critical field strength where electrokinetic and convective transport forces balance each other, resulting in no net movement of charged solutes towards the membrane. This is the ideal field strength, and the beneficial effects of cross-flow velocity can be superimposed on this effect. However, there is danger of electrolysis at the electrodes, gas being generated, and a pH change of the product streams proportional to the applied voltage (although gas generation at the membrane surface has been suggested as a way to increase flux). Corrosion of electrodes and high power cost has inhibited the commercial practice of this technique.

Pulsed electric fields, where the field is switched on and off at regular intervals, has an even greater benefit on flux with less of the problems of continuously applied electric fields (Bowen 1991; Brors and Kroner 1992; Robinson et al. 1993; Wakeman and Tarleton 1987).

6.F.
SUMMARY: MEMBRANE FOULING

Some methods of retarding fouling are effective and are used commercially, but others—electric fields, intermittent jets, acoustic methods, and pulsatile flows (other than backpulsing)—are still in the laboratory stage with scaleup and economic issues yet to be resolved. With high-flux UF and MF systems, it may be necessary to operate at suboptimum conditions initially to minimize long-term fouling. This means operating at low pressures or below a certain "critical flux" or above a critical shear stress/flux ratio to control the rate at which particles come to the membrane surface in relation to the rate at which they are removed by shear stresses. It may be that each feed-membrane combination has a particular *flux* (rather than pressures or cross-flow velocities) above which there will be rapid fouling.

These concepts require "slow-start" procedures when introducing feed into the membrane modules. Startup conditions are critical, since even a momentary excursion above the critical flux results in irreversible fouling; e.g., a slow start resulted in 13–26% higher long-term flux than abrupt startup procedures during ceramic MF of raw sugar solutions (Dornier et al. 1995). Maintaining a high permeate backpressure (at startup and throughout the cycle), ramping up the motor of the pump to avoid undue pressure bursts, and arriving at the final (pre-determined) TMP several minutes or even hours after startup allow operation below the critical flux and minimize certain types of fouling. The phenomenon of critical flux and critical shear stress/flux ratio is now being elucidated in the scientific literature (e.g., Field et al. 1995; Gesan et al. 1995). Operating below this flux may result in little or no flux decline. This critical flux has to be determined in each case experimentally. These concepts have been known intuitively (and practiced) for many years, except it was usually expressed in terms of operating at or below a certain critical *pressure* to minimize fouling.

However, the key economic factor in membrane technology is not flux per se, but productivity, which is expressed in terms of V_P, the volume of permeate produced per cycle, i.e., between cleanings:

$$V_P(\text{L}) = J_{av}(\text{L/m}^2 \cdot \text{h}) \cdot A(\text{m}^2) \cdot t(\text{h}) \tag{6.9}$$

where J_{av} is the average flux in the operating period t. Overall productivity could be higher if operated above the "critical" flux, even if it means a decline in flux during operation and more frequent cleaning cycles (e.g., every 12 h instead of every 24 h). This has to be evaluated on a case-by-case basis.

6.G.
CLEANING MEMBRANES

Cleaning is the removal of foreign material from the surface and body of the membrane and associated equipment. The vast majority of the literature over the past 2 decades has focused on "fouling" rather than *cleaning*, even though what appears to be a fouling problem may really be a *cleaning* problem. Considerable progress has been made in this period on understanding the interactions between the foulants (soil), the membrane, and the operating conditions during cleaning (Dychdala 1993). The frequency of cleaning is a critical economic factor, since it has a profound effect on the operating life of a membrane. Indeed, it may be more meaningful to specify membrane lifetime on the basis of number of cleaning cycles rather than in terms of operating time.

Cleaning and sanitizing membranes is desirable for several reasons: (1) laws and regulations may demand it in certain applications (e.g., the food and biotechnology industries), (2) reduction of microorganisms to prevent contamination of the product streams, and (3) process optimization—it may be better to take time off for cleaning and restoring the flux, rather than continuing with a fouled membrane with a low flux.

There are three meanings that we can attach to a "clean" membrane:

- A *physically* clean membrane is one that is free from visible impurities or foreign matter.

- A *chemically clean* membrane has "all" foulants and impurities removed.
- A *biologically clean*, or *sanitized*, membrane is free of all viable microorganisms (Dychdala 1993).

Most plant operators consider a membrane to be "clean" when the previous, or "new membrane," water flux has been restored. Actually, it may not be possible to obtain the initial water flux (i.e., when the membrane was new); water flux (even when only water is processed) typically drops to a stable value after a few runs. In some cases, the importance of restoring the previous water flux can be overemphasized and may not be necessary or practical. The important criterion is that the previous *process* flux should be restored.

Rejection of individual components of a feedstream respond differently to cleaning. Gradual adsorption of feed components on the membrane during processing can increase the rejection of a partially rejected component. This will obviously be improved by cleaning, at least in the initial stages of postcleaning operation. However, rejection of highly rejected or poorly rejected molecules rarely improves upon cleaning.

In order to clean a membrane, three types of energy inputs are required:

- *Chemical energy*, in the form of detergents or cleaners: Cleaning is essentially a physicochemical reaction. The cleaner serves to solubilize or disperse the foulant or soil.
- *Thermal energy*, in the form of heat: Due to the nature of the action of chemical cleaners, its efficiency will increase with temperature. In general, the hotter the solution, the more effective is the cleaning treatment.
- *Mechanical energy*, in the form of high velocities in pipelines, and perhaps actual scouring when chemical cleansers are not effective enough

The effect of these energies on cleaning efficiency is affected by a fourth factor: time of cleaning (Dychdala 1993).

The general principles of cleaning and sanitation are just as applicable to membranes as to other contact surfaces. The nature of the surface influences the rate of deposition of the foulant/soil, but it is the nature of the soil and its interaction with the cleaner that controls the cleaning process. One key difference from other contact materials is that membranes have pores and are made of relatively active surfaces. This can result in considerable absorption and adsorption of compounds from the feedstream. It is also important to remember that cleaning of the permeate side is just as important as the feed/retentate side, especially if the permeate stream contains the desired product. The type of foulant will also determine the type of cleaner that should be used: one that solubilizes the soil or disperses it. This is important because some soluble components that are small enough to pass through membrane pores could precipitate in the pores during rinsing.

The most important rule of cleaning is never to let the membrane dry out after contact with the process stream. At the end of the run, the process stream should be immediately followed by a water rinse of the entire membrane plant, including tanks, pipelines, pumps, etc., until the exit water appears clean.

6.G.1.
IMPORTANT FACTORS DURING CLEANING

- Membrane materials and chemistry: This determines a membrane's ability to withstand the action of the chemical cleaners. Cellulose acetate is the least tolerant to extremes of pH and temperature, although occasional and/or brief exposure of cellulose acetate to acid and alkaline cleaners is practiced. Polyamides (which form the active separation layer of many thin-film composite membranes) are very sensitive to chlorine. Ceramic/inorganic membranes can withstand high acid, alkali, and chlorine concentrations at high temperatures, as well as a wide variety of chemicals. However, although the membrane material itself may be resistant to these conditions, due consideration should be given to the gaskets, epoxy materials, plasticizers, overwraps, etc., that are part of the membrane element/housing/module.

- Fluid mechanics: Cleaning solutions should be pumped through the system under turbulent flow conditions, generally at Reynolds numbers greater than 2100 (preferably 3000–5000, which typically means linear velocities of 1.5–2 m/sec in tubular modules). If these high Reynolds numbers are not possible, e.g., with spiral modules, the maximum possible linear velocity in the feed/retentate channel should be used during cleaning, especially if surface fouling is the main problem. In addition, the pressures should be as low as possible, but consistent with the pressure drop required to maintain high flow rates. High transmembrane pressures merely serve to refoul the membrane with the foulant. Hollow fibers used for MF and UF generally operate in laminar flow, but they operate at high linear velocities and high shear rates at the membrane surface, which facilitates cleaning.

  If pore fouling has occurred or if the permeate side of the module is to be cleaned adequately, moderately high transmembrane pressures should be utilized during cleaning, but only after surface soil has been removed. Care should be taken when sending aggressive cleaning solutions, especially those containing chlorine, through associated equipment such as conductivity cells and rotameters with plastic floats.

  Backflushing (Section 6.E.2.) can be practiced with membrane configurations where the membrane is a self-supporting structure with no danger of peeling off from the membrane backing or support layer, e.g., hollow fibers and inorganic membranes. Backflushing as a cleaning technique has been found to be helpful with colloidal and particulate fouling, but relatively ineffective with macromolecular or small soluble solutes. This backflushing during cleaning is not to be confused with "periodic backwashing" practiced with some membrane units during microfiltration to enhance flux (Section 6.E.2.).

Air entrapment and foaming should be minimized during cleaning cycles, especially with enzyme cleaners, because it may cause inactivation of the enzyme.

- Time: In general, most chemical cleaners complete their action within 30–60 min. Prolonged cleaning after the optimum time may actually cause refouling of the membrane due to the filtration effect. It is better to use several cleaning steps successively with fresh cleaning solutions. Thirty minutes is optimum with most chlorine-containing cleaning solutions. Enzyme cleaners may require 60–90 min, or perhaps longer, depending on the soil and enzyme activity.

- Temperature: Chemical reaction rates double with a 10°C increase in temperature. Thus, the temperature of the cleaning solution should be as high as possible, consistent with temperature limitations of the membrane/module. For cleaning lipids, fats, or oils, the temperature of the cleaning solution and rinse water should be higher than the melting point of the fat. Temperatures of 55–60°C should suffice.

- Water quality: The quality of the water used for cleaning is an oft-neglected factor. Ideally, soft water should be used. If hard water is used, sufficient amounts of complexing agents ("builders") must be added. Typical examples of builders are polyphosphates (which are not recommended for environmental reasons), phosphonates, gluconates (which are not as effective as the others), citrates, EDTA, and NTA (the last two are water conditioning agents).

Of special concern are the levels of iron, manganese, alumina, and silica in the water. If silicates are less than 5 ppm, then iron should be less than 0.5 ppm and manganese less than 0.2 ppm. If silicates are in the 30–50 ppm range, then iron should be less than 0.05 ppm and manganese less than 0.02 ppm. A dirty yellow color on the membrane surface is an indication of high iron in the wash water. It may be the main reason for long-term decrease in UF membrane performance (Armishaw 1982).

Silicates usually occur in a microcolloidal, high-polymer form, which can be retained by most membranes and can severely foul them. Since silicates cannot be easily removed from water, it is important that hardness is sufficiently reduced or complexed by the addition of builders. The same is true for alumina, which should be below 0.05 ppm. Chloride salts (e.g., NaCl) should be kept as low as possible, especially with acid cleaners since it will otherwise promote corrosion of metallic components.

From a microbiological point of view, total bacterial count should be of drinking water quality ($<100/\text{mL}$) and *E. coli* should be zero. Chlorine should be as low as possible (zero in some cases) for cellulose acetate, polyamide, and thin-film composite membranes.

For these reasons, permeate from reverse osmosis is preferred for flushing, cleaning, and diafiltration operations, since this is often the best quality water in a plant. Table 6.1 lists some specifications of water for cleaning.

- pH: The pH of the cleaning solution or rinse water depends on the type of foulant and membrane. Alkaline cleaners (up to pH 12, if compatible with the membrane) containing NaOH or KOH are particularly effective for organics and proteins. They act by solubilizing these materials. Carbonates (Na_2CO_3, K_2CO_3) aid cleaning through their pH regulation properties. Polyphosphates are added to act as dispersants and also help to solubilize carbonates and emulsify fats. (Remember, no silicates should be present, not even the water-soluble kind.) High pH may saponify fats and make them easier to remove, but if any calcium or magnesium is present, it can react with the soap and cause problems.

    Acidic cleaners (pH 1.5–2.8) are primarily used to combat inorganic salt fouling. Phosphoric acid is the least aggressive in terms of handling and for the membranes per se (except for some alumina membranes, where they should not be used). In addition, it has its own detergent action because of the phosphate groups. Nitric acid is very corrosive but effective. Citric acid is especially good with iron deposits. Blends of acids may be particularly effective. Sulfuric and hydrochloric acids should not be used since they are especially hard on stainless steel components. Sometimes these acids can inhibit the removal of protein.

    The pH of the cleaning solution during the cleaning cycle should always be monitored to ensure that it is within the recommended limits of the membrane. With the many prepackaged cleaning formulations now readily available on the market, there is a tendency to just mix the ingredients and pump it through the system. Considering the tremendous damage that can be done during the cleaning cycle, it would be wise to at least measure the pH of the cleaning solution prior to use, if not actually monitoring it continuously on-line during the cleaning cycle. (Keep in mind that the effect of pH depends on the temperature and time of exposure. For example, acidic solutions can be used with CA membranes, but for short periods at low temperatures).

    The pH of the last cleaning solution that comes in contact with the membrane appears to have an influence on the water flux measured after the cleaning (Bragulla and Lintner 1986). With polymeric membranes, water flux is higher if the last cleaning solution is alkaline. An acid rinse often leads to low flux. On the other hand, inorganic membranes (e.g., zirconia) show higher water fluxes with a last acid rinse. In fact, the positive effects apparently last long enough with inorganic membranes to result in an improved flux with acid whey. With polymeric membranes, however, the effect of the alkaline cleaner (which may be to "open" the pores) does not last long enough to be effective with the process feed.

6.G.2.
TYPICAL FOULANTS AND SOILS

- Protein: In general, proteins are most soluble at high and low pH and least soluble at pH 4–5 (their isoelectric point). High pH is preferred for protein foulants, not only because proteins are slightly more soluble than at low pH, but also because of possible "peptization" (hydrolysis) of the protein, which expedites cleaning.
- Fats, oils, and grease: Fatty deposits are especially difficult to remove, except at high temperatures or with organic solvents. Some fats such as tallow or lard may be initially unresponsive to high temperatures (Figure 6.27). Fatty deposits have a greater affinity for hydrophobic synthetic polymers than for hydrophilic polymers or inorganic materials. Fats are easiest to remove from glass, followed by stainless steel, acrylics, polyethylene, polyvinyl chloride, and polysulfone, the latter being the most troublesome among the common membrane materials. Fats and oils will actually be less of a problem with hydrophilic membranes than hydrophobic membranes; e.g., it is usually easier to clean fatty deposits off a CA membrane, but this advantage is

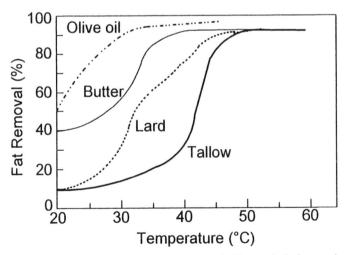

Figure 6.27. *Effect of water temperature on removal of fats and oils from polysulfone membrane (adapted from Bragulla and Lintner 1986).*

counterbalanced by CA's limited temperature and pH. If excess calcium ions are present, i.e., more calcium than can be handled by chelating agents in the cleaning solution, then calcium soaps may be formed with caustic cleaners. These form insoluble deposits that can plug up membrane modules.

- Carbohydrates: Low molecular weight sugars are readily soluble in water and thus need no special cleaners. Starches, polysaccharides, fiber, and pectin materials may need some special treatment. It has been found that the order of cleaning is critical with pectin foulants, as shown in Figure 6.28.

- Salts: These can come not only from the process fluid, but also from water and even from additives such as emulsifiers. Acids and sequestering agents such as EDTA are used to dissolve salt foulants. Citrates are particularly effective due to their combined detergent and sequestering activities.

6.G.3.
CLEANING CHEMICALS

Several companies sell chemical cleaning compounds specifically for membranes, among them Ecolab-Klenzade, Diversey, H. B. Fuller's Monarch Division, and Pfizer. Many membrane manufacturers also supply their own cleaning reagents, either as powders or liquids. Powders are preferred because of ease of transportation, formulation, cost, and flexibility, although liquids are easier to dose.

Cleaning solutions should be used only after due consideration of the type of soil and the membrane. Final selection of the membrane/module should be based as much on cleanability and sanitation as flux and rejection properties. Clean-in-place (CIP) procedures should take into account the order of usage of chemicals. For example, with milk fouling, if alkaline cleaning alone is insufficient, a sequence of alkaline–acid–alkaline works best. On the other hand, with cheese whey, since fouling is primarily caused by the interaction or deposition of salts on the membrane, the first step should be an acid cleaner, followed by an alkaline cleaner. For pectin fouling, alkaline alone may be better (see Figure 6.28).

Table 6.4 lists typical cleaning reagents and their modes of action. Chlorine is an extremely effective membrane cleaner, especially when pore fouling occurs. It is not recommended for polyamide surfaces (e.g., thin-film composites), but it can be used at levels of 50 ppm for a few minutes with CA membranes and at 200–400 ppm with PES and many other polymeric membranes. It

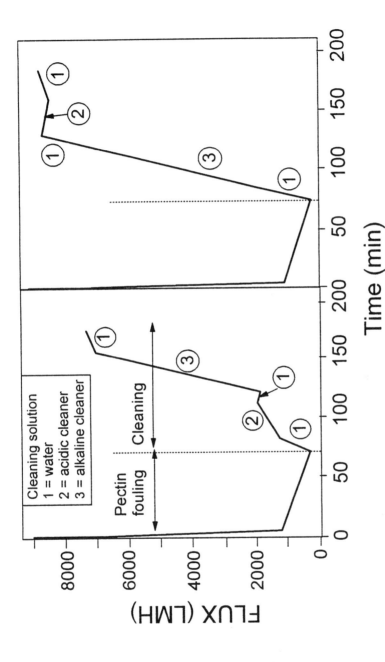

Figure 6.28. Effect of acidic (Ultrasil 75) and alkaline (Ultrasil 11) cleaning solutions on cleaning a polysulfone membrane after fouling by pectin (adapted from Bragulla and Lintner 1986).

Table 6.4. Typical cleaning reagents and their modes of action.

Foulant	Reagent	Time and Temperature	Mode of Action
Fats and oils, proteins, polysaccharides, bacteria	0.5N NaOH with 200 ppm Cl_2	30–60 min 25–55°C	Hydrolysis and oxidation
DNA, mineral salts	0.1M–0.5M acid (acetic, citric, nitric)	30–60 min 25–35°C	Solubilization
Fats, oils, biopolymers, proteins	0.1% SDS, 0.1% Triton X-100	30 min–overnight 25–55°C	Wetting, emulsifying, suspending, dispersing
Cell fragments, fats, oils, proteins	Enzyme detergents	30 min–overnight 30–40°C	Catalytic breakdown (proteolysis)
DNA	0.5% DNAase	30 min–overnight 30–40°C	Enyme hydrolysis
Fats, oils, and grease	20–50% ethanol	30–60 min 25–50°C	Solubilization

Source: Adapted from Pall Filtron (1995).

works by a combination of hydrolysis and oxidation of the soil. With polymeric membranes, it apparently "opens" the pores sufficiently to allow the foulant to be flushed out at high pressure with the chlorine solution or quickly with water. With inorganic membranes, chlorine at 1000–2000 ppm is especially effective.

Of course, in all cases, the pH of the chlorine solution should be alkaline (pH 10–11) to minimize corrosion. Due care should be taken regarding the effect of chlorine on other wetted parts of the system. The chlorine solution must be freshly made. Household bleach can be used: 4 ml of 5% bleach in 1 L of 0.5 N NaOH will give a 200 ppm solution. However, with severely fouled membranes, the chlorine will rapidly be depleted. Thus, it is important to measure and dose chlorine into the cleaning solution continuously during cleaning, if necessary, to ensure the most effective and rapid cleaning.

Caustic solutions (e.g., NaOH) act by breaking bonds between the membrane surface and the fouling material. At high enough concentrations and temperatures, it could cause "peptization" of proteins and thus help solubilize them. However, using plain NaOH may not be effective since it will be difficult to control pH during the cleaning. Buffers and "builders" may be needed in the cleaning solution. Sodium carbonate, soda ash, or phosphates are good builders and buffering compounds, provided they are completely solubilized before being used to clean the membrane system. (Silicates should not be used under any circumstances.) Using builders and sequestering agents can actually reduce the pH needed for efficient cleaning.

Acids are a good way to get rid of inorganic salts. Citric acid combines acidity with detergency and chelating ability and is preferred for salt fouling, especially salts of Ca, Mg, Mn, and Fe. HCl is very corrosive and should be avoided.

Enzymes are effective for troublesome fouling problems, especially with proteins. They are slow to act, however, and expensive and are generally used as a last resort. Enzyme use should be followed by a chemical cleaning, since enzymes are proteins that may foul the membrane if left as a residue.

Organic solvents, e.g., ethanol solutions, can be used to remove hydrophobic materials such as fats, oils, and grease, depending on the compatibility of the membrane to the solvent. Chlorinated hydrocarbons and petroleum solvents should not be used on membranes used for aqueous applications.

Surfactants are valuable components of a cleaning solution. These are compounds that possess both hydrophilic and hydrophobic functional groups. Their main purpose in a cleaning solution is to enhance wettability and rinsability, improve contact between foulants and cleaning agents, and ultimately reduce water usage and cleaning time. They can be anionic, cationic, or nonionic. Examples of anionic wetting agents are soaps and alkylsulphonates. The best example of a cationic surfactant is quartenary ammonium compounds (QACs), which have a formula of NH_4^+-R, where R is an alkyl group. QACs are not recommended with some membranes such as polyamide-based thin-film composites. These membranes have a large proportion of free carboxyl groups on the membrane surface, which react with the positively charged cationic detergent. The resulting large number of alkyl groups on the membrane surface changes it from hydrophilic to hydrophobic, resulting in a large loss of water flux even after rinsing. QAC also interacts with stainless steel and glass. It is difficult to rinse out with water and has been found to inhibit dairy cultures. Excessive foaming is also a problem.

Nonionics, such as phenol compounds and ethylene oxide, are amphylotes that can have a synergistic effect on anionic and cationic surfactants. The nonionics possess hydrophilic groups (SO_3, NH_4, CHO, CH_2OH) and hydrophobic regions within the molecule. Nonionics also adsorb to glass and stainless steel, causing possible degradation of the membrane and/or module components over the long term (e.g., inducing stress cracking by dissolving the plasticizers out of PS). This is the main problem using commercial laundry detergents as membrane cleaners. Great care must be exercised when using surfactants. They themselves can be adsorbed by the membrane, causing more problems.

6.G.4.
SANITIZERS

The purpose of a sanitizer is to reduce microbial concentration down to acceptable levels. Sterilization, on the other hand, infers the inactivation of all microorganisms such that they cannot grow or reproduce. Microorganisms are ubiquitous and multiply rapidly. It is not unknown for certain molds to utilize the membrane material itself as carbon or nitrogen sources for growth. However, it should be remembered that sanitizing or sterilization without prior cleaning is of no value.

The first step in a cleaning/sanitizing program, which is the removal of loose soil by rinsing, will usually remove about 90% of all microorganisms. The next step, using chemical cleaners, can remove 99.9%, while sanitizers can reduce microbial counts by 99.999%. Unlike inorganic membranes (which can be steam sterilized), the best that can be obtained with chemical sanitizers is a 5-log cycle reduction in microbial activity.

Each type of microorganism responds differently to sanitizers. A good sanitizer should (a) kill many microorganisms of different types (broad spectrum), (b) be effective at low concentrations, (c) be active even in presence of soil, (d) have a lasting effect but be easy to remove, (e) not be toxic to the user, (f) not be corrosive to equipment, and (g) not leave harmful residues in the food product.

The most common disinfectants are chlorine, hydrogen peroxide, and peracetic acid. It may be advisable to routinely change the sanitizer to minimize chances of building up a class of resistant microorganisms.

- Chlorine (Cl_2) is a universal disinfectant, and its effectiveness depends on concentration, time, and pH. For disinfection of water, a residual Cl_2 level of 0.5 mg/L is used. Fouling of water intake lines, heat exchangers, sand filters, etc., in a water treatment scheme can be minimized with this level of free residual Cl_2. Residual chlorine refers to the total amount of chlorine (combined and free chlorine) remaining in the water at the time of measurement. Combined available chlorine refers to one or more of the family of chlorine–ammonia compounds, called chloramines, resulting from the reaction of chlorine with ammonia compounds already present in water. Free available chlorine is actually either hypochlorous acid or hypochlorite ion or a mixture of the two, depending on pH and temperature. Free chlorine is usually present after sufficient chlorine has been added to satisfy the demand of ammonium ions present.

  Chlorine is most commonly available as hypochlorites of calcium and sodium or chlorine gas.

$$Cl_2 + H_2O \rightarrow HOCl + H^+ + Cl^- \qquad (6.10)$$

  Hypochlorite (NaOCl) decomposes into $NaCl^+$ oxygen. It is a broad spectrum agent available in liquid form, it is fast acting, and it is economical. However, below pH 9, NaOCl generates Cl_2 gas. This is corrosive, even for stainless steel, and is inherently unstable in the presence of organic compounds. It is extremely aggressive, especially to polyamide membranes. Odor and flavor of food become affected by residual chlorine.

In case it is necessary to remove residual chlorine from tapwater (e.g., for use with polyamide or thin-film composite membranes), the following can be done:

Pass through activated carbon:

$$C + 2Cl_2 + 2H_2O \rightarrow 4HCl + CO_2 \qquad (6.11)$$

Treat with metabisulfite:

$$NaHSO_3 + HOCl \rightarrow HCl + NaHSO_4 \qquad (6.12)$$

In theory, 1.37 lb of sodium metabisulfite will neutralize 1 lb of free chlorine; in practice, 3 lb are used.

- Iodophores (iodine + surfactants) are broad spectrum agents effective at low concentrations. However, it stains plastic components, rinses poorly, and is aggressive to polymeric membranes and to the support materials. It has been known to cause stress cracking with polysulfone and polyamide membranes. It is generally not recommended as a sanitizer.
- Quartenary ammonium compounds are generally noncorrosive but may reduce flux due to adhesion to surfaces. QACs are selective in activity but rinse poorly. They are bacteriostatic (not bactericidal) at low concentrations in the product. They destabilize foam in beer, so absolutely should not used in the beer industry. It is best if QACs do not come in contact with any membrane.
- Metabisulfite, at slightly acidic pH, releases SO_2, which is a reducing agent. It is nonaggressive to membranes and leaves no harmful residues. It is active against bacteria, yeasts, and molds. It is good for storage of membranes, especially for polyamides since it is a reducing agent. However, its mode of action is slow, and there are suggestions that it may be corrosive at acidic pHs.
- Hydrogen peroxide is a broad spectrum disinfectant and relatively noncorrosive and leaves no harmful residues since it decomposes into oxygen and water. It is slow acting and is less effective at low temperatures, and it is incompatible with some polyamide membranes.
- Per(oxy)acetic acid is a mixture of H_2O_2 and acetic acid and is a broad spectrum agent active on bacteria, viruses, mold, and spores. It releases oxygen, which is active on microorganisms. This is useful since the active ingredient can get to the permeate channel of RO membranes, since oxygen gas will go through RO membranes. It is noncorrosive to stainless steel, leaves no harmful residues, is fast acting at low temperatures, and is both sporicidal and virucidal. It can be used on cellulose acetate, polysulfone, and inorganic membranes but generally

not on polyamides, unless the manufacturer specifically approves it. However, there are reports from the field of its use at low temperatures. It is nonfoaming and has a low biochemical oxygen demand/chemical oxygen demand (BOD/COD) content.

• Formaldehyde is generally used for storage at concentrations of 0.5–1%. However, great care should be taken to ensure that all organics have been removed and the membrane is absolutely clean. Otherwise, aldehydes form insoluble complexes with proteins.

In summary, membrane lifetime will be affected by the cleaner and procedures used. Certain applications may require compliance with regulatory agencies, especially in processing of food and pharmaceuticals. These requirements are to ensure that ultrafiltered or microfiltered products are not subsequently contaminated by extractables or residues of the cleaners. It is not too farfetched to state that, in many applications, selection of the membrane/module depends heavily on how easy it is to clean and sanitize.

REFERENCES

AKHTAR, S., HAWES, C., DUDLEY, L., REED, I. and STRATFORD, P. 1995. *J. Membrane Sci.* 107: 209.

ARMISHAW, R. F. 1982. *N.Z. J. Dairy Sci. Technol.* 17: 213–228.

ARROYO, G. and FONADE, C. 1993. *J. Membrane Sci.* 80: 117.

BAKER, R. J., FANE, A. G., FELL, C. J. D. and YOO, B. H. 1985. *Desalination* 53: 81.

BELFORT, G., DAVIS, R. H. and ZYDNEY, A. L. 1994. *J. Membrane Sci.* 96: 1.

BERTRAM, C. D., HOOGLAND, M. R., LI, H., ODELL, R. A. and FANE, A. G. 1993. *J. Membrane Sci.* 84: 279.

BHAVE, R. R. 1995. Personal communication. USFilter, Warrendale, PA.

BOWEN, W. R. 1991. In *Chromatographic and Membrane Processes in Biotechnology.* C. A. Costa and J. S. Cabral (eds.), Kluwer Academic Publishers, The Netherlands. p. 207.

BRAGULLA, S. and LINTNER, K. 1986. *Sonderdruck aus Alimenta* 5: 111–116.

BRORS, A. and KRONER, K. H. 1992. In *Proc., 9th International Biotechnology Symp.* M. Ladisch and A. Bose (eds.) American Chemical Society, Washington, DC. p. 254.

BRINK, L. E. S. and ROMJIN, D. J. 1990. *Desalination* 78: 209.

BSI. 1996. Company literature. Eden Praire, MN.

BUSBY, T. F. and INGHAM, K. C. 1980. *J. Biochem. Biophys. Methods.* 2: 191.

CABRAL, J. M. S., CASALE, B. and COONEY, C. L. 1985. *Biotechnol. Lett.* 7: 749.

CHAMCHONG, M. and NOOMHORM, A. 1991. *J. Food Process Engr.* 14: 21.

CHERYAN, M. 1986. *Ultrafiltration Handbook.* Technamic, Lancaster, PA.

CHERYAN, M. and CHIANG, B. H. 1984. In *Engineering and Food, Volume 1.* B. McKenna (ed.), Applied Science Publishers, London. p. 191.

CHERYAN, M. and MERIN, U. 1980. In *Ultrafiltration Membranes and Applications.* A. R. Cooper (ed.), Plenum, New York. p. 619.

DAUFIN, G., MICHEL, F. and MERIN, U. 1992. *Aust. J. Dairy Technol.* 47: 7.

DEFRISE, D. and GEKAS, V. 1988. *Process Biochem.* 23: 105.

DILLMAN, W. J. and MILLER, J. F. 1973. *J. Colloid Interface Sci.* 44: 221.

DORNIER, M., PETERMANN, R. and DECLOUX, M. 1995. *J. Food Engineering* 24: 213.

DYCHDALA, G. 1993. *The Chemistry of Membrane Cleaning.* EcoLab-Klenzade technical bulletin.

FANE, A. G., FELL, C. J. D. and SUKI, A. 1982. Presented at the *Symposium on Membranes and Membrane Processes*, Perigia, Italy.

FANE, A. G., FELL, C. J. D. and SUKI, A. 1983. *J. Membrane Sci.* 16: 195.

FIELD, R. 1996. In *Industrial Membrane Separation Technology* K. Scott and R. Hughes (eds.), Blackie Academic, London, U.K. p. 67.

FIELD, R. W., WU, D., HOWELL, J. A. and GUPTA, B. B. 1995. *J. Membrane Sci.* 100: 259.

GEKAS, V. and ZHANG, W. 1989. *Process Biochem.* 24: 159.

GEPPERT, G. and THIELEMANN, H. 1983. *Chem. Techn. (Germany)* 35 (10): 517.

GESAN, G., DAUFIN, G. and MERIN, U. 1995. *J. Membrane Sci.* 104: 271.

GOURLEY, L., BRITTEN, M., GUTHIER, S. F. and POULIOT, Y. 1994. *J. Membrane Sci.* 97: 283.

GUPTA, B. B., BLANPAIN, P. and JAFFRIN, M. Y. 1992. *J. Membrane Sci.* 70: 257.

GUPTA, B. B., DING, L. H. and JAFFRIN, M. Y. 1985. In *Progress in Artificial Organs.* Y. Nose, C. Kjellstrand and P. Ivanovich (eds.), ISAO Press, Cleveland, OH. p. 891.

GUPTA, B. B., HOWELL, J. A., WU, D. and FIELD, R. W. 1995. *J. Membrane Sci.* 99: 31.

HANEMAAIJER, H. 1988. *I2-Procestechnolgie (Neth.)* 4 (1): 15.

HARRIS, T. A. J., REUBEN, B. G., COX, D. J., VAID, A. K. and CARVELL, J. 1988. *J. Chem. Technol. Biotechnol.* 42: 19.

HAYES, J. F., DUNKERLEY, J. A., MULLER, L. L. and GRIFFIN, A. T. 1974. *Aust. J. Dairy Technol.* 29: 132.

HENRY, J. D. and ALLRED, R. C. 1972. *Dev. Indust. Microbiol.* 13: 177.

HENRY, J. D., LAWLER, L. F. and KUO, C. H. A. 1977. *AIChE J.* 36: 907.

HERMIA, J. 1982. *Trans. I. Chem. E.* 60: 183.

HIDDINK, J., DEBOER, R. and NOOY, P. F. C. 1981. *Milchwiss.* 36: 11.

HILDEBRANDT, J. R. 1991. In *Chromatographic and Membrane Processes in Biotechnology.* C. A. Costa and J. S. Cabral (eds.), Kluwer Academic Publishers, The Netherlands. p. 363.

HODGINS, L. T. and SAMUELSON, E. 1990. U.S. Patent 4,906,379.

JAFFRIN, M. Y., DING, L. H., COUVREUR, C. and KHARI, P., 1997. *J. Membrane Sci.* 124: 233.

JAGANNADH, S. N. and MURALIDHARA, H. S. 1996. *Ind. Eng. Chem. Res.* 35: 1133.

JONSSON, A. S. 1993. *J. Membrane Sci.* 79: 93.

JUCKER, C. and CLARK, M. M. 1994. *J. Membrane Sci.* 97: 37.

KAI, M., ISHII, K., HONDA, Z., MIYANO, T. and TAMADA, M. 1989. In *Advances in Reverse Osmosis and Ultrafiltration.* T. Matsuura and S. Sourirajan (eds.), National Research Council, Ottawa. p. 15.

KENNEDY, T. J., MERSON, R. L. and McCoy, B. J. 1974. *Chem. Eng. Sci.* 29: 1927.

KHORAKIWALA, K. H., CHERYAN, M. and MEHAIA, M. A. 1986. *Biotechnol Bioeng. Symp. Ser.* 15: 249.

KIM, K. J., CHEN, V. and FANE, A. G. 1993. *J. Colloid. Interface Sci.* 155: 347.

KIM, K. J., FANE, A. G. and FELL, C. J. D. 1988. *Desalination.* 70: 229.

KIM, K. J., FANE, A. G. and FELL, C. J. D. 1989. *J. Membrane Sci.* 43: 187.

KIM, K. J., FANE, A. G., FELL, C. J. D. and JOY, D. C. 1992. *J. Membrane Sci.* 68: 79.

KLOOSTERMAN, J., VAN WASSENAAR, P. D., SLATER, K. H. and BAKSTEEN, H. 1988. *Bioprocess Engr.* 3: 181.

KO, M. K. AND PELLEGRINO, J. J. 1992. *J. Membrane Sci.* 74: 141.

KUO, K. P. and CHERYAN, M. 1983. *J. Food Sci.* 48: 1113.

KRONER, K. H., HUMMEL, W., VOLKEL, J. and KULA, M.-R. 1986. In *Membranes and Membrane Processes.* E. Drioli and M. Nakagaki (eds.), Plenum Press, New York. p. 223.

KRONER, K. H., SCHUTTE, H., HUSTEDT, H. and KULA, M.-R. 1984. *Process Biochem.* 19 (April): 67.

LAINE, J.-M., HAGSTROM, J. P., CLARK, M. M. and MALLEVIALLE, J. 1989. *J. Amer. Water Works Association* 81 (November): 61.

LE, M. S., SPARK, L. B. and WARD, P. S. 1984. *J. Membrane Sci.* 21: 219.

LEVY, P. F. and SHEEHAN, J. J. 1991. *BioPharm.* 4(4): 24.

LOCKLEY, A. K., WHITE, W. J. P. and HALL, G. M. 1988. *Intern. J. Food Sci. Technol.* 23: 11.

MACKLEY, M. R. and SHERMAN, N. E. 1992. *Chem. Eng. Sci.* 47: 3067.

MARSHALL, A. D., MUNRO, P. A. and TRAGARDH, G. 1993. *Desalination* 91: 65.

MATTHIASSON, E. and SIVIK, B. 1980. *Desalination* 35: 59.

MATSUMOTO, K., KATSUYAMA, S. and OHYA, H. 1987. *J. Ferment. Technol.* 65: 77.

MCDONOGH, R. M., WELSH, K., FANE, A. G. and FELL, C. J. D. 1988. *Desalination* 70: 251.

MCGREGOR, W. C., WEAVER, J. F. and TANSEY, S. P. 1988. *Biotechnol. Bioeng.* 31: 385.

MERIN, U. and CHERYAN, M. 1980. *J. Food Process. Preserv.* 4: 183.

MERIN, U., GORDIN, S. and TANNY, G. B. 1983. *N.Z. J. Dairy Sci. Technol.* 18: 153.

MICHAELS, S. L. 1994. *BioPharm.* 7(8): 38.

MILISIC, V. and BERSILLON, J. L. 1986. *Filtration & Separation* 23 (Nov.): 347.

MILLER, K. D., WIETZIL, S. and RODGERS, V. G. J. 1993. *J. Membrane Sci.* 76: 77.

NICHOLS, D. J. and CHERYAN, M. 1981. *J. Food Sci.* 46: 357.

NILSSON, J. L. 1990. *J. Membrane Sci.* 52: 121.

OLDANI, M. and SCHOCK, G. 1989. *J. Membrane Sci.* 43: 243.

PADILLA, O. I. and MCLELLAN, M. R. 1993. *J. Food Sci.* 58: 369.

PALL FILTRON. 1995. Company literature. Northborough, MA.

PATEL, P. N., MEHAIA, M. A. and CHERYAN, M. 1987. *J. Biotechnol.* 5: 1.

PERSSON, K. M., CAPANNELLI, G., BOTTINO, A. and TRAGARDH, G. 1993. *J. Membrane Sci.* 76: 61.

PIOT, M., MAUBOIS, J.-L., SCHAEGIS, P., VEYRE, R. and LUCCIONI, L. 1988. *Le Lait* 64: 102.

PITT, A. M. 1987. *J. Parenteral Sci. Technol.* 41: 110.

PORTER, M. C. and MICHAELS, A. S. 1971. *CHEMTECH* 1: 440.

RADLETT, P. J. 1972. *J. Appl. Chem. Biotechnol.* 22: 495.

RANE, K. D. and CHERYAN, M. 1996. Stillage processing with ceramic membranes (unpublished data). University of Illinois, Urbana.

REED, I. M., DUDELY, L. Y. and GUTMAN, R. G. 1987. *Proc. 4th Eur. Congr. Biotechnol.* 2: 573.

REIHANIAN, H., ROBERTSON, C. R. and MICHAELS, A. S. 1983. *J. Membrane Sci.* 16: 237.

ROBINSON, C. W., SIEGEL, M. H., CONDEMINE, A., FEE, C., FAHIDY, T. Z. and GLICK, B. R. 1993. *J. Membrane Sci.* 80: 209.

RODGERS, V. G. J. and SPARKS, H. E. 1993. *J. Membrane Sci.* 78: 163.

ROGERS, P. L., LEE, K. J. and TRIBE, D. E. 1980. *Process Biochem.* 15 (Aug.–Sept.): 7.

ROLCHIGO, P. 1995. Personal communication. Membrex Inc., Fairfield, NJ.

SAEED, M. and CHERYAN, M. 1989. *J. Agric. Food Chem.* 37: 1270.

SAGLAM, N. 1995. Ph.D. thesis, University of Illinois, Urbana.

SHELDON, J. M., REED, I. M. and HAWES, C. R. 1991. *J. Membrane Sci.* 62: 87.

SPIAZZI, E., LENOIR, J. and GRANGEON, A. 1993. *J. Membrane Sci.* 80: 49.

SUKI, A., FANE, A. G. and FELL, C. J. D. 1984. *J. Membrane Sci.* 21: 269.

VRADIS, I. and FLOROS, J. D. 1995. In *Food Process Design and Evaluation.* R. K. Singh (ed.), Technomic, Lancaster, PA. p. 1.

WAKEMAN, R. J. and TARLETON, E. S. 1987. *Chem. Eng. Sci.* 42: 829.

WAKEMAN, R. J. and TARLETON, E. S. 1991. *Desalination* 83: 35.

WENTEN, I. G. 1995. *Filtration & Separation* 32(3): 253.

WINZLER, H. B. and BELFORT, G. 1993. *J. Membrane Sci.* 80: 35.

YAMAGIWA, K., KOBAYASHI, H., OHKAWA, A. and ONODERA, M. 1993. *J. Chem. Eng. Japan.* 26: 13.

CHAPTER 7

Process Design

7.A.
PHYSICS OF THE MEMBRANE PROCESS

Microfiltration is primarily used for clarifying liquids, while ultrafiltration is used for fractionating solutions. If we wish to quantitatively estimate the relative degree of purification in a microfiltration (MF) or ultrafiltration (UF) process or, conversely, to calculate the amount of membrane processing required to attain a certain degree of concentration, separation, or purification, we can use simple mathematical models. This is based on the assumption that the probability of a particle (i.e., a component of the feed solution) passing through the membrane is highest (i.e., probability $= 1$) for solutes with 0% rejection. Rejection (R) at any point in the process is defined as

$$R = 1 - \frac{C_P}{C_R} \qquad (7.1)$$

where C_P is the solute concentration in the permeate and C_R is the solute concentration in the retentate. Conversely, the probability is zero for solutes that are completely (i.e., 100%) rejected by the membrane. The data are presented in terms of volume concentration ratio (VCR):

$$\text{VCR} = \frac{\text{Initial feed volume } (V_0)}{\text{Retentate volume } (V_R)} \qquad (7.2)$$

The weight concentration ratio (WCR) is expressed in the same manner, except with weights rather than volume. VCR and WCR are also frequently referred to in the literature as "concentration factor" (X). Sometimes the data is also presented in terms of percent volume reduction or percent water removed, where

$$\text{Percent volume reduction} = \frac{V_0 - V_R}{V_0} \times 100 = \frac{V_P}{V_0} \times 100 \qquad (7.3)$$

where V_P is volume of permeate. Assuming the size of the potential measurement error is proportional to the size of the observation and the probability is constant throughout UF and subsequent diafiltration processes, the following expression was derived (Cheryan 1986):

$$C_R = C_0(\text{VCR})^R \qquad (7.4)$$

Equation (7.4) shows that the concentration of a solute at any time or stage of membrane processing is a function of both the volume reduction and the value of R, where R is expressed by Equation (7.1). In other words, if the probability that a solute will go through the membrane is 1, then it implies it is a freely permeable solute that will not be rejected by the membrane, and its concentration on either side of the membrane will be equal. Or from Equation (7.1), $C_P = C_R$, and it has a rejection of zero. Similarly, if the probability is zero, the solute will not go through the membrane, and we will measure a rejection $R = 1$.

7.A.1.
EXAMPLE

Typical data from the fractionation of cheese whey is given in Table 7.1. The first row is VCR (or X) = 1, showing the composition of the cheese whey used as feed. A sample calculation for VCR = 20 is given below, using Equation (7.4):

- for protein, $R = 1$, $C_R = 0.4(20)^1 = 8.0\%$
- for lactose, $R = 0$, $C_R = 5(20)^0 = 5.0\%$
- for NPN, $R = 0$, $C_R = 0.2(20)^0 = 0.2\%$
- for ash, $R = 0.2$, $C_R = 0.68(20)^{0.2} = 1.24\%$

Figure 7.1 shows the behavior of cheese whey during UF. Straight lines should be obtained when plotting data as C versus VCR, if rejection is 0 or 1. Rectangular hyperbolas will be obtained for all other R values. Deviations from expected behavior will suggest the following could be occurring: (a) solute adsorption by the membrane; (b) change in rejection during UF, which is quite common, especially at higher retentate concentrations; or (c) volume exclusion effects of the solute becoming significant, which becomes important at high solute concentrations.

The total solids at VCR 20 is 14.44%. Thus, the composition of the retentate solids is as follows, on a dry basis:

- protein = 8/14.44 = 55.4%
- lactose = 5/14.44 = 34.6%
- NPN = 0.2/14.44 = 1.3%
- ash = 1.24/14.44 = 8.6%

Table 7.1. Ultrafiltration of cheese whey. Basis: feed is 100 L of whey with the composition shown in the first row. Rejections (R) of protein = 1, lactose and NPN = 0, ash = 0.2. Density of feed, retentate, and permeate = 1.0 g/ml.

Volume (L)	VCR	Volume Reduction (%)	Protein %w/v	Protein kg	Lactose %w/v	Lactose kg	NPN* %w/v	NPN* kg	Ash %w/v	Ash kg	Total Solids %w/v	Total Solids kg	Protein Content of Solids (% d.b.)
100	1	0	0.4	0.4	5.0	5.0	0.2	0.20	0.68	0.68	6.28	6.28	6.4
50	2	50	0.8	0.4	5.0	2.5	0.2	0.10	0.78	0.39	6.78	3.39	11.8
25	4	75	1.2	0.4	5.0	1.25	0.2	0.05	0.90	0.22	7.30	1.98	16.4
10	10	90	4.0	0.4	5.0	0.50	0.2	0.02	1.08	0.11	10.28	1.03	38.9
5	20	95	8.0	0.4	5.0	0.25	0.2	0.01	1.24	0.06	14.44	0.72	55.4

*Nonprotein nitrogen
Concentration of solutes (%w/v) calculated using Equation (7.4)
Mass of solute (kg) calculated with Equation (7.5)

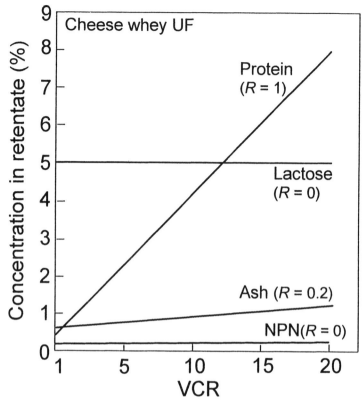

Figure 7.1. *Effect of volume concentration ratio (VCR) on retentate composition. Example is ultrafiltration of cheese whey. Rejection (R) of protein = l, lactose = 0, nonprotein nitrogen (NPN) = 0, and ash components = 0.2.*

Figure 7.2 is a plot of protein, ash and lactose contents on a dry basis as a function of VCR and percent volume reduction. Table 7.1 also shows the mass of each solute (M_S) in the retentate at each VCR where:

$$M_S = C_R V_R \tag{7.5}$$

Another important parameter is the yield (Y) of a component, which is the fraction of that component in the original feed recovered in the final retentate:

$$Y = \frac{C_R V_R}{C_0 V_0} \tag{7.6}$$

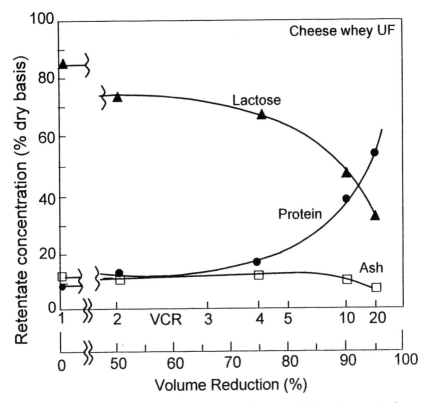

Figure 7.2. *Composition of retentate solids as a function of VCR and percent volume reduction during ultrafiltration of cheese whey.*

From Equations (7.2), (7.4), and (7.6):

$$Y = (VCR)^{R-1} \tag{7.7}$$

Equation (7.7) states that the yield of a particular component is an exponential function of the decreasing volume of the feed in the system. This has been experimentally confirmed in a large number of cases. Figure 7.3 is a plot of the yield of protein and lactose during the ultrafiltration of cheese whey expressed as a function of VCR and percent volume reduction. The significance of Equation (7.7) must be reemphasized. In an ideal system [i.e., no membrane absorption, solutes completely soluble and permeable ($R = 0$) and no solute interactions], a 50% reduction in volume (VCR 2) implies removal of 50% of the feed as permeate, together with 50% of the freely permeable solutes, and

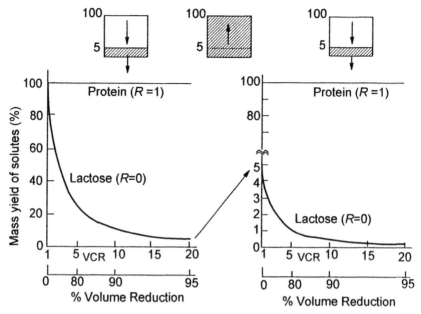

Figure 7.3. Yield of protein and lactose as a function of VCR during discontinuous diafiltration. left: solute remaining in retentate during first stage UF; right: solute remaining during second-stage UF.

90% removal of the solution by permeation (VCR 10) implies that only 10% of the permeable solute mass will remain in the retentate.

There is another way of expressing the behavior of individual solutes. From Equation (7.4):

$$\frac{C_R}{C_0} = \left(\frac{V_0}{V_R}\right)^R \tag{7.8}$$

$$\log(\text{SCR}) = R\log(\text{VCR}) \tag{7.9}$$

where SCR is the solute concentration ratio (C_R/C_0). Equation (7.9) enables us to calculate rejection using only retentate data, as opposed to Equation (7.1), which is based on both permeate and retentate data. Figure 7.4 is a plot of Equation (7.9) at different values of R.

7.B.
MODES OF OPERATION

Although, as indicated in Table 7.1 and Figures 7.2 and 7.3, a considerable purification of the protein can be done by direct ultrafiltration, the flux will drop

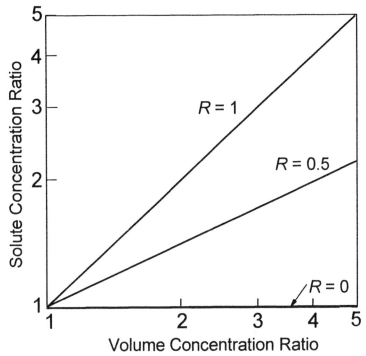

Figure 7.4. *Relationship between solute concentration ratio (SCR) and VCR as a function of rejection (R).*

to uneconomically low values, and the pumping power required will rise due to increase in viscosity of the retentate. Thus, in order to efficiently effect the maximum purification of a retained solute, diafiltration (DF) will have to be done. DF refers to the process of adding water to the retentate and continuing the elimination of membrane-permeating species along with the water. DF can be conducted under either one of two modes: discontinuous or continuous diafiltration. A schematic of the two modes is shown in Figure 7.5.

7.B.1.
DISCONTINUOUS DIAFILTRATION

Discontinuous diafiltration (DD) refers to operations where permeable solutes are cleared from the retentate by volume reduction, followed by redilution with water and reultrafiltration in repetitive steps. Each time the reultrafiltration step is done, Equations (7.4) and (7.6) can be applied with C_0 being the diluted concentration at the start of the reultrafiltration. For example, Table 7.2 shows typical data when the VCR 20 retentate from Table 7.1 was subjected to a DD process by diluting with a volume of water equal to the final retentate

Table 7.2. Discontinuous diafiltration of the VCR 20 retentate of Table 7.1. An equal volume of water (5 L) was added to the retentate and reultrafiltered to 5 L. Same assumptions as in Table 7.1.

Description	Volume (L)	Protein		Lactose		NPN*		Ash		Total Solids % w/v
		% w/v	% d.b.	% w/v	% d.b.	% w/v	% d.b.	% w/v	% d.b.	
VCR20 retentate diluted	10	4.0	55.4	2.5	34.6	0.1	1.4	0.62	8.6	7.22
After reultrafiltration to VCR 20	5	8.0	70.7	2.5	22.1	0.1	0.9	0.71	6.3	11.31

*Nonprotein nitrogen

Table 7.3. Discontinuous diafiltration of VCR 20 retentate of Table 7.1. The retentate was diluted back to the original volume of the cheese whey (100 L) and reultrafiltered to 20-fold concentration factor. Same assumptions as in Table 7.1.

	Volume	Concentration in Retentate (% w/v)					Protein
Description	(L)	Protein	Lactose	NPN*	Ash	Total Solids	(% d.b.)
VCR20 diluted to original whey volume	100	0.4	0.25	0.01	0.062	0.722	55.4
Re-UF 2X	50	0.8	0.25	0.01	0.071	1.13	70.7
Re-UF 10X	10	4.0	0.25	0.01	0.098	4.36	91.7
Re-UF 20X	5	8.0	0.25	0.01	0.113	8.37	95.5

*Nonprotein nitrogen

volume and UF back to VCR 20. Of course, a further purification could have been effected by diluting the VCR 20 retentate in Table 7.1 all the way back to the original volume. This would result in the same protein concentration as VCR 1 in the first stage, but one-twentieth of the lactose and nonprotein nitrogen (NPN) concentration and one-fourth of the original ash concentration (Table 7.3). Reultrafiltration of this diluted retentate will result in the composition shown in Table 7.3 and Figure 7.3. Thus, by a two-stage VCR 20 DD

DISCONTINUOUS DIAFILTRATION (DD)

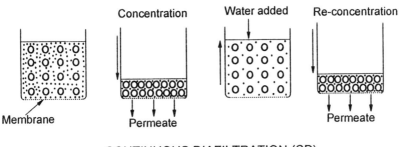

CONTINUOUS DIAFILTRATION (CD)

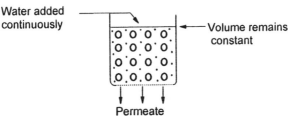

Figure 7.5. *Schematic illustration of discontinuous (top) and continuous diafiltration (bottom).*

process, the protein content of whey solids can be increased from 6.6% to the far more desirable 95% (dry basis).

When an equal volume reduction is given in each stage, solute concentration in the final retentate (C_R) can be calculated by combining Equations (7.4) and (7.6) for a multistage process:

$$C_R = C_0(\text{VCR})^{1+n(R-1)} \tag{7.10}$$

where C_0 is the solute concentration in the *original feedstream* (i.e., at the start of the first stage of DD), n is the number of stages, and R is the average rejection of that solute. If VCRs and/or rejection are not the same in each stage, then the following equation can be used:

$$C_R = C_0(\text{VCR})_1^R(\text{VCR})_2^{R-1} \tag{7.11}$$

where $(\text{VCR})_1$ is the volume concentration ratio in the first stage and $(\text{VCR})_2$ is the second stage VCR.

7.B.2.
CONTINUOUS DIAFILTRATION

Continuous diafiltration (CD) involves adding water at the appropriate pH and temperature to the feed tank at the same rate as the permeate flux, thus keeping feed volume constant during processing (Figure 7.5). Permeable solutes are removed at the same rate as the flux. This mode of DF is particularly useful if the concentration of the retained solute is too high to permit effective UF or DD operations for purification. CD data are usually presented in terms of volumes permeated or diluted (V_D) where

$$V_D = \frac{\text{Volume of liquid permeated } (V_P)}{\text{Initial feed volume } (V_0)} \tag{7.12}$$

An analysis similar to the one used for deriving the equations for UF and DD would give

$$\ln(C_0/C_R) = V_D(1 - R) \tag{7.13}$$

The yield (Y), defined as C_R/C_0, is the fraction of the original solute remaining in the retentate:

$$Y = e^{-V_D(1-R)} \tag{7.14}$$

$$\text{Fraction of solute removed} = 1 - Y = 1 - e^{-V_D(1-R)} \tag{7.15}$$

Table 7.4. Continuous diafiltration of cheese whey. Same basis and assumptions as Table 7.1.

V_D	Volume (V_P), L	Protein % w/v	Lactose % w/v	NPN* % w/v	Ash % w/v	Total Solids % w/v	Protein in Solids (% d.b.)
0	0	0.4	5.0	0.2	0.68	6.28	6.4
1	100	0.4	1.84	0.074	0.306	2.62	15.3
2	200	0.4	0.68	0.027	0.14	1.244	32.1
3	300	0.4	0.25	0.01	0.062	0.72	55.4
6	600	0.4	0.012	0.001	0.006	0.419	95.5

*Nonprotein nitrogen
Concentration of solutes calculated using Equation (7.15)

The concentration of solute in the retentate after V_D volumes of continuous DF is

$$C_R = C_0 e^{-V_D(1-R)} \qquad (7.16)$$

Figure 7.6 is a plot of Equation (7.15) for different values of R. Note that it takes much more processing (i.e., more water to be added) with CD to obtain a certain elimination of a solute than with DD. Table 7.4 shows the composition of cheese whey subjected to continuous diafiltration. Comparing Tables 7.1 and 7.4, a removal of 95 L from a 100-L initial volume of cheese whey by UF resulted in a 55% [dry basis (d.b.)] protein retentate, while it took the permeation of 300 L to effect the same degree of purification by CD. Similarly, to obtain a 95% protein powder, it would require two VCR 20 stages by DD, i.e., a total of $95 + 95 = 190$ L of permeate. In contrast, CD would require about 600 L to obtain the 95% protein powder.

Not only is there considerably more solvent to be handled in the CD operation, but it also results in a more dilute permeate, which may be an important consideration in its handling and disposal. Flux during CD should be higher than the average flux during DD, but it may not compensate for the extra processing that is required by CD. In addition, the final retentate in CD is dilute, and thus an additional concentration step will be required. The optimum may well be a combination of UF–CD–UF, as discussed in Section 7.D.

It should be remembered that, as far as solute elimination from a certain volume of feed is concerned, UF concentration is the limiting form of diafiltration when no DF fluid is used in the process. For these combined UF–CD processes, a convenient parameter to plot the data is V_C, which is defined by combining VCR and V_D (Rajagopalan and Cheryan 1991):

$$V_C = V_D - \frac{1}{\text{VCR}} \qquad (7.17)$$

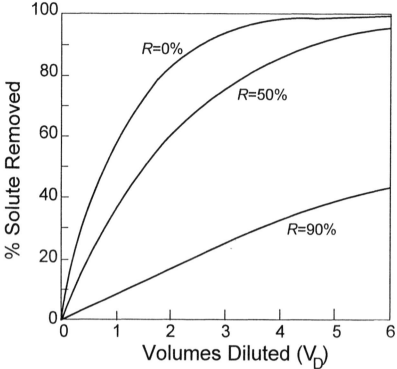

Figure 7.6. Theoretical plots of solute removal by continuous diafiltration (CD) for different values of solute rejection.

This parameter serves to bridge the gap between the two processes. V_C is -1 at the start of the process. During UF, V_D is zero and V_C ranges from -1 at VCR 1 to zero at VCR $= \infty$. This is shown schematically in Figure 7.7 for solutes with 0% and 100% rejection. During UF, V_C starts at -1 and can go to an unlimited value. The use of V_C is limited to processes carrying out sequential UF concentration and DF in that order. However, if a series of UF concentration and DF operations are to be carried out, V_D can be redefined more generally as

$$V'_D = \frac{V_F}{V_R} \tag{7.18}$$

where V_R is the volume at which the DF is carried out and V_F is the volume of diafiltration liquid added. V'_D can be used in V_C [Equation (7.17)] to provide a continuous scale for any sequence of UF and DF operations. Section 7.C.6. discusses DF control modes in continuous operations.

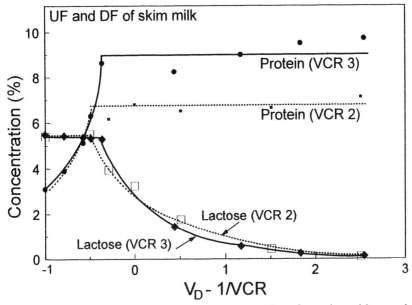

Figure 7.7. *UF and CD of skim milk. The concentration of protein and lactose is shown as a function of the new parameter V_C [Equation (7.17)]: full lines: skim milk was UF to 3 × before CD; broken lines: skim milk was UF to 2 × before CD (adapted from Rajagopalan and Cheryan 1991).*

7.B.3.
DIALYSIS ULTRAFILTRATION

The UF and DF methods discussed so far rely on the convective elimination of permeable microsolutes. The rate at which microsolutes are removed through the membrane (J_S) depends on its concentration (C_S) and flux of water (J_W):

$$J_S = C_S \cdot J_W \tag{7.19}$$

In addition, if $R = 0$,

$$(C_S)_R = (C_S)_P \tag{7.20}$$

where the subscripts P and R refer to permeate and retentate, respectively. The flux J_W is controlled by transmembrane pressure or mass transfer considerations. To enhance J_S, the membrane module can be operated in a dialysis mode, as shown in Figure 7.8. Instead of allowing the permeate merely to flow out of the module freely, water is pumped into the permeate side of the membrane (usually at three to ten times greater flow rate than the flux), while keeping the net pressure in the permeate channel lower than the feed side pressure. Due to

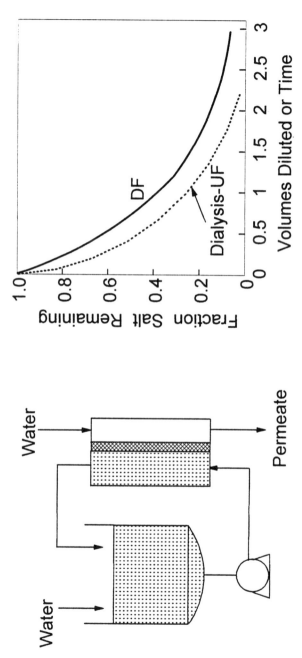

Figure 7.8. Dialysis UF: left: operating principle; right: comparison of rates at which a solute with R = 0 is removed by conventional CD and by dialysis UF.

dilution by the fresh water, the concentration of microsolutes in the permeate is now lower than the feed side, in proportion to the relative flow rates of permeate and dialysis water. Or,

$$C_{SR} > C_{SP} \tag{7.21}$$

This will establish a concentration gradient from the retentate side to the permeate side, which will enhance the normal permeation of the permeable solutes. Or,

$$J_S = C_S \cdot J_W + D_S(C_{SR} - C_{SP}) \tag{7.22}$$

where D_S is the diffusion coefficient of the permeable solute. As shown in Figure 7.8, this will result in less time for the diafiltration or less volumes diluted (although the total amount of water will be greater). To minimize the amount of fresh water needed for dialysis UF, the permeate (which will be much more dilute with dialysis UF) can be sent to a reverse osmosis (RO) system, which will simultaneously concentrate the solute while producing the water needed for dialysis UF.

Dialysis UF has proven useful in a variety of applications, e.g., producing low-alcohol beer and wine [using RO or nanofiltration (NF) membranes], desalting gelatin (Dutre and Tragardh 1995), and for blood purification (hemodiafiltration), which uses much lower pressures and operates at lower flux (Chang and Lee 1988; Sigdell 1982; Villarroel et al. 1977). In fact, if the pressures are equalized on the retentate and permeate side, J_W is zero, and microsolute transport will occur only by dialysis. This could improve the separation of similarly sized molecules. Dialysis UF can also be used for ion substitution or solute exchange, e.g., permeable solute A in the feed can be exchanged for another solute B by pumping in a concentrated solution of B on the permeate side instead of water.

7.C.
BATCH VERSUS CONTINUOUS OPERATION

There are several system approaches that one can use to design and lay out a MF or UF plant. They are

- Batch operation
- Single-pass processing
- Feed and bleed
- Multistage recycle operation

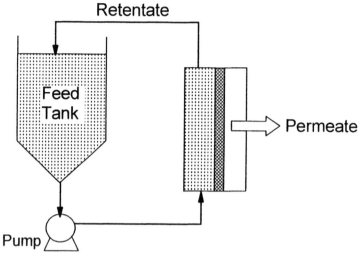

Figure 7.9. *Batch filtration with full recycle of retentate.*

7.C.1.
BATCH OPERATION

There are several variations of this operation. The simplest one that is most common with laboratory- and pilot-scale units is shown in Figure 7.9. Heat exchangers are usually included in the recycle loop to control temperature (not shown in Figures 7.9–7.14). One pump can be used for both feed and recirculation. The retentate is returned to the feed tank for recycling through the module. It is the fastest method of concentrating a given amount of material, and it will also require the minimum membrane area. The batch with partial recirculation of the retentate (Figure 7.10) is used when a continuous feedstream needs to be processed and other storage tankage may not be available. In addition, a smaller diameter return piping can be used, there should be less foaming in the tank, and there is a small reduction in recirculating pump energy. When feed is added continuously at the same rate as the flux, it is referred to as "batch with recirculating loop, topped off." This is done when the permeate is the required product, e.g., fruit juice clarification.

Batch operations and their modifications are used where the permeate is the required product, e.g., fruit juice and effluent treatment where the concentrated retentate is to be hauled away. They are not suitable for applications where the retentate is the desired product if it will be damaged due to prolonged exposure to heat and shear. The residence time of each particle within the system is the longest in batch processing, all other factors being equal. As a first approximation, the average flux (J_{av}) during concentration can be estimated as:

$$J_{av} = 0.5(J_i + J_f) \tag{7.23}$$

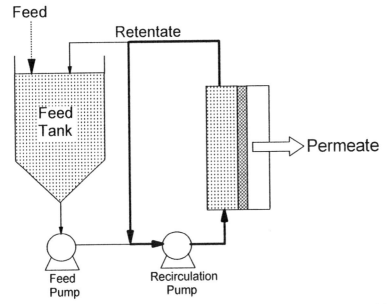

Figure 7.10. *Batch ultrafiltration with partial recycle of retentate. If feed is added continuously at the same rate as the flux, this is the "batch with topped off" mode.*

where J_i is the flux at VCR1 and J_f is the flux at the highest concentration factor.

7.C.2.
SINGLE PASS

Figure 7.11 shows this type of operation, involving no recycling of the retentate. Normally, in most MF and UF applications, the recirculation flow rate (Q) is much higher than the flux through a module. A single pass through the module will result in a very low volume of permeate and small X factor (i.e., a low recovery) unless a very large membrane area is used. Thus, the single-pass operation is generally restricted to situations where concentration polarization effects are negligible and high flow rates are not required. Typical applications may be hollow fiber bioreactors, water treatment, pyrogen removal, etc. Residence time of particles is least with this system.

7.C.3.
FEED-AND-BLEED

This mode of operation, shown schematically in Figure 7.12, is the most common method used for continuous full-scale operation. It essentially merges the batch and the single-pass operations. Two pumps may be required: the feed pump provides the transmembrane or system pressure, while the recirculation pump provides the cross-flow. At startup, the feed pump is used to

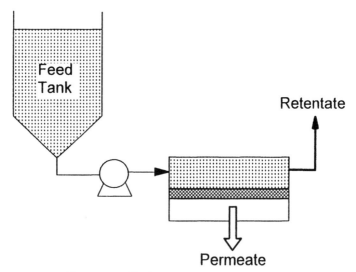

Figure 7.11. *Single-pass continuous filtration.*

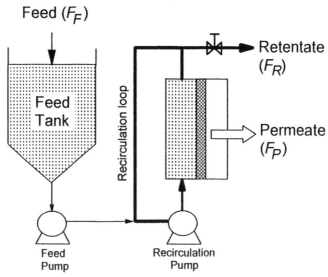

Figure 7.12. *Feed-and-bleed operation.*

fill the recirculation/module loop, after which the recirculation pump is started. After a few seconds to stabilize the pressures and flow rates, retentate is bled off the recirculation loop at a flow rate (F_R) such that

$$X = \frac{F_F}{F_R} = \frac{F_R + F_P}{F_R} \tag{7.24}$$

where F_F, F_R, and F_P are volumetric flow rates of feed, retentate (concentrate), and permeate, respectively. The feed will flow into the loop at the same rate as the permeate flow rate plus concentrate flow rate. Permeate flow rate is given by

$$F_P(\text{L/h}) = J(\text{L/m}^2 \cdot \text{h}) \cdot A(\text{m}^2) \tag{7.25}$$

The advantage here is that the final concentration is immediately available as the feed is pumped into the loop. In fact, the feed tank shown in Figure 7.12 is not really necessary for this operation, and this savings in tankage is another advantage. The disadvantage here is that the process loop is operating continuously at a concentration factor equivalent to the final concentration of a batch system. The flux, therefore, is lower than the average flux in a batch mode, and the membrane area required is correspondingly higher.

For design purposes, the following equations are useful:

$$F_F = F_R + F_P \tag{7.26}$$

$$F_F C_F = F_R C_R + F_P C_P \tag{7.27}$$

where C_F, C_R, and C_P are concentrations of a particular solute in the feed, concentrate, and permeate, respectively.

Continuous operation is preferred with large-scale operations, e.g., with 100 m^2 of membrane area or more. Since the liquid in the recycle loop is well mixed, it assumes the mixing pattern of a continuous, stirred-tank reactor (CSTR), i.e., there will be a wide distribution of residence times. If fouling occurs, F_P in Equation (7.24) will decrease with time, necessitating a corresponding decrease in F_R to keep the X value constant during operation.

7.C.4.
MULTISTAGE OPERATIONS

In order to overcome the low flux disadvantage of the feed-and-bleed operation, and yet to maintain its continuous nature, large-scale continuous plants are usually staged, using several individual feed-and-bleed stages. The stages are operated in series as far as the concentrate flow is concerned but in parallel with respect to permeate flow. A schematic of a three-stage, multistage recycle plant is shown in Figure 7.13. Only the final stage (in this case, the third stage) is operating at the highest concentration and lowest flux, while the other stages

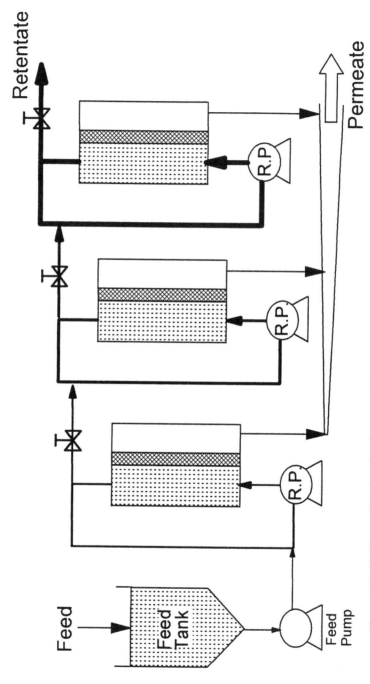

Figure 7.13. Multistage filtration showing several feed-and-bleed systems connected in series. RP is recirculating pump.

are operating at lower concentrations with consequently higher flux. Thus, the total membrane area should be less than a single-stage feed-and-bleed operation and should approach that required for a batch process. Usually, the economic optimum is three to seven stages since increasing the number of stages increases the cost of controls, valves, piping, and fittings faster than the decrease in membrane area. The residence time, volume holdup, and tankage required is much less than for the same duty in a batch operation. In addition, such continuous systems can be operated in a sanitary manner over a 24-h cycle, with a single 1- to 3-h cleaning and sanitizing break per 24 h being adequate.

In contrast, due to microbiological or product stability considerations, the batch process is usually restricted to an 8- to 16-hour operational cycle. Recovering the product, cleaning, sanitizing, and refilling the feed tank may take another 2–4 h. The batch plant may be in actual operation for only one to three cycles per 24-h day. Hence, although the average flux rate may be higher and membrane area lower with a batch plant, the total volume throughput on a daily basis could be higher in a multistage continuous plant. The greater the number of stages, the narrower is the residence time distribution of particles, all other factors being equal.

7.C.5.
EXAMPLE

Calculate the surface area required to produce a fivefold concentration of skim milk in a hollow fiber UF system. The $5\times$ retentate (total solids = 20%) has a flow rate of 1000 L/hour. Assume the following conditions: temperature $= 50°C$, flow velocity $= 1.1$ m/sec. The flux versus concentration data are given in Figure 4.20. Calculate the membrane area required for (1) batch operation, (2) feed-and-bleed operation, and (3) multistage recycle operation. For part (3), estimate the optimum number of stages.

The data in Figure 4.20 at 50°C and velocity of 1.1 m/sec can be expressed in terms of VCR or X value as

$$J = 35 - 25.75 \log(\text{VCR}) \qquad (7.28)$$

1. Batch: From Equation (7.23), average flux is estimated as 26 liters per square meter per hour (LMH). From Equation (7.24):

$$F_F = 5 \times 1000 = 5000 \text{ L/h}$$

From Equation (7.26):

$$F_P = 5000 - 1000 = 4000 \text{ L/h}$$

From Equation (7.25):

$$A = \frac{4000}{26} = 153.8\,\text{m}^2$$

2. Feed-and-bleed, single stage: Since the system will be operating at the final VCR 5 concentration, $J_f = 17$ LMH. Thus, membrane area $= 4000/17 = 235.3\,\text{m}^2$.
3. Multistage recycle operation: This is essentially done by trial and error to minimize the area required. We also know that, regardless of the number of stages that will be used, the last stage will be operating at VCR 5, i.e., at a flux of 17 LMH. A computer program could be written that uses the flux–VCR relationship [Equation (7.28)] with different combinations of VCR in each stage to estimate the area in each stage and minimize the total area by iterative calculations. The final result would probably indicate using unequal areas in each stage, i.e., more area in the initial stages. However, it may be desirable to have the same area in each stage to minimize the cost of manufacture and fabrication. Even within this equal-area limitation, due consideration has to be given to the number of modules and housings within each stage, to minimize the excess area in each stage (this is discussed later, Section 7.G.1.). In exceptional circumstances (e.g., very high viscosity), the last stage could be different, e.g., special pumps, modules with larger feed channel dimensions, less residence time, etc.

In this example, using unequal areas resulted in only 1–2% lower area than assuming equal areas, and so the equal area case is shown in Table 7.5, with the corresponding VCRs and fluxes. For example, for a three-stage system, flux of the first stage at VCR 1.5 is 30.5 LMH [from Equation (7.28)]. From Equation (7.26):

$$F_{P1} = 5000 - 5000/1.5 = 1666.7\,\text{L/h}$$

From Equation (7.25):

$$A_1 = 1667.7/30.5 = 54.6\,\text{m}^2$$

At VCR 2.5, flux is 24.75 LMH. A material balance around the second stage will give

$$F_{P2} = (5000/1.5) - 5000/2.5 = 1333.3\,\text{L/h}$$

$$A_2 = 1333.3/24.75 = 53.7\,\text{m}^2$$

The flux at VCR 5 is 17 LMH. Thus,

$$F_{P3} = (5000/2.5) - 5000/5 = 1000\,\text{L/h}$$

$$A_3 = 1000/17 = 58.8\,\text{m}^2$$

Table 7.5. Calculation of membrane area for multistage feed-and-bleed operation for the example in 7.C.5.

Process	Stage 1		Stage 2		Stage 3		Stage 4		Stage 5		Area (m²)
	VCR	J	VCR	J	VCR	J	VCR	J	VCR	J	
Feed-and-bleed											
One-stage	5	17	—	—	—	—	—	—	—	—	235
Two-stage	2	27.3	5	17	—	—	—	—	—	—	180
Three-stage	1.5	30.5	2.5	24.75	5	17	—	—	—	—	167
Four-stage	1.35	31.5	1.9	27.5	2.9	23.2	5	17	—	—	162
Five-stage	1.25	32.5	1.6	30	2.25	26	3.25	21.75	5	17	158
Batch	1–5	26									154

J is in LMH

The total area is $54.6 + 53.7 + 58.8 = 167.1$ m^2. Continued iterations with different VCRs in the first and second stages could have refined the estimates for equal areas.

Similar calculations for two-, four-, and five-stage processes are shown in Table 7.5. A considerable savings in membrane area can be achieved using multiple stages instead of a single stage. In this particular case, going beyond four stages shows only a marginal decrease in total membrane area. Considering that additional piping, instrumentation, controls, and pumps are needed, it may not even be economical to go beyond three stages in this particular example.

7.C.6.
CONTROL METHODS

The manner in which the system is controlled and DF is conducted depends on the mode and type of operation. *Constant pressure* mode is the simplest way of running a plant. This means keeping the average of the inlet and outlet pressures (i.e., the transmembrane pressure) at a fixed value. If flux decreases because of fouling or concentration, the flow rate can be increased by increasing the pressure drop (e.g., opening the inlet and outlet valves) but keeping the average pressure the same. This works well in a batch operation, but in a feed-and-bleed system, a decrease in permeate flow will mean a reduction in retentate flow, thus lowering the overall capacity of the plant.

Constant retentate flow is more difficult to achieve in UF and MF, since it relies on compensating for drop in flux by increasing the transmembrane pressure. Flux is, at best, a weak function of pressure in most UF and MF applications, and thus this mode is seldom preferred. A variation of this is *constant permeate flow* (i.e., *constant flux*), which works best if the flux is deliberately kept low (below the "critical" flux; see Section 6.F.)

The most common method of control of continuous plants is *constant X-factor*. From Equation (7.24), it can be seen that this becomes a matter of measuring the feed flow rate and retentate flow rate and using control valves at the inlet and outlet of the module(s). Large automated plants would use a programmable logic controller (PLC) and flow ratio controller to keep the VCR or X value constant. This mode is best if pressure is not a control variable and if flux declines during operation.

Diafiltration is conducted in a batch plant simply by adding water to the feed tank. With continuous feed-and-bleed multistage systems, either counter current or parallel flow CD can be done (Merry 1996). Parallel flow involves the addition of DF water in one or more stages of the plant (Figure 7.14). Since each stage is well mixed, the added water essentially leaves as permeate from each stage. Compared to batch DF, parallel flow may require 35% more area and water and will result in a more dilute permeate.

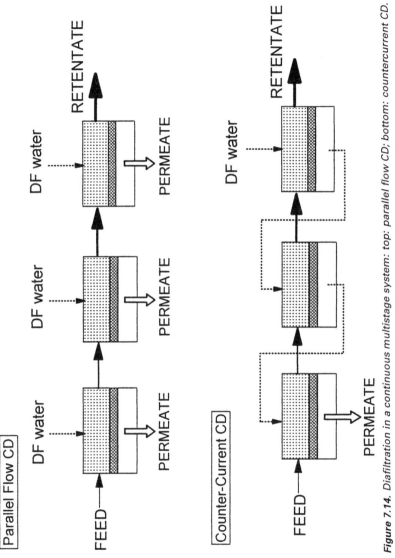

Figure 7.14. Diafiltration in a continuous multistage system: top: parallel flow CD; bottom: countercurrent CD.

317

In countercurrent DF, fresh water is introduced only in the last stage of the plant (Figure 7.14). The permeate from this stage, which is now more dilute and of greater volume than without DF, is used as the DF "water" for the preceding stage. This is repeated for all preceding stages until the required degree of purification has been achieved in the final retentate. Permeate leaves the plant only from the first stage. Compared to parallel flow, countercurrent DF results in less dilute permeate, requires less water for the same degree of purification, but requires greater membrane area/time. Compared to batch DF, countercurrent DF will require 35% more area and 40% less water and will result in a more concentrated permeate.

7.D.
MINIMUM PROCESS TIME

The time required for a given ultrafiltration operation will depend on (a) whether discontinuous or continuous diafiltration is chosen, (b) the concentration of the retained solids, (c) the total volume to be processed, and (d) the membrane area. The most important is concentration of the solute, in those systems that are mass-transfer limited and obey the semilogarithmic relationship:

$$J = k \ln C_G / C_B \tag{7.29}$$

In a batch operation, the time of processing (t) is given by

$$t = \frac{V_P}{J \cdot A} = \frac{V_0 - V_R}{J \cdot A} \tag{7.30}$$

Flux determines the time for processing and/or the membrane area required and thus directly controls capital costs. For the same amount of retained solids in the feed, low concentration results in higher flux and shorter processing time. However, low feed concentration also implies much larger volumes are to be processed, requiring longer processing times. Thus, there should be an optimum solids concentration that will result in the minimum process time. The process decision to be made is between operating at low C_B/high flux/large volume to be permeated versus high C_B/low flux/low permeation volume.

To address this problem, Ng et al. (1976) obtained the following expression for the optimum macromolecule concentration (C_B^*) that will minimize the time of processing for a constant volume diafiltration process:

$$C_B^* = C_G / e \tag{7.31}$$

Thus, one can predict an optimum solids concentration to conduct the diafiltration operation for a completely rejected molecule that will result in the minimum process time, knowing only the value of the gel concentration, C_G. Table 4.2 shows C_G values of several solutes.

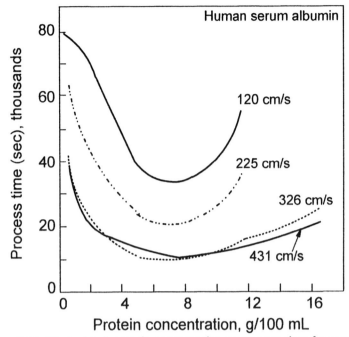

Figure 7.15. *Effect of bulk protein concentration on process time for continuous diafiltration of human serum albumin: pH = 6.8, 22°C, 25 psig. The variable is shear rate per unit length (adapted from: Ng et al. 1976).*

As an example of the use of this concept, the literature (e.g., Table 4.2) suggests C_G values of 24–28 g/100 mL for human serum albumin. According to Equation (7.31), C_B^* is 8–10 g/100 mL, which is close to what was experimentally obtained (Figure 7.15).

As shown in Equation (7.31), C_B^* is independent of operating parameters such as pressure, temperature, and shear rate. Figure 7.15 confirms this is true for shear rate. However, if C_G is really only the osmotic equivalence of the applied pressure, then there probably will be an optimum pressure also.

As mentioned in Section 7.B.2., the combined process of concentration + diafiltration can simultaneously concentrate and purify a macromolecule. In the food industry, for example, such a combination could be used to manufacture protein isolates with a high protein and low sugar and/or salt content (Section 8.J.). If removal of undesirable low molecular weight permeable solutes is the primary goal, it is convenient to define the purity (P) of the retained species in the retentate at any time as

$$P = \frac{C_B}{C_B + C_{SR}} \qquad (7.32)$$

where C is the concentration, the subscript B is the retained species that has to be purified, S refers to freely permeable solutes, and R refers to retentate. During direct UF or MF, the ratio of the concentration of the retained components to the permeable components increases. However, owing to the rapid drop in flux during concentration as shown by Equation (7.29), high degrees of purification require DF.

Thus, there are two competing factors to be considered in the design of a membrane-based purification process: (1) Flux decreases during the membrane concentration operation, but the volume to be processed is low; (2) flux remains high during CD, but this must be balanced against the larger process volume required to effect the same degree of purification. The process objective can be to minimize membrane area/time or to minimize the quantity of DF water required.

To determine the optimum combination of UF and CD, we need a functional relationship between time, flux, purity, and concentration of the retained solute during UF and CD. From Equation (7.30), for a solute that is 100% rejected:

$$t_{\mathrm{UF}} = \frac{V_0(1 - 1/\mathrm{VCR})}{(J_{\mathrm{av}})_{\mathrm{UF}} \cdot A} = \frac{V_0(1 - C_{B0}/C_B)}{(J_{\mathrm{av}})_{\mathrm{UF}} \cdot A} \tag{7.33}$$

$$t_{\mathrm{CD}} = \frac{V_0 C_{B0}(-\ln C_B - \ln\{[1 - P]/PC_{S0}\})}{(J_{\mathrm{av}})_{\mathrm{CD}} \cdot A} \tag{7.34}$$

where C_{B0} is the initial concentration of the retained component, C_{S0} is the initial concentration of the permeable solute that has to be removed, C_B is the concentration of the retained component at any time t, (J_{av}) is the average flux during the operation, P is purity defined by Equation (7.32), and V_0 is the initial volume of the feed. Differentiating Equation (7.34) and setting $dt/dC_B = 0$ would give the relationship between C_B^* and the parameters P, C_{S0}, C_G and the mass transfer coefficient k for continuous DF (Asbi and Cheryan 1992).

These equations were used to determine the time required to purify cheese whey solids from an initial 6.6% protein (d.b.) to a final 95% protein (d.b.). The case study was for an initial volume (V_0) of 100 L and for a membrane area (A) of 1 m^2. For DD, the cheese whey was concentrated to VCR 20 before adding water and reultrafiltering. UF time was calculated from Equation (7.33) using flux data of Cheryan and Kuo (1984). For CD, the initial flux was used as $(J_{\mathrm{av}})_{\mathrm{CD}}$ in Equation (7.34).

As shown in Figure 7.16, discontinuous diafiltration requires much less time than CD. However, CD was conducted at a low initial protein concentration. The situation could be improved if CD was performed at higher initial protein concentration; e.g., if the whey was preconcentrated 5$\times$ to 2% protein, the CD time would reduce by 75%. In fact, in most applications, the concentration of the retained solute in the feed is much lower than C_B^*. To reach the optimum

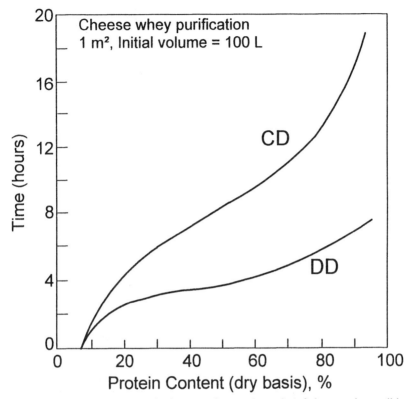

Figure 7.16. *Time required to obtain a certain protein purity of cheese whey solids. Flux and rejection data were obtained from Cheryan and Kuo (1984). Initial volume (V_0) was 100 L, membrane area was 1 m². For discontinuous diafiltration, the cheese whey was concentrated to a VCR of 20 (55% protein) before adding water to the feed tank and reultrafiltering (adapted from Asbi and Cheryan 1992).*

concentration, the feed will have to be preconcentrated by UF or MF. In that case, the time for concentration by MF or UF must also be considered, which will change C_B^*. The total time for the combined process is

$$t = t_{\text{UF}} + t_{\text{CD}} \tag{7.35}$$

From Equations (7.33) and (7.34),

$$t = \frac{V_0(1 - C_{B0}/C_B)}{(J_{\text{av}})_{\text{UF}} \cdot A} + \frac{V_0 C_{B0}(-\ln C_B - \ln\{[1 - P]/PC_{S0}\}}{(J_{\text{av}})_{\text{CD}} \cdot A} \tag{7.36}$$

The minimum time can be obtained by differentiating Equation (7.36) and setting $dt/dC_B = 0$. An example of its use is shown for the UF of soybean

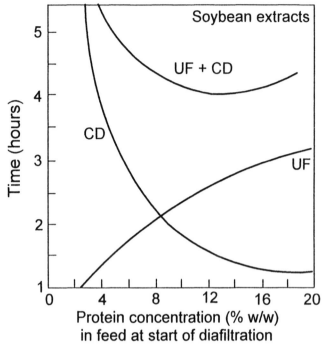

Figure 7.17. *Time required for ultrafiltration and continuous diafiltration of soybean water extracts. Initial protein concentration in feed = 2.6%. Initial concentration of permeable species = 1.6%. Initial protein purity = 61.9%. Final protein purity = 95%. For UF, the time is to concentrate the feed from its initial value of 2.6% protein to the indicated value on the graph. For CD, the time shown is for continuous diafiltration of the UF-concentrated retentate to obtain a final protein purity of 95% (adapted from Asbi and Cheryan 1992).*

water extracts. Flux data were obtained from Nichols and Cheryan (1981). A trial-and-error solution of the differentiated form of Equation (7.36) gave $C_B^* = 13.1\%$ (Asbi and Cheryan 1992). The actual times for the separate operation of UF and CD calculated using the individual equations for t_{UF} and t_{CD} are shown in Figure 7.17. If CD commenced at the initial protein concentration of 2.6%, $t_{UF} = 0$ and $t_{CD} = 6.3$ h (for an initial volume of 100 L and unit membrane area). If the feed is preconcentrated by UF, t_{UF} increases while t_{CD} decreases (since a lesser volume has to be diafiltered). The minimum in the curve occurs at 13.0%, which is close to the C_B^* value obtained from Equation (7.36).

When both UF and CD times are included in the optimization, the optimum value of C_B^* is lower than when CD alone is considered. For example, with the soybean water extracts example, C_B^* for the combined process is 13% rather than the 18.6% obtained for CD only. Similarly, with cottage cheese whey,

C_B^* for CD alone is 15.4% versus 10.2% for the combined process (Asbi and Cheryan 1992).

The model derived here could also be used to minimize the volume of permeate generated (e.g., in situations where the diafiltration solvent is limited in availability and/or where waste disposal must be kept to a minimum). The total permeate generated is

$$(V_P)_{total} = (V_P)_{UF} + (V_P)_{CD} \tag{7.37}$$

$$(V_P)_{UF} = V_0 \, (1 - C_{B0}/C_B) \tag{7.38}$$

$$(V_P)_{C_D} = V_R \, V_D = V_0 \, C_{B0} \, V_D/C_B \tag{7.39}$$

$$(V_P)_{total} = V_0[1 - (1 - V_D)/\text{VCR}] \tag{7.40}$$

Table 7.6 summarizes the results of the optimization study for two cases. For soybean extracts being purified from an initial 61.9% protein to 95% protein, the combined process is marginally better in terms of lowest permeate volume and lowest process time compared to DD. With cottage cheese whey, which required a much greater amount of purification to increase protein content from 6.6% to 95.5%, the combined process was clearly superior in minimizing both permeate volume and process time.

There is some evidence in the literature that the film theory/gel model [Equation (7.29)] may not be valid under all conditions likely to be encountered during UF of macromolecules such as proteins. Jaffrin (1990) observed that, in the purification of plasma, flux was a function of both ethanol and albumin concentration, increasing with increase in alcohol concentration, especially when

Table 7.6. Process time and permeate volume for various diafiltration options in batch mode. Initial volume = 100 L, membrane area = 1 m^2.

Feed	Operation	Permeate Volume (L)	Process Time (h)
Soybean extracts	CD at VCR 1	246	6.29
	DD (VCR 5/VCR 2.5)	120	4.43*
	DD (VCR 7.14/VCR 1.67)	126	4.18*
	Combined UF + CD (UF until VCR 5)	97	4.1
Cheese whey	CD at VCR 1	600	18.5
	DD (VCR 20/VCR 20)	190	7.64*
	Combined UF + CD (UF until VCR 25.5)	106	5.18

*Does not include time between UF stages for adding diafiltration water to feed tank
Source: Data from Asbi and Cheryan (1992)

albumin concentration was low. Peri et al. (1973) and Rajagopalan and Cheryan (1991) observed that the removal of lactose and salts from skim milk by DF increased the flux. Higher flux during DF was observed for UF of perilla extracts (Lin et al. 1989) and glutamic acid (Kuo and Chiang 1987). In some cases, flux changed from being mass transfer–controlled initially to pressure-controlled at the end of the DF runs.

7.E.
FRACTIONATION OF MACROMOLECULES

The examples considered so far were purification of a macromolecule, i.e., removal of freely permeable low molecular weight solutes from a solution containing proteins. It is also possible to fractionate macromolecules by a judicious choice of membranes and operating parameters. In this section we will consider the fractionation of a mixture of three proteins X, Y, and Z with the following molecular weights: $X = 3000$; $Y = 40,000$; and $Z = 500,000$, each at a concentration of 1.0% w/v.

To separate these proteins, a two-stage fractionation will have to be employed. The first stage should consist of a membrane of about 20,000 molecular weight cut-off (MWCO) (designated 20 K), and the second stage can contain a 100,000 MWCO membrane (100 K). Table 7.7 shows typical rejection characteristics of each membrane. Figure 7.18 is a schematic of the process. If the first stage of UF using the 20 K membrane is done to a VCR of 10, it will result in the following composition of the reject and permeate streams, using Equation (7.4):

$$C_{RX} = 1(10)^{0.1} = 1.26\%$$

$$C_{RY} = 1(10)^{0.95} = 8.91\%$$

$$C_{RZ} = 1(10)^{0.99} = 9.77\%$$

To determine the composition of the permeate stream, a material balance should be done on each component. Assuming an initial volume of 100 L,

Table 7.7. Rejections of proteins with the two membranes.

Protein	MW	Rejection (20 K membrane)	Rejection (100 K membrane)
X	3,000	10	5
Y	40,000	95	10
Z	500,000	99	98

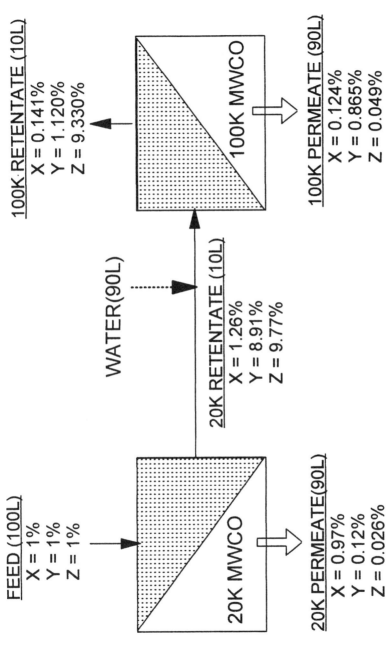

Figure 7.18. Schematic process diagram for fractionation of three proteins: X (MW = 3000), Y (MW = 40,000), and Z (MW = 400,000).

from Equation (7.27):

$$100(0.01) = 90(C_{PX}) + 10(0.0126)$$

$$C_{PX} = 0.97\%$$

Similarly,

$$C_{PY} = 0.12\%$$

$$C_{PZ} = 0.026\%$$

The retentate (10 L) is then diluted with water back to the original volume (100 L) and sent through the 100 K membrane module for a 10-fold reduction in volume. The composition of the reject and permeate streams are as follows:

Component	100 K Reject Stream	100 K Permeate Stream
X	0.141%	0.124%
Y	1.12%	0.865%
Z	9.33%	0.049%

The final composition on a dry weight basis is shown in Table 7.8. Each of the three final product streams could be purified further using the appropriate membranes and then concentrated using tight membranes.

7.F.
ENERGY REQUIREMENTS

Several energy inputs are required in UF and MF, as shown in Figure 7.19:

$$E = E_{\text{Thermal}} + E_F + E_Q \tag{7.41}$$

Table 7.8. Composition (% d.b.) of final protein fractions after two-stage purification by UF.

Protein	Feed	Permeate from 20 K Membrane	Permeate from 100 K Membrane	Retentate from 100 K Membrane
X	33.3	86.9	11.9	1.3
Y	33.3	10.8	80.4	10.6
Z	33.3	2.3	4.7	88.1

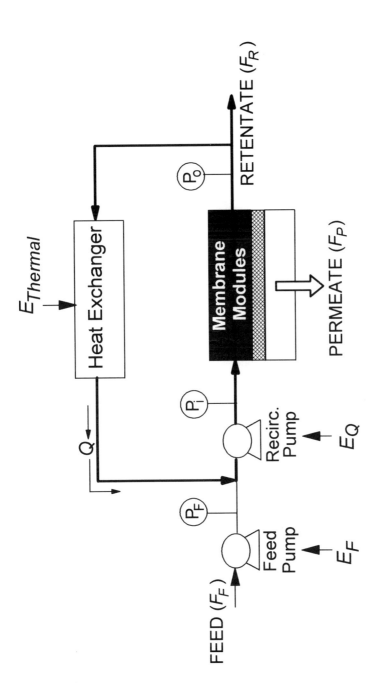

Figure 7.19. Energy inputs for a typical membrane operation. P_i and P_o are inlet and outlet pressures across the membrane module. P_F is the operating pressure of the feed pump. F_F, F_P, and F_R are volumetric flow rates of feed, permeate and retentate, respectively. Q is the recirculation rate.

FEED (F_F)

Feed Pump

E_F

P_F

Recirc. Pump

E_Q

P_i

Q

Heat Exchanger

$E_{Thermal}$

P_o

RETENTATE (F_R)

Membrane Modules

PERMEATE (F_P)

where $E_{Thermal}$ is the thermal energy (either heating or cooling as required) to maintain the process fluid at the required temperature, E_F is the pressure energy required to feed the system at the proper transmembrane pressure, and E_Q is the recirculation energy for maintaining the required flow rate or turbulence. The thermal energy input is usually provided through a heat exchanger in the recirculating loop in each stage. Simple heat transfer equations can be used for estimating $E_{Thermal}$, but the heat input by the pumps should also be taken into account.

E_F and E_Q represent pumping energy:

$$E_F = P_F \cdot F_F / \eta \tag{7.42}$$

$$E_Q = \Delta P \cdot Q / \eta \tag{7.43}$$

where P_F is feed pump pressure, F_F is feed flow rate, Q is recirculation (flow) rate, ΔP is pressure drop across the module ($P_i - P_o$), and η is pump efficiency (usually 0.5–0.85).

For single-pump systems, where the recirculation and the pressure energy are provided by the same pump:

$$(E_F + E_Q) = P_i(Q + F_F) \tag{7.44}$$

For batch systems, with no continuous feed, $F_F = 0$.

In most UF and MF applications, $F_F \ll Q$. For example, in the case study cited in Section 7.C.5. of a feed-and-bleed system of 235 m² area operating at VCR 5, the recirculation rate (Q) for 4″ spiral membranes will be ~260,000 lb/h, and 830,000 lb/h for 1.1-mm hollow fibers. In contrast, $F_F = 5000$ lb/h and $F_P = 4000$ lb/h. Thus, $E_F \ll E_Q$, and E_F can be neglected as a first approximation.

The relationship between Q and ΔP is given by the general relationship

$$\Delta P \propto (Q)^n \tag{7.45}$$

where $n = 1$ for laminar flow and $n = 1.5$–2 for turbulent flow. Thus, plots of ΔP versus Q should be linear for laminar flow modules and hyperbolic for turbulent flow modules. An example of such plots is in Figure 5.37.

Combining Equations (7.43) and (7.45),
For laminar flow:

$$E_Q \propto (Q)^2 \tag{7.46}$$

For turbulent flow:

$$E_Q \propto (Q)^{2.5-2.85} \tag{7.47}$$

We had earlier seen from Equations (4.35) and (4.36) that increases in cross-flow rate would be more beneficial for turbulent flow systems due to the higher exponent on the Reynolds Number term. However, Equations (7.46) and (7.47) indicate that the power consumption will also increase much faster for turbulent flow systems.

To compare energy consumption of different membrane systems, it is important to express the data on a common basis. There are two ways to do this:

1. Unit permeate volume basis: Since a certain amount of permeation or reduction in volume is the ultimate aim of UF or MF, the energy required per unit volume permeate is expressed as

$$E_Q = \frac{\Delta P \cdot Q}{J \cdot A \cdot \eta} \qquad (7.48)$$

This above definition can be rewritten in order to express energy consumption in recognizable energy units, such as kJ/m^3 or kWh/gal:

$$E_Q = C \frac{\Delta P \cdot Q}{J \cdot A \cdot \eta} \qquad (7.49)$$

Table 7.9 lists values of the conversion factor C, depending on the units used for other parameters.

2. Unit membrane area basis: This is useful to compare relative energy consumption of equipment without regard to the flux:

$$E_Q = C \frac{\Delta P \cdot Q}{A \cdot \eta} \qquad (7.50)$$

Table 7.9. Values of conversion factor C in Equation (7.49). Pump efficiency (η) has not been included in the C value.

ΔP	Q	J	A	E_Q	C in Equation (7.49)
bar	m^3/h	LMH	m^2	kJ/m^3	99,988
bar	m^3/h	LMH	m^2	kWh/m^3	27.8
kg_f/cm^2	L/min	LMH	m^2	kJ/m^3	5882
kg_f/cm^2	L/min	LMH	m^2	kWh/m^3	1.634
kPa	L/h	LMH	m^2	kJ/m^3	1
kPa	L/h	LMH	m^2	kWh/m^3	2.78×10^{-4}
psi	gpm	GFD	ft^2	kWh/gal	0.01044

Table 7.10. Values of conversion factor C in Equation (7.50). Pump efficiency (η) has not been included in the C value.

Units in Equation (7.50)				C in
ΔP	Q	A	E_Q	Equation (7.50)
bar	m^3/h	m^2	kW/m^2	2.777×10^{-2}
kg_f/cm^2	L/min	m^2	kW/m^2	1.634×10^{-3}
kPa	L/h	m^2	kW/m^2	2.78×10^{-7}
psi	gpm	ft^2	kW/ft^2	4.351×10^{-4}

The units would be kJ/m^2 or kW/m^2 or kW/ft^2. Table 7.10 lists values of C for Equation (7.50). Annual pumping costs could be obtained as follows:

$$\text{Pumping cost/year} = E_Q(kW/m^2) \times \text{membrane area in plant } (m^2)$$

$$\times \text{operating time (hours/year)}$$

$$\times \text{power cost (\$/kWh)} \tag{7.51}$$

In Equations (7.48)–(7.51), the total ΔP and/or total Q, as well as the total membrane area, in all modules in a plant must be considered.

7.F.1.
EXAMPLE

Calculate the energy consumption for the example in Section 7.C.5. for batch UF operation. Compare energy consumption for spirals, hollow fiber, and tubular modules. Assume a pump efficiency of 50%, an operating time of 8000 hours per year, and a power cost of $0.05/kWh.

We will first consider the case of hollow fibers with characteristics shown in the table below. We had earlier estimated the average flux in the batch operation to be 26 LMH (Section 7.C.5.). For the removal of 4000 L/h of permeate, the membrane area required for the batch operation was 153.8 m². Thus, the number of modules = 153.8/1.4 = 109.8, which means 110 modules should be used. The actual membrane area in the plant will be $110 \times 1.4 = 154$ m². In all likelihood, these modules will be placed parallel to each other (e.g., see Figure 7.20). Thus, the total ΔP will be 0.9 kg_f/cm^2 and total flow rate = $110 \times 38 = 4180$ L/min. From Equation (7.49) and Table 7.9:

$$E_Q = 5882\frac{(0.9)(4180)}{(26)(154)(0.5)} = 11{,}053 \text{ kJ/m}^3$$

	Hollow Fibers	Tubular	Spiral
Channel height, mm	1.1	12.5	1.1
Membrane area per module, m^2	1.4	1.7	6
Average flux (J), LMH	26[1]	45[2]	17[1]
Cross-flow rate (Q), L/min	38	265	76
Pressure drop (ΔP), kg/cm^2	0.9	2	1
E_Q			
(kJ/m^3)	11×10^3	81×10^3	8.8×10^3
(kWh/m^3)	3.1	23	2.4
(kW/m^2)	0.08	1.0	0.041
Number of modules	110	53	40
Pumping cost ($/year)	4928	36,700	3970

[1] Cheryan and Chiang (1984).
[2] DeBoer and Hiddink (1980).

From Equation (7.50) and Table 7.10:

$$E_Q = 1.634 \times 10^{-3} \frac{(0.9)(4180)}{(154)(0.5)} = 0.08 \, \text{kW/m}^2$$

From Equation (7.51):

Annual pumping cost $= 0.08 \times 154 \times 8000 \times 0.05 = \$4928$ per year

Similar calculations for tubular and spiral-wound modules are shown in the table above. Recirculation pumping energy (kW/m^2) is lowest with spirals and highest with tubular modules—a 24-fold difference in this example. However, when compared on a cost basis, the difference is less because average flux of the tubular module is 2.6 times higher.

Table 7.11 is a comparison of energy consumption and performance of several types of membrane modules processing cheese whey. When comparing the data, the following points should be noted:

1. The application was the concentration of cheese whey by ultrafiltration to result in a 35% protein (d.b.) solids. This usually meant processing to VCR 10. The flux data is the average flux determined as volume of permeate/time of processing, except for the Abcor spiral-wound data, which is initial flux.
2. Although a common feed appears to have been used, the source of the cheese whey is quite different. Cheryan and Kuo used acidified cottage cheese whey with hollow fibers and spiral-wound modules. Hiddink et al. used Gouda whey that had been decalcified. Native cottage cheese whey was apparently

Table 7.11. Comparison of energy consumption for various UF modules processing cheese whey.

Module	Manufacturer	Membrane	Whey	Flux (LMH)	Pressure (bar)	Energy (kWh/m³)	Reference
Spiral	Abcor/Koch	HFM-100	Acid	38	4.8	2.88	Abcor/Koch (1981)
	Osmonics	SEPA-20 K	Cottage cheese (pH 3)	12	3.35	1.64	Cheryan and Kuo (1984)
Hollow fibers	Romicon	PM-10	Cottage cheese (pH 3)	28	1.55	3.30	Cheryan and Kuo (1984)
Plate and frame	Rhône-Poulenc	IRIS 3038	Gouda whey (pH 6.6)	51	3	4.88	Hiddink et al. (1981)
	DDS	GR6P	Sulfuric acid (pH 4.5)	72	4.5	5.25	Matthews et al. (1978)
Tubular	Wafilin	WFA500	Gouda whey (pH 6.6)	74	5	8.11	Hiddink et al. (1981)

Energy calculated using Equation (7.49) assuming pump efficiency of 50%, except for DDS data of Matthews et al. (1978), which was from current drawn by pump motor

Table 7.12. Operating economy of UF plants processing whole milk for cheese manufacture.

	Plate and Frame	Hollow Fibers
Membrane	DDS GR61PP	Romicon PM-10
Total area (m^2)	189	203
Operating temperature (°C)	50	50
Feed flow rate (L/h)	7,550	6,000
VCR	4.9	5.0
Average flux (LMH)	32	24
Electricity (kWh/m^3 permeate)	19.7	10
Water (L/m^3 feed)	40	30

Source: Adapted from Knudsen and Braun (1985)

used with Abcor spirals, while Matthews et al. used sulfuric acid casein whey. The rate of fouling would be quite different with each whey, and this factor could partially explain some of the differences in the flux data.

3. The transmembrane pressures (TMP) used are quite different, and since whey UF is expected to be pressure-controlled in this application, this could also have a significant effect on flux.

Despite these limitations in the data, Table 7.11 again shows that spirals use the least energy while the large diameter tubular modules use the most. Table 7.12 is a comparison of energy consumption for operating plants that were using either plate or hollow fiber systems for milk UF. However, even though the energy consumption of the plate system was double that of the hollow fibers, the overall cost of production of the plant using the plate system was lower because of the lower cost of membrane replacement (Knudsen and Braun 1985). This emphasizes the importance of considering all costs when comparing module designs or operating strategies, as discussed in the next section.

Another point to consider when comparing energy costs of membrane technology with other technologies (e.g., evaporation) is the relative cost of the energy source. For example, steam cost in 1995 in the United States averaged \$5/1000 lb steam (\$11/ton) and electricity for industrial customers averaged \$0.05/kWh. However, on a unit *energy* basis, the cost of steam = \$5.15 per million Btu and electricity = \$14.60/million Btu (1 Btu $\cong$ 1 kJ). Thus, electric power (the energy source for membrane plants) is three times more expensive than steam (the prime energy source for evaporators). With these costs, a membrane process must use 65% less energy to compete with a steam-based dewatering process, if energy alone is the criterion of choice. If the relative cost of electrical energy was higher, e.g., \$0.10/kWh versus \$5/1000 lb steam, then the membrane process would have to use 75% less energy.

7.G.
COSTS AND PROCESS ECONOMICS

7.G.1.
ARRAYS AND CONFIGURATIONS

Membrane modules can be arranged in the flow path in one of two ways:

1. In parallel, with each membrane module/element arranged in parallel to each other, as shown in Figure 7.20
2. In series, with several individual elements in the same flow path as shown in Figure 7.21

Combinations of series and parallel flow are common, depending on the pump characteristics and cost of the modules. The examples shown in these figures are for eight modules.

The nature of the feed–pressure relationship also has a bearing on the configuration or arrangement of modules. If the feed displays little or no polarization, e.g., with characteristics shown in Figure 7.22 (left), the best possible TMP is the highest value (90 psi in this example). The ideal configuration will be the parallel array (Figure 7.20). If we further assume that the recommended recirculation flow is 300 gallons per minute (gpm) per module and it results in a pressure drop of 20 psi per module, then all modules in the system will need the conditions shown for module #1 in Figure 7.22 (left). The total flow will be $8 \times 300 = 2400$ gpm, and the inlet pressure to all modules will be 100 psi and outlet pressure 80 psi in order to meet the two pressure requirements (TMP $= 90$ psi, $\Delta P = 20$ psi). In general, this combination shown in Figure 7.20 will result in the lowest pressure drop and highest flux of all possible arrays for this type of feed. However, the power consumption and housing cost will be the highest. The pump will need a capacity of 2400 gpm at 100 psi, for a total power consumption of 280 hp (assuming overall η of 50%).

Stacking more than one element in series in the same flow path (as shown in Figure 7.21) will result in downstream elements 2, 3, and 4 being at a lower effective TMP than the first one (Figure 7.22, left). The flux in each downstream module will be correspondingly lower, requiring more total modules and housings. Pump capacity will be lower, however, since we now need only $2 \times 300 = 600$ gpm at 100 psi. The economic criteria to be decided is between the lower cost of the pump versus the higher cost of more modules.

On the other hand, if the system is highly polarized with the characteristics shown in Figure 7.22 (right), it would not make any difference in terms of membrane area whether the series or parallel array is used since flux in all modules is the same. It does make a difference in terms of pump capacity, however. In parallel, all modules could run at inlet and outlet pressures of 40

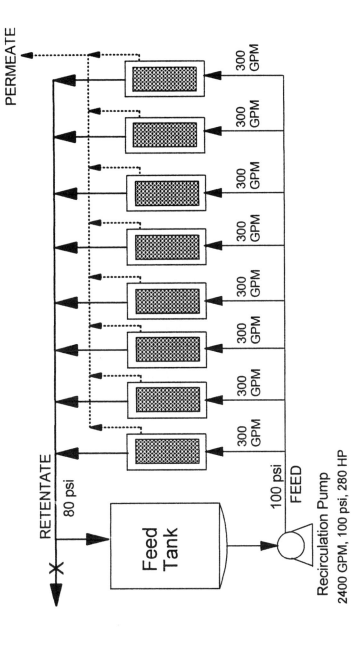

Figure 7.20. Parallel array of modules in a plant. The assumptions are that each module requires 300 gpm, a pressure drop of 20 psi, and a transmembrane pressure of 90 psi.

335

Table 7.13. *Housing cost and pump power for high-polarization feed with flux characteristics shown in Figure 7.22 (right). Number of modules = 8, cost of each module = $1200, cost of 1-long housing = $1350, 4-long housing = $1750.*

| | Type of Housing | |
	1-Long	4-Long
Number of modules	8	8
Number of housings	8	2
Cost of modules ($)	9,600	9,600
Cost of housing ($)	10,800	3,500
Total cost ($)	20,400	13,100
Total flow (gpm)	2,400	600
Inlet pressure (psi)	40	100
Pump power, η = 50% (hp)	112	70

Costs based on 8″ × 40″ spiral modules and housings provided by Osmonics, Inc., Minnetonka, MN in 1995

and 20 psi, resulting in a pump of 2400 gpm at 40 psi with 112 hp (Table 7.13). In contrast, two trains of four modules each will require 600 gpm at 100 psi for 70 hp.

Most membrane modules (inorganics, tubular, hollow fibers) are sold with their housing as an integral unit. The major exception is the spiral membrane

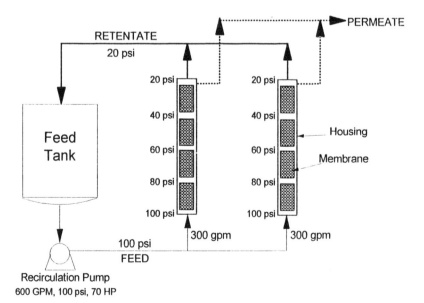

Figure 7.21. *Series array of modules in a plant. Same assumptions as Figure 7.20.*

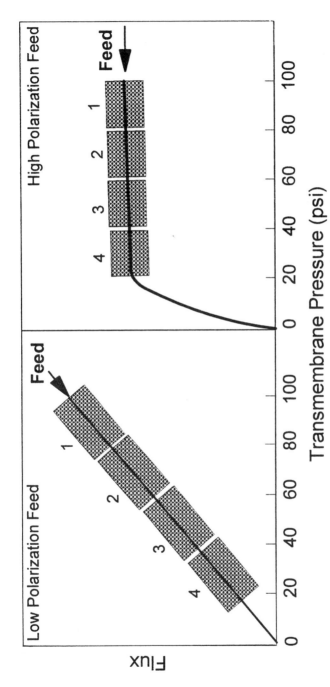

Figure 7.22. Flux–pressure behavior: left: low-polarization feed; right: highly polarized feed.

Table 7.14. Housing cost and power consumption for low-polarization feed shown in Figure 7.22 (left). Other assumptions same as Table 7.13.

	Housing Type			
	1-Long	2-Long	3-Long	4-Long
Number of modules	8	9.2 (10)*	10.6 (11)*	12.8 (13)*
Cost of modules ($)	9,600	12,000	13,200	15,600
Number of housings	8	5	4**	4†
Cost of housings ($)	10,800	7,250	6,250	8,350
Total cost ($)	20,400	19,250	19,450	23,950
Total flow (gpm)	2,400	1,500	1,200	1,200
Inlet pressure (psi)	100	100	100	100
Pump power (hp)	280	175	140	140

*Calculated number of modules and actual number (in parentheses) of modules needed
**Three 3-long housings and one 2-long housing
†Three 4-long housings and one 1-long housing

where the membrane element and housings are available as separate units. A stainless steel (SS) housing with the appropriate end-caps can cost as much or more than the membrane element. For example, a single-element (1-long) 304SS housing with end-caps for 8″ spirals costs about $1350, and a 4-long housing with end-caps is $1750 (highly polished sanitary 316SS housings would be $2800 and $3300, respectively). In this case, it would obviously be cheaper to use a series array with several elements per housing, as shown in Table 7.13 (at least for highly polarized feeds).

However, if the feed is low in polarization, there will probably be an optimum number of modules that can be placed in series. Table 7.14 shows calculations for the case study shown in Figure 7.22 (left). Single-element housings will require the least number of modules but the most housings and pump power. With two elements per housing, the average flux is lower, requiring 9.2 modules (which actually means 10 modules) and five housings. Three-element housings require a total of 11 modules; the modules can be placed in three 3-element housings and one double-element housing in four parallel paths. The 4-element housings will require a total of 13 membrane elements placed in three 4-element housings and one single-element housing, also in four parallel paths.

The lowest total cost is with double-element housings. However, since pump power increases in proportion to the number of parallel flow paths, the optimum could also be the 3-element housing combination, depending on the cost of the pump. If membrane replacement cost is also considered, the optimum may be double-element housings or even single-element housings, depending on the degree of polarization.

Table 7.15. Cost of various MF and UF configurations.

Module Type	Total System* ($/m^2)	Membrane Replacement ($/m^2)	Energy Consumption (W/m^2)	Key Factor
Spiral wound	300	40–100	40–130	Prefiltration
Hollow fibers	1000	600	80–700	Low pressures
Tubular (0.5″)	1300–2000	300–500	700–2000	High flow rate
				Large floor space
Inorganic	2200–6000	500–3000	400–2000	Long life
				High flow rate

*Includes housings, fittings, controls and first set of membranes. Does not include tankage, prefilter, or CIP system

7.G.2.
SYSTEM COST

Table 7.15 is a budget cost for various module configurations. With very large spiral systems (> 1000 m^2) using industrial (rather than sanitary) fittings, piping and pumps, and simple automation and controls, the system cost could go down to $250/m^2. Membrane replacement could be $30–40/m^2 for cellulose acetate (CA), polyethersulfone (PES), and polyvinylidene fluoride (PVDF) membranes in 8″ or larger modules. Smaller systems using 4″ modules with highly polished pharmaceutical and sanitary components with extensive automation and computer controls could reach $450/m^2. Keep in mind that spirals and ceramics need only to have membrane elements replaced and not the housings. Hollow fiber and tubular replacements come with new polymeric housings (except for PCI's removable core design).

At the other end of the cost scale are the inorganic membranes. The first set of membranes are purchased together with the housings. If the membrane elements need to be replaced, they can be removed from the housing, unless they are welded or glued to the end-caps of the housing.

Also shown in Table 7.15 is the key factor to take into consideration with a particular module design. With spirals, good prefiltration is required, perhaps to 1/10–1/50th the spacer thickness. With tubular and inorganic modules, high flow rates and high floor space should be factored in to the decision-making process.

Table 7.16 shows calculations for capital and operating costs for an application using CeraMem ceramic membranes processing ethanol stillage (see Section 8.D.2. for details). In this case, each ceramic membrane module has 120 ft^2 of area and requires 300 gpm of cross-flow with a pressure drop of 20 psi. The membrane cost (less housing) is $45/ft^2, and the system (including the first set of membranes) is $210/ft^2. The feed rate was 1400 gpm and the required

Table 7.16. Cost calculations for feed-and-bleed ceramic membrane system.

	Single Stage	Multistage Feed-and-Bleed			
		Stage 1	Stage 2	Stage 3	Total
Feed flow (gpm)	1,400	1,400	368	236	
Concentration factor (X)	7.0	3.8	5.93	7.0	
Retentate flow (gpm)	200	368	236	200	
Permeate flow (gpm)	1,200	1,032	132	36	1,200
Average flux (GFD)	52.9	131.7	103.53	52.9	
Membrane area (ft^2)	32,641	11,279	1841	982	14,102
Number of modules	272	94	15	8	118
System cost ($, million)	6.85	2.369	0.386	0.206	2.96
Depreciation ($/year)	769,389	265,867	43,386	23,138	332,391
Membrane replacement ($/year)	293,767	101,513	16,566	8,834	126,913
Power ($/year)	567,949	196,258	32,027	17,080	245,365
Cleaning ($/year)	22,849	7,895	1,288	687	9,871
Labor, maintenance, interest ($/year)	685,455	236,864	20,614	20,614	296,131
Operating cost ($ million/year)	2.34	0.808	0.132	0.070	1.011

concentration factor (X) was 7. Flux could be correlated with concentration factor (X) as follows:

$$J[\text{gallons per square foot per day (GFD)}] = 133.4 - 6.84 \times 10^{-4}(X)^6$$

The calculations are shown below, and final results are shown in Table 7.16.

1. Single-stage feed-and-bleed:
   - From Equation (7.24): retentate flow rate (F_R) = 1400/7 = 200 gpm
   - From Equation (7.26): permeate flow rate
     (F_P) = 1400 − 200 = 1200 gpm = 1,728,000 gal/day (gpd)
   - From Equation (7.25): Membrane area (A) = 1,728,000 gpd/52.9
     GFD = 32,641 ft^2
   - Number of ceramic modules = 32,641/120 = 272
   - System cost = $210/ft^2 × 32,641 ft^2 = $6.85 million
   - Energy/power cost: From Equation (7.50) and Table 7.10,

$$\text{Energy consumption} = 4.351 \times 10^{-4}(20)(300)/(120)(0.5)$$

$$= 0.0435\,\text{kW/ft}^2$$

From Equation (7.51):

$$\text{Energy usage} = 0.0435\,\text{kW/ft}^2 \times 32641\,\text{ft}^2 \times 8000\,\text{h/year}$$

$$\times\ \$0.05/\text{kWh} = \$567,949/\text{year}$$

- Capital charge: This provides for depreciation, insurance, etc., and is calculated annually at about 10–20% of capital cost, based on a depreciation over 7- to 14-year periods. Membrane costs are not normally included in this category and are considered instead as a separate operating expense. In this example, system cost less membranes = $210 − 45 = $165/ft^2. Using a 7-year straight line method of depreciation:

$$\text{Depreciation} = \$165 \times 32641/7 = \$769,381/\text{year}$$

- Membrane replacement costs vary widely as mentioned in Chapter 5. Most polymeric membranes are guaranteed for 1 year under normal operating conditions expected in MF and UF that require a daily cleaning. The exception may be certain water treatment applications, which have milder operating conditions and (perhaps) a lower frequency of cleaning, which could have a higher operating life. Seasonal use, e.g., fruit juice, could get 2-year warranties. Ceramic and inorganic membranes, however, should get much longer operating lifetimes. Depending on how much experience the company has with a particular application, 5–10 years of guaranteed life are not uncommon. However, for business reasons, the warranty of ceramic membranes sometimes may not be more than 2 years (even though the user and manufacturer know intuitively that higher lifetimes can be expected). Long lifetimes may be enough to move this category of costs from the operating expense category to the depreciation/capital charge category. For this example, we use a 5-year life.

$$\text{Membrane replacement cost} = \$45/\text{ft}^2 \times 32,641\,\text{ft}^2/5\text{years}$$

$$= \$293,767/\text{year}$$

- Maintenance, labor, and interest: Small plants (<400,000 L per day of feed) require about one person per shift, while large plants (>2 million L/day of feed) require about one person per 500,000 L/day of capacity per shift. Labor cost can be reduced by automation, but that increases the capital cost. Labor and maintenance are charged at a nominal rate of 2–5% installed cost. In this example:

$$\text{Maintenance costs } @3\% \text{ of capital cost} = \$205,636/\text{year}$$

$$\text{Interest } @7\% \text{ of fixed capital cost} = \$479,819/\text{year}$$

- Cleaning cost: If conventional detergents and sanitizers are used, chemicals cost will be $1–10/m^2/\text{year}$ ($0.1–1/\text{ft}^2/\text{year}$), being lower for plants with low holdup and which need only one brief cleaning cycle per day using NaOH and/or chlorine. It will be on the high side if two or more cleanings are needed with combinations of acid, alkali, and enzymes. If a separate clean-in-place (CIP) system (pump, tank, heat exchangers, controls, etc.) is needed for the membrane system, another $20,000–$50,000 should be added to the capital cost. In this particular example, only chemicals are considered.

$$\text{Cleaning cost} = \$0.7/\text{ft}^2/\text{year} \times 32641 \text{ ft}^2 = \$22,849/\text{year}$$

- The total operating cost in this example for the single-stage feed-and-bleed is $2.34 million per year. This comes to be $3.48 per 1000 gallons ($0.92/m^3$) of feed. The high cost of the ceramic system makes depreciation the single biggest cost. If spiral membranes could have been used instead (assuming the flux was the same), capital costs would be one-tenth, and membrane replacement and power would be half the cost shown in Table 7.16.

2. Multistage feed and bleed: The optimum number of stages and X values in each stage would be determined as shown in Section 7.C.5. In this partic-ular case, it is better to use unequal areas in each stage since it results in significantly lower total area than using equal areas, because of the flux–X relationship (13% less in the case of a three-stage system). The rest of the calculations would follow the same procedure as for the single stage. Using three stages reduces membrane area and operating cost by 57% compared to a single-stage system.

7.H.
SUMMARY

Figure 7.23 summarizes all the factors to be considered in the process design of an MF or UF system. Not specifically included in the diagram is the cleaning mode. Cleaning requires conditions different than the process mode; e.g., cleaning is best at higher velocity and lower pressure, with flux that is usually much higher during cleaning. Thus, it is difficult to find one pump that can serve

PROCESS DESIGN

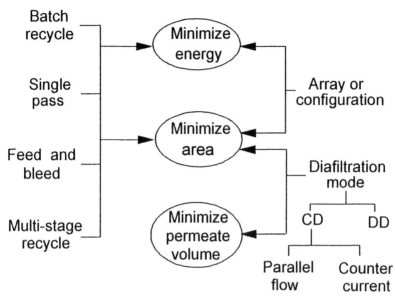

Figure 7.23. *Important factors for process design of a membrane system.*

for both processing and for cleaning; in fact, a separate CIP system is usually needed for large continuous systems. Many original equipment manufacturers (OEMs) and integrated membrane companies have computer programs that can perform these calculations rapidly and arrive at the minimum capital and operating cost. However, this assumes that you have prior knowledge of or experience with many of the design factors shown in Figure 7.23.

REFERENCES

ABCOR/KOCH. 1981. Product Bulletins FD/DB-80-2, GN/DB-81-3, Wilmington, MA.

ASBI, B. A. and CHERYAN, M. 1992. *Desalination* 86: 49.

CHANG, Y. L. and LEE, C. J. 1988. *J. Membrane Sci.* 39: 99.

CHERYAN, M. 1986. *Ultrafiltration Handbook*, Technomic, Lancaster, PA.

CHERYAN, M. and CHIANG, B. H. 1984. In *Engineering and Food, Vol. 1*, B. M. McKenna (ed.), Applied Science Pub., London, U.K. p. 191.

CHERYAN, M. and KUO, K. P. 1984. *J. Dairy Sci.* 67: 1406.

DEBOER, R. and HIDDINK, J. 1980. *Desalination* 35: 169.

DUTRE, B. and TRAGARDH, G. 1995. *J. Food Engr.* 25: 233.

HIDDINK, J., DEBOER, R. and NOOY, P. F. C. 1981. *Milchwiss.* 35: 657.

JAFFRIN, M. Y. 1990. Presented at *4ᵗʰ Symposium on Protein Purification Technologies*, Clermont-Ferrand, France.

KNUDSEN, A. and BRAUN, H. 1985. *Report No. 14*, Danish Government Dairy Research Institute.

KUO, W. H. and CHIANG, B. H. 1987. *J. Food Sci.* 52: 1401.

LIN, S. S., CHIANG, B. H. and HWANG, L. S. 1989. *J. Food Engr.* 9: 21.

MATTHEWS, M. E., DOUGHTY, R. K. and HUGHES, I. R. 1978. *N.Z. J. Dairy Sci. Technol.* 13: 37.

MERRY, A. J. 1996. In *Industrial Membrane Separation Technology*, K. Scott and R. Hughes (eds.), Blackie Academic, London, U.K.

NG, P., LUNDBLAD, J. and MITRA, G. 1976. *Separation Sci.* 11: 499.

NICHOLS, D. J. and CHERYAN, M. 1981. *J. Food Process. Preservation.* 5: 104.

PERI, C., POMPEI, C. and ROSSI, F. 1973. *J. Food Sci.* 38: 135.

RAJAGOPALAN, N. and CHERYAN, M. 1991. *J. Dairy Sci.* 74: 2435.

SIGDELL, J. E. 1982. *J. Art. Org.* 5: 361.

VILLARROEL, F., KLEIN, E. and HOLLAND, F. 1977. *Trans. Am. Artif. Intern. Organs.* 23: 225.

CHAPTER 8

Applications

Industrial ultrafiltration was initially developed primarily for the treatment of wastewaters and sewage to remove particulate and macromolecular materials. Its applicability has now widened considerably to include such diverse fields as water treatment, chemicals processing, food processing, and biotechnology. This chapter will focus on selected applications of ultrafiltration (UF) and microfiltration (MF) that demonstrate the breadth and scope of this separation technique.

8.A.
ELECTROCOAT PAINT

This is perhaps the first large industrial application of ultrafiltration. Electrodeposition of paint was introduced in the mid-1960s and the first UF system on an electrocoat (E-coat) paint line was installed in 1970. The piece to be painted (made of the appropriately conductive material) is immersed in an aqueous solution of paint and a voltage applied between the piece to be coated and the paint tank or an electrode. There are two types of electrocoating processes: anodic electrodeposition, where the piece to be painted is positively charged, and cathodic deposition, where the piece to be painted is the cathode; i.e., it is negatively charged. After electrodeposition of the primer coat, the piece is lifted out from the tank, and the undeposited, loosely adhering paint is washed off in a spray of water, and the part is then cured in an oven.

In the anodic method, the piece being coated is active, which causes a small amount of the piece to dissolve into the paint. These dissolved metal ions reduce the corrosion resistance of the paint. The advantage of the cathodic process is that the part being coated is passive and not affected by the reaction process. Increased corrosion resistance is the main reason why most automakers have switched to the cathodic process.

Although electrophoretic painting was rapidly accepted for priming metal products such as automobile parts, appliances, and office furniture, one of its disadvantages has been the "drag-out" loss of undeposited paint. This drag-out is removed by spraying the piece with deionized water (Figure 8.1). The diluted

345

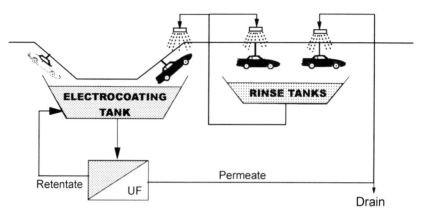

Figure 8.1. *Schematic of electrocoat painting plant for automobiles using UF. An optional deionized water rinse at the end of the line is not shown.*

drag-out could not be reused in the main dip tank, since it would upset the emulsion stability of the paint. Emulsion stability is affected by a delicate balance between the water content and the concentration of paint particles. Adding the diluted drag-out back to the paint tank is obviously not recommended. In addition, the drag-out could contain ionic impurities, such as chromic acid and zinc phosphate from pretreatment stages and neutral metal atoms from the workpiece dissolving in the anodic paint film. Thus, the drag-out would have to be treated if it had to be dumped into the sewer, an expensive and uneconomical option. The logic for incorporating an ultrafiltration plant is obvious.

An ideal electrocoating system would keep the main dip tank in constant chemical balance and would not discharge any paint particles. The best solution is to have the UF unit operate directly on the main dip tank to provide just enough clean permeate for the drag-out rinsing. The UF unit also has to operate on a low recovery (i.e., low ratio of permeate to retentate flow) to avoid upsetting the chemical balance of the paint; typically, the paint solids are increased by less than 1%. The permeate flows through a closed-loop, countercurrent rinse system (Figure 8.1). It is used sparingly to wash the workpiece, and the diluted drag-out is returned to the main paint dip tank. A small amount of permeate— perhaps less than 10% of the total amount—is periodically bled off in order to maintain the correct ionic balance in the paint tank.

Successful operation of ultrafilters in the paint recovery field requires careful selection of membrane and operating parameters. A high fluid velocity or shear rate is required to minimize concentration polarization and fouling effects. For anodic paints that tend to operate at 12–15% solids, it is best to use membranes that are negatively charged. One example is the Romicon XM-50 membrane, made of acrylic vinyl copolymer; typical data showing the flux versus long-term operation are in Figure 8.2. To maintain high flow rates at reasonable pressure

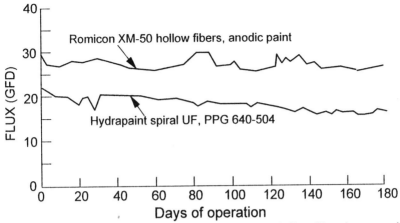

Figure 8.2. *Ultrafiltration of electrocoat paint. Romicon hollow fiber data was obtained with PPG anodic electrocoat paint (adapted from Breslau et al. 1980). Spiral UF startup data were obtained with 8" modules of 250 ft², containing a hydrophilic polyolefin membrane. The paint was PPG 640-504 paint with 21.5% solids. The spiral was operated for 6 months on line with no cleaning (adapted from Hydranautics 1996).*

drops, the shorter and larger diameter fibers (HF15-43-XM50) are preferred. The 0.5″ Ultra-Cor VII multitube modules from Koch/Abcor are designed with 25 liters per square meter per hour (LMH) at 25°C. Anodic UF systems are relatively easy to operate: membrane cleaning is infrequent (perhaps one to two times a year), resulting in long membrane lifetimes of 3 years or more.

With cathodic paints, however, the XM50 membrane was found to foul very rapidly (Figure 8.3). This was probably due to charge interactions between the positively charged paint and the negatively charged XM50 membrane. A much better performance was obtained with positively charged membranes (designated "X_1 and X_2" in Figure 8.3). Average fluxes are lower for the cathodic paint, but typical fluxes are 25–70 LMH using the appropriate hollow fiber module and operating conditions (Breslau et al. 1980). Spiral (with large spacers) and 0.5″ tubular modules of hydrophilic polyolefin membrane or polyvinylidine fluoride (PVDF) are also used with both anionic and cationic paints. Average fluxes of 15–60 LMH are possible, depending on the paint (Figure 8.2). Membrane life with cathodic paint is about 2 years with one to three cleanings per year.

Figure 8.4 shows a typical hollow fiber installation for the recovery of paint. A similar installation in Adelaide, Australia, consisted of 40 HF26.5-45-CXM cartridges for a total of 100 m² membrane area, designed to produce about 2800 L of permeate per hour, using a 37-kW centrifugal recirculation pump. The prefilter was a 50-μm cartridge filter. That plant produced enough permeate to rinse about 300 small cars per day (*FiltratioNews* 1984).

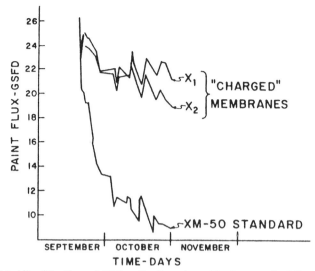

Figure 8.3. Ultrafiltration of PPG cathodic paint with the standard Romicon XM-50 module and the charged CXM hollow fiber module, designated as X_1 and X_2 (adapted from Breslau et al. 1980).

It is commonly believed that electrodeposition of paint would not be economical today without the use of ultrafiltration. UF helps to maintain the required paint bath control, recovers more than 90% of the paint drag-out, and substantially reduces the load on wastewater treatment. Thousands of such units are presently in use, primarily in the automotive and appliance fields. The savings in paint alone pay for the operation (payback periods of 1 year or less are

Figure 8.4. View of a UF paint operation with hollow fibers at the Chrysler assembly plant in Windsor, Canada (Source: Romicon, 1988, with permission).

common), not to mention the savings in sewage treatment and deionized water costs, since the only fresh water demand is for the final rinse.

8.B.
THE DAIRY INDUSTRY

Among the food industries, dairy applications probably account for the largest share of installed membrane capacity. This is partially a reflection of the universality of milk in the human diet, as well as the need to develop new methods of processing to arrest the static or declining consumption in milk and milk products in many dairying countries. Figures 8.5 and 8.6 are general schematics of the possible applications of UF and MF in the dairy industry. Reverse Osmosis (RO) of cheese whey, a by-product of cheese manufacture, was the first successful commercial application in 1971 (not shown in Figure 8.5; see Cheryan and Alvarez 1995). UF is more widely used and is well established in the dairy industry. The largest applications are the fractionation of cheese whey and, to a lesser extent, the preconcentration of milk for cheese manufacture. Current estimates are that over 300,000 m^2 of membrane area are installed in the dairy industry worldwide. Present usage is weighted towards RO and UF, although there are expectations of significant growth in MF (Cheryan and Alvarez 1995).

To successfully process a complex material, such as milk and its by-products with membranes, requires an understanding of the interactions between individual components of milk under different processing conditions. Diverse factors

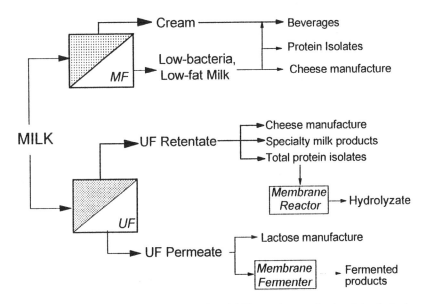

Figure 8.5. *Processing options with UF and MF membranes in the dairy industry (adapted from Cheryan and Alvarez 1995).*

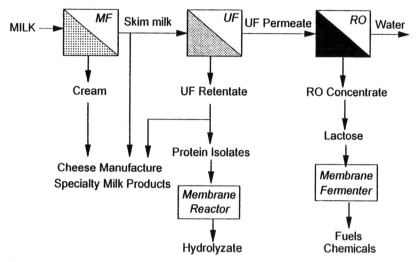

Figure 8.6. Sequential membrane processing in the dairy industry (adapted from Cheryan and Alvarez 1995).

impose additional limitations, the most important being that, as a food product, the membrane process must follow certain sanitary guidelines, which imposes further limitations on the selection of the membrane and other materials of contact.

Table 8.1 lists the gross composition of milk and some manufactured products. It is important to realize that

1. There can be significant variations in raw milk over a fairly wide range, depending on the breed of cow, season, pretreatments received, etc. In the case of cheese whey, the type of cheese and the pretreatments the whey received have a tremendous influence on the performance of the UF and MF units, as discussed in Chapter 6.

Table 8.1. Composition of selected dairy products.

Component	Milk	Quarg	Cottage Cheese	Feta Cheese	Mozarella Cheese	Cheddar Cheese	Cheese Whey
Total solids	12.5	18.0	20.3	44.8	45.8	63.3	6.3
Protein (N × 6.38)	3.3	12.6	17.3	14.2	19.4	24.9	0.6
Fat	3.5	0.4	0.4	21.3	21.6	33.1	0.1
Lactose	4.9	4.0	1.9	4.1	2.2	1.3	5.0
Ash	0.7	0.8	0.7	5.2	2.6	3.9	0.6

Sources: Data from USDA (1978) and Renner and Abd El-Salam (1991)

2. The functional, physical, and chemical properties of the final UF product will also significantly influence the processing parameters.
3. Most food products, especially milk-based ingredients, have certain defined standards of identity and rather strict regulations concerning their processing and handling. The same holds true for the cleaning and sanitizing aspects of the process.

The size distribution of milk components is shown in Figure 8.7. Lactose and soluble salts pass through UF membranes, while protein, fat, and some of the insoluble or bound salts are retained. Polyethersulfone membranes are most popular in the dairy industry, even though they foul more than cellulosic membranes. However, polyethersulfone (PES), polysulfone (PS), and PVDF have the advantage that higher temperatures can be used (generally 50–55°C

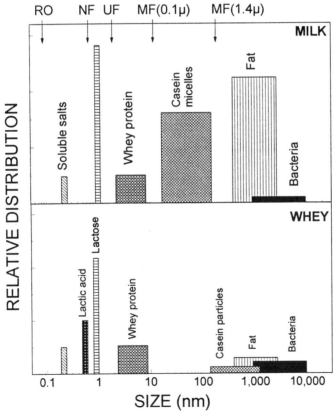

Figure 8.7. Size distribution of components in milk (top) and whey (bottom), showing possible applications of membrane technology.

to keep the viscosity low, flux high, and the growth of microorganisms under control), they can better tolerate extremes of pH (thus making cleaning much easier) and chlorine (for sanitation and cleaning). It is interesting that milk tends to foul membranes much less than cheese whey, despite its higher concentration of solids (Table 8.1).

8.B.1.
FLUID MILK AND FERMENTED PRODUCTS

Typical data for UF of milk were shown previously in Figures 4.10, 4.11, 4.15, 4.19, 4.20, 4.23, 4.26, and 4.29. Permeate flux increases as the pressure increases, but becomes independent of pressure at a level that is a function of the recirculation velocity of the retentate and solids content. Concentration polarization is quite severe with milk, and thus it is best to operate at the highest velocity possible. In most cases, maximum total solids obtainable with UF is about 38–42%. Some specific applications are discussed below.

There is considerable potential in the manufacture of specialty milk-based beverages. Milk is the major source of calcium and a significant source of high-quality protein and other nutrients. Removal of fat and cholesterol by skimming results in poorer sensory qualities. UF has been used to compensate for the poor taste and to maximize the concentration of desirable protein and calcium (Cheryan and Alvarez 1995). Since calcium in milk is mostly in the insoluble or bound form or is associated with impermeable casein micelles, concentrating skim milk by UF will simultaneously concentrate the calcium and protein, while leaving the concentration of sodium, potassium, and lactose unchanged. Fat can be added back to improve sensory quality if desired. The taste and texture of the UF milk is superior to regular skim milk and better than even skim milk with added milk solids. This product was commercialized in Australia in the late 1980s.

When used for the production of yogurt, UF milk gave a product with better viscosity and firmness than yogurt made from evaporated, milk powder-added, or RO-concentrated milks, although organoleptic scores were lower than the others, caused, most likely, by the loss of lactose and salts (Abrahamsen and Homen 1980).

"Ymer" has also been manufactured from UF milk. After 2× concentration by UF, cream is added if necessary, followed by homogenization and pasteurization. The concentrate is acidified by the addition of microbial culture. UF Ymer is generally much better in overall quality; furthermore, a 15–20% increase in yield is obtained because of the inclusion of whey proteins (Pal and Cheryan 1987).

On-farm ultrafiltration of milk has been studied in the United States and France, with a view to reducing transportation and refrigeration costs. While several studies indicate that it is technically feasible for large dairy herds and if the concentrated milk is used for manufacture of cheese, regulatory and marketing problems would have to be resolved. Stability of ultrafiltered milk is satisfactory with proper pretreatment (Kosikowski 1986), e.g., "thermalization" at 65–70°C for 10–20 sec to minimize lipase activity, and if health and safety requirements are met on the farm. One large dairy farm in the southwestern United States is implementing this idea in 1997 and shipping the milk to other areas. On the other hand, if milk could be processed and packaged on-site at dairy farms, it will enable producers to ship value-added finished products rather than raw milk. This will open up new opportunities for UF on the dairy farm.

Total milk protein isolates are usually manufactured by coprecipitation using a combination of heat, acid, and/or calcium salts. This generally results in low protein solubility, which restricts its use as a functional food ingredient. UF of milk, in combination with diafiltration (DF), can produce 90% protein isolates from milk with lactose concentrations less than 0.1% (Rajagopalan and Cheryan 1991). However, even exhaustive diafiltration of milk will not reduce the ash content below 8% on a protein basis, probably because these components (primarily calcium salts) are insoluble or associated with casein micelles and, thus, also get rejected. The ash concentration could be decreased further by lowering the pH during UF or DF, which would solubilize some of the salts and allow passage through the membrane.

8.B.2.
CHEESE MANUFACTURE

The principal use of milk ultrafiltration in the dairy industry today is the manufacture of cheese. The pioneering work of Maubois et al. (1969) that resulted in the "MMV" process for the manufacture of cheese promised to revolutionize the dairy industry. From a membrane technologist's point of view, cheese manufacture can be defined as a fractionation process whereby protein (casein) and fat are concentrated in the curd, while lactose, soluble proteins, minerals, and other minor components are lost in the whey. Thus, one can readily see the application of UF (Figure 8.8). First, milk is preconcentrated to a protein/fat/solids level normally found in cheese (Table 8.1). Second, this "pre-cheese" is then converted to cheese by conventional cheesemaking methods. This basic process can have several variations, depending on the cheese. Some of the UF cheesemaking options are shown in Figure 8.9.

The benefits of using UF in cheesemaking are (Cheryan and Alvarez 1995)

- There is an increase in yield of cheese of 10–30% because of the inclusion of the whey proteins and associated bound water. This is most readily apparent with soft and semisoft cheeses.
- The amount of enzyme (rennet) required is lower. There is conflicting scientific evidence in the literature, which is about evenly divided between those who found slower clotting times with UF milk and those who observed faster coagulation with UF milk (Mehaia and Cheryan 1983). In some cases, however, it has reduced enzyme requirements by 70–85%.

TRADITIONAL PROCESS

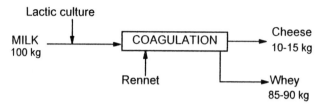

UF PROCESS (MMV)

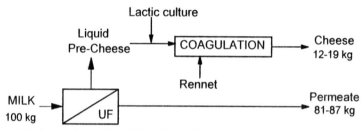

Figure 8.8. Comparison of traditional and UF methods of manufacturing cheese (adapted from Maubois 1989; Maubois et al. 1969).

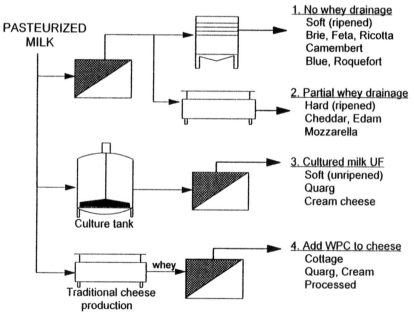

Figure 8.9. Possible options for incorporating UF in the manufacture of various types of cheese (adapted from Koch Membrane Systems product literature).

- There is reduced volume of milk to handle, reducing the number of cheesemaking vats and better utilization of plant space.
- Little or no whey is produced because most of the water and lactose has been already removed. However, there is the UF permeate to dispose but it is relatively "sweet" (less acidic) and clear, it contains no protein, and it is high in fermentable carbohydrate (lactose) and soluble nonprotein nitrogen. Thus, the permeate can presumably be treated or disposed of more easily than whey.
- There is uniformity in the quality of the product. Protein level can be standardized and the product suffers less from seasonal variations.
- It has the ability to use a continuous and automated process, which improves quality control, especially of moisture in the final product. This also results in improved sanitation and environmental conditions.

However, there are some technical difficulties with this process:

- The increase in viscosity is dramatic when the protein content of milk exceeds 12–14%. This may make it more difficult to mix the starter culture and rennet, which could cause texture problems in the cheese. This can be partially overcome by addition of NaCl, citrate, or acid before UF.
- The buffer capacity of the retentate is increased, so even if there is a higher lactic acid production, pH may not drop to the desired value.
- Recirculation of retentate, even at the relatively low pressures encountered in UF, may lead to a partial homogenization of fat and reduce texture of hard cheeses. In addition, exposure of retentate to heat and air may damage whey proteins, which may affect water content and texture of the cheese.
- Semihard and hard cheeses have a low moisture content and, consequently, a high solids content. A true "pre-cheese" cannot be made for these products, and some wheying off will occur; however, even a partial preconcentration can have significant benefits.

Some specific examples of UF cheese are presented below.

For *feta* cheese production the advantages of ultrafiltration are especially remarkable (Figure 8.10). Feta cheese, also called white cheese, originated in southeastern Europe and Asia (Table 8.1). It has a white color, high salinity, and sharp taste. There are two routes: in the low-concentration UF process, milk is preconcentrated to about 2.5×, and the final product is exactly the same as traditional feta, except for the inclusion of whey proteins, which makes the cheese smoother in texture than the traditional product. The heat treatment denatures the whey protein, giving smoothness to the cheese, as well as higher

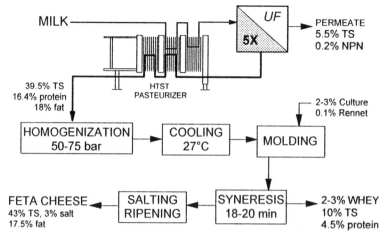

Figure 8.10. Manufacture of feta cheese by UF (adapted from Alfa-Laval literature).

moisture. In addition, cheese yield is 15% higher. In the high-concentration process, whole milk is preconcentrated 5×. Yield improvements are 25–30% and there is little or no whey drainage, but the product structure is different. In addition to the higher yield, much less rennet is used, and fat losses (normally in the whey) are eliminated or reduced. Automation has made long operating times of 20 h per day possible, and the payback period is usually only 9–18 months. Other white cheeses from Turkey, Egypt (*Domiati, Kariesh*), Greece (*Teleme*) and South America (*Queso fresco*) have also been studied (Renner and Abd El-Salaam 1991).

Quarg is an unripened, soft, smooth curd cheese produced from skim milk, similar to the *shrikhand* popular in the Indian subcontinent and *labneh* in the Middle East. To minimize possible bitterness due to high residual minerals in the product, UF should be done at a low pH (which will solubilize the minerals and allow them to permeate). A typical flow sheet is shown in Figure 8.11. The pasteurized skim milk is cooled to 35°C and incubated with starter culture until the pH drops to about 4.7. The acidified milk is then warmed to about 50°C for UF to volume concentration ratio (VCR) 3–4 [17–18% total solids (TS)], cooled to 4°C, the cream and spices added, and packaged. Yields can be increased 25–40% by using ultrafiltered milk. It is interesting that UF is done at the isoelectric point of the milk proteins, where the proteins are unstable and precipitate out of the milk system. Apparently, UF is still possible if a very high velocity/shear stress is maintained in the module during UF, to prevent the deposition of the isoelectric proteins on the membrane. Thin-channel designs would be particularly applicable in such an operation (Patel et al. 1986).

Camembert is another variety of soft cheese with a special surface ripening. The internal part remains soft, mainly because of the migration of lactic acid,

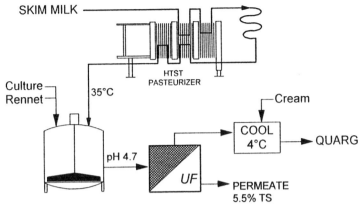

Figure 8.11. *Manufacture of quarg by ultrafiltration (adapted from Maubois 1989).*

along with minerals, to the surface. This could explain the need to control the mineral content of the retentate in order to ensure good quality in the final product. UF is done to about 5×, the concentrate is heated to 80°C for 5 min, and the liquid pre-cheese is filled into cheese forms where starter culture and rennet are added. Coagulation takes place in the containers. After ripening, the cheese shows properties that are similar to the traditional product. Whey proteins do not significantly affect the texture of the cheese, and so its inclusion has no detrimental effect, and in fact, it helps to increase the yield by 20%.

Goat's milk cheese is another interesting application. Because goats produce very little, if any, milk in the winter, cheesemakers keep frozen acid curd from goat's milk obtained in the late summer for use in the winter. However, this causes the curd to become oxidized and acquire off-flavors. UF retentate of goat's milk, however, can be successfully stored with few problems, thus enabling this industry to continue production of goat's cheese throughout the year. The UF product is comparable in taste to the best traditional cheeses. Yields were increased by about 12–20%, rennet usage was reduced by 80%, and the disposal of the sweet, neutral whey was much less of a problem than with traditional processes.

Mozzarella cheese was another challenge for UF (Figure 8.12) because whey proteins affect stretchability and meltability of the cheese. When cheese is made from low-concentrate retentates, it had similar stretching and melting characteristics as the regular mozzarella. Most UF processes for mozzarella include reduction of calcium by acidification of the milk to about pH 5.8 before ultrafiltration, followed by diafiltration of the retentate. This is considered essential for the development of stretching properties of the cheese. Homogenization of milk is known to reduce the stretchability of the cheese; therefore, the fat content is adjusted by adding cream after the skim milk is ultrafiltered. High whey protein in the mozzarella results in poorer melting quality. Denaturated

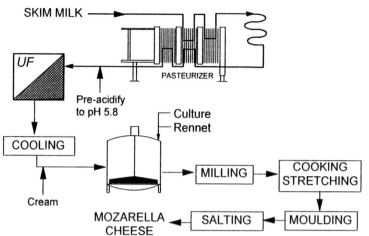

Figure 8.12. Manufacture of mozzarella cheese by ultrafiltration (adapted from Maubois 1989).

whey proteins allow the entrapment of moisture, lowering the melting quality; complexes between whey protein and casein are also likely to be responsible. At least one plant was installed in the United States in the mid-1980s (Cheryan 1986) but has since closed down.

Cheddar by UF has long been considered the ultimate challenge for the cheesemaker. Most semihard cheeses require a relatively long period of maturation (aging) in order to achieve the desired texture and flavor. The proteolytic activity of the coagulant is a key factor in ripening. Slow proteolysis reported in early studies of UF hard and semihard cheeses was caused by the low levels of residual rennet. Therefore, more residual rennet is required in the UF cheese to obtain the same rate of ripening.

There are numerous publications on the problems of making cheddar cheese by ultrafiltration (Cheryan and Alvarez 1995). In order to obtain a pre-cheese of the same total solids content (62%), whole milk must be concentrated 9×. This is not possible with most common UF equipment. Some expulsion of moisture during curd handling (in addition to water loss by UF) is necessary to reach that solids level. In the late 1980s, APV and CSIRO in Australia marketed a process shown in Figure 8.13. Whole milk was concentrated about 5×. The retentate (38–40% TS) was split into two streams. About 10% was repasteurized and fermented with starter culture. This was then mixed with the rest of the retentate, and the enzyme is added in a continuous coagulator. At the end of the coagulator, a knife cut the curd, which was cooked in syneresis drums. The curd–whey mixture was pumped to a draining and matting conveyor and to a standard cheddaring system. Permeate from the UF plant was sent to an RO plant to produce sterile water for cleaning and diafiltration.

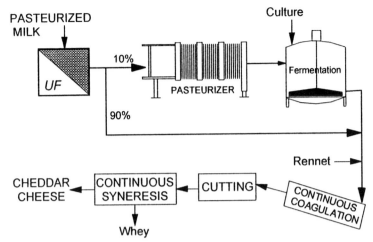

Figure 8.13. *Manufacture of cheddar cheese by ultrafiltration using the APV-Sirocurd process (adapted from Jameson, 1987 and Parrot, 1990).*

The effect of whey proteins on the properties of UF cheeses is still a matter of debate. Undenaturated whey proteins appear to be resistant to the action of rennet, starter, and plasmin. Whey proteins could act as an inert filler, giving the cheese a smoother texture or have a diluting effect and limit the accessibility of enzymes to the casein micelles. Cheddar cheese yields were not as high as with soft cheeses. If $5\times$ milk concentrate was used, inclusion of whey proteins increased yield by 6–8%, assuming no losses of casein and fat.

After two commercial attempts in 1986 and 1990 in Australia and the United States, the APV-Sirocurd process was abandoned, perhaps because of problems related more to the quality of the cheese product rather than the performance of the UF system.

Processed cheese is another commercial success. It is a blended product with part of the cheddar cheese replaced by a "cheese base," which is a cultured UF retentate. After standardization of milk to 3.8% fat, milk is pasteurized and ultrafiltered to about 30% solids, followed by diafiltration to reduce lactose sufficient to assure a final pH of 5.2 after culturing. UF may then be continued until the retentate is about 40% TS, 21% fat, 16.7% protein, and 1.1% lactose. Salt and culture are then added. When the pH of the base has reached about 6.4, it is concentrated by evaporation at low temperature to the desired solids level (usually 60% TS). The base is then blended with regular cheddar cheese for flavor, body, and appearance. Overall yields are about 16% higher.

Even though the profit potential of incorporating UF in the cheese manufacturing process has been known for a long time (Table 8.2), UF cheesemaking has not lived up to its early promise. Numerous varieties of commercial and

Table 8.2. Profit potential for cheese made by ultrafiltration (basis: 100,000 kg of milk/day).

	Feta	Mozzarella	Cheddar	Queso Fresco
Cheese yield (kg)				
Traditional process	13,700	9,930	10,360	11,432
UF process	17,800	11,750	12,290	14,824
Extra yield per day (%)	30	18	18	30
Cost of cheese (DKr*/day)	12	20	22	15
Extra income (DKr/day)	49,200	36,400	42,460	50,880
Extra costs per day (DKr):				
Milk fat	20,450	10,030	17,070	12,246
Electric power for UF	330	1,200	1,320	432
Steam for cheese base	—	—	1,000	—
Membrane operating costs	700	1,600	2,600	700
Total extra costs per day (DKr)	21,480	12,830	21,990	13,378
Income minus costs (DKr/day)	27,720	23,570	20,470	37,502
Profit				
DKr/year	7,207,200	6,128,200	5,322,200	4,875,260
% of traditional process	17	12	9	22

*DKr = Danish Krones
Source: Pasilac Company (1982)

ethnic cheeses have been studied, but except for a few soft cheeses, there are few commercial successes.

8.B.3.
MILK MICROFILTRATION

MF in dairy applications became a practical unit operation with the development of inorganic/ceramic membranes. It is interesting that inorganic membranes also suffer from typical fouling problems caused by interactions between the membrane, protein, and/or minerals (Fauquant et al. 1988; Pafylias et al. 1996; Piot et al. 1987; van der Horst and Hanemaaijer 1990; Vetier et al. 1988), yet these membranes are being successfully used for separation of milk components.

The main applications of MF in milk processing are fat separation, bacterial removal, and caseinate concentration (Figure 8.5). Casein micelles (milk protein in suspension) are smaller than the bacteria and fat (Figure 8.7). In theory, a membrane with pores of 0.2 μm should be able to separate the fat and bacteria from the rest of the milk components, provided the membrane had a fairly uniform and narrow pore size distribution and the appropriate physicochemical properties that minimized fouling. However, attempts with polymeric membranes operated in the conventional cross-flow manner failed, because a "dynamic," or secondary, membrane of the polarized particles quickly caused MF membranes to behave as UF membranes; i.e., the caseins and whey proteins

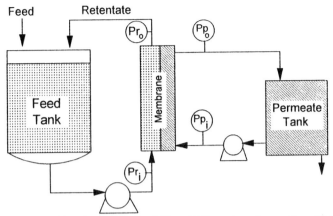

Figure 8.14. *Microfiltration using the CPF or UTP modes of operation. The feed is pumped through retentate channels at high velocity creating a pressure drop (Pr_i – Pr_o). The permeate is pumped through the shell side of the module with the same pressure drop as the retentate (Pp_i – Pp_o). The transmembrane pressure throughout the length of the module is uniform ($P_T = Pr_i - Pp_i = Pr_o - Pp_o$) (adapted from Cheryan and Alvarez 1995).*

were also being rejected. To minimize the formation of this secondary membrane, the units had to be operated at extremely high velocities (e.g., over 6 m/sec) and low transmembrane pressure (less than 5 psig/0.3 bar/40 kPa) to minimize compaction of the polarized layer.

This concept has been put into commercial practice by Alfa(Tetra)-Laval and USFilter as uniform transmembrane pressure (UTP) or co-current permeate flow (CPF), as described in Section 6.E.3. Membralox tubular ceramic membranes with 1.4-μ pores are operated in a double loop constant-pressure operation as shown in Figure 8.14. Because of the uniform and low-pressure profiles (e.g., see Figure 6.23), fouling is low and flux is very high. Fluxes of 700–1000 LMH can be achieved at 10× (Figure 8.15), usually for periods of over 6 h at low pressures of 3–6 psi. If needed, the transmembrane pressure can be increased slightly during the run to compensate for fouling. Very high pressures are unlikely to improve flux and may aggravate fouling, as shown in Figure 8.16. It is interesting that milk with some fat in it gave higher and more stable flux than skim milk; e.g., compare the CPF/UTP data shown in Figure 8.15 with the CPF/UTP data for skim milk flux obtained by Pafylias et al. (1996). Perhaps this was because skim milk has a greater tendency to foam during pumping.

Bacterial retention is 99% with the microbial load usually found in milk (Olesen and Jensen 1988; Pafylias et al. 1996; van der Horst and Hanemaaijier 1990), and fat is also substantially rejected. On the other hand, there is no significant change in the concentration of other components, so the permeate is essentially bacteria-free skim milk. Interestingly, Pafylias et al. (1996) and

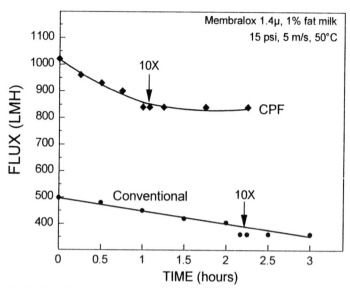

Figure 8.15. Microfiltration of 1% fat milk in the conventional and CPF/UTP modes. The system was operated in a batch concentration mode until 10× retentate concentration and then operated in total recycle mode to simulate feed and bleed (adapted from Pafylias et al. 1996).

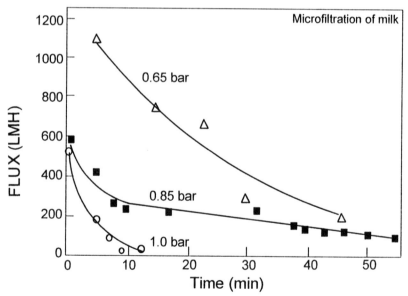

Figure 8.16. Microfiltration of milk: effect of transmembrane pressure on flux (adapted from Piot et al. 1987).

van der Horst and Hanemaaijier (1990) successfully microfiltered milk using high velocities with ceramic membranes in the conventional mode (i.e., without using UTP/CPF), although flux was 50% lower.

This process is known as "Bactocatch" in the dairy industry. It became commercial in 1989 to produce more stable pasteurized and refrigerated milk products. It could also be useful in subtropical and tropical countries, where inadequate refrigeration and transportation facilities result in high microbial loads in the milk coming in to dairy plants. A Bactocatch system on the receiving dock to lower the microbial load can significantly improve the quality of the milk products in these countries.

Enriched casein fractions (i.e., separated from whey proteins and other soluble milk components without isoelectric precipitation) can be produced using $0.2\text{-}\mu\text{m}$ MF membranes (Fauquant et al. 1988). With $0.1\text{-}\mu$ ceramic membranes, a flux of 90 LMH was obtained with 70–80% whey protein transmission and total casein retention (Le Berre and Daufin 1996). In addition, β-casein can be isolated from casein micelles if the temperature is lowered to less than 5°C, which causes β-casein to dissociate from the micelle and be removed in the permeate (a loose UF membrane of $>100,000$ molecular weight cut-off (MWCO) can also be used to isolate β-casein from milk). This protein has potential biological activity useful in pharmotherapeutic applications (Maubois 1991).

8.B.4.
CHEESE WHEY ULTRAFILTRATION

A good example of the successful application of membrane technology, and UF in particular, is the processing of cheese whey. Whey is a by-product of the cheese industry. It is the liquid fraction that is drained from the curd during the manufacture of cheese. Typically, every 100 kg of milk will give about 10–20 kg of cheese, depending on the variety, and about 80–90 kg of liquid whey (Figure 8.8). Its disposal is a major problem for the dairy industry as reflected by its composition (Table 8.1). It has a low solids content and a very unfavorable lactose:protein ratio, which makes it difficult to utilize as is. The biological oxygen demand (BOD) is 32,000 to 60,000 ppm, which creates a very severe disposal problem.

Despite continuing efforts to find uses for the whey, either as is or in dry form, or its major components (high-quality protein and lactose), it is estimated that as much as 40–50% of the whey produced is disposed of as sewage, with the rest being used primarily for animal feed or human food. World production is estimated at 80–130 million tons per year with the United States producing about 30 million tons per year. Since cheese consumption is increasing around the world, we can assume that the whey disposal problem is getting worse. This explains why membrane technology has attracted the attention of cheese and whey producers: the appropriate membrane can simultaneously fractionate, purify, and concentrate whey components (Figure 8.7), thus enhancing their

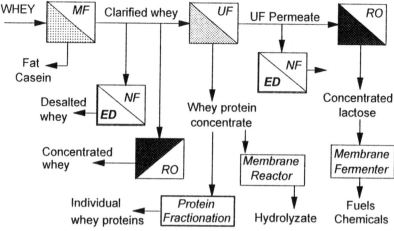

Figure 8.17. *Membrane processing of cheese whey.*

utilization and reducing the pollution problem. Figure 8.17 shows a general schematic of possible applications of membranes in whey treatment.

There are two kinds of cheese whey: sweet whey (resulting from cheeses that are produced by enzyme action, with a pH above 5.6) and acid whey (which results from cheeses made by acidification, with a pH of about 4.6). The latter generally has twice the amount of calcium phosphate and more lactic acid. Calcium phosphate is more soluble at low pH and low temperature. Thus, process temperature and pH are selected based on the behavior of calcium salts, in addition to considerations of viscosity, protein denaturation, and microbial growth.

Cheese whey is a notoriously bad fouling stream, as discussed earlier in Chapter 6. It appears to foul hydrophobic membranes (e.g., polysulfone) much more than hydrophilic membranes (e.g., cellulose acetate), primarily because of interactions between protein and the membrane. Mineral salts, primarily calcium phosphate, also play a very important role by themselves; e.g., they can precipitate on the membrane or within the pores, and/or they can enhance protein fouling by interacting with the membrane, perhaps forming "bridges" between the membrane and proteins. Salt and protein deposition have also been implicated in fouling of ceramic membranes by whey (Aimar et al. 1988).

Even the coagulant used in cheesemaking can have an effect on membrane fouling. *Mucor pusillus* rennet gave the highest permeate flux during UF of whey, while the traditionally used calf rennet had the worst effect (Tong et al. 1988). It is interesting that whey from cheese made with calf rennet, when subsequently treated with *Mucor pusillus* rennet, resulted in increased flux almost to the level of the whey obtained with the *Mucor pusillus* coagulant.

The pH should be far away from the isoelectric point of the proteins (Figures 6.1 and 6.16), and the temperature should be far away from 30°C (Figure 6.19). There is an optimum combination of pressure and velocity to minimize fouling (Figure 6.19). The appropriate pretreatment is important for high flux, e.g., heat treatment at 55°C for 30–90 min (Figure 6.17); lowering pH to increase solubility of the salts so that they would remain soluble (Figure 6.1); removal of calcium (Figure 6.16); pretreatment with citric acid and EDTA at pH 2.5 to sequester the calcium (Patocka and Jelen 1987) or addition of calcium chloride after cooling to 0–5°C, followed by pH adjustment to 7.3; heating to 50°C; and centrifugation (Kim et al. 1989). However, even though these pretreatments increased average flux, they could affect the functionality of the whey protein concentrates. A sequential combination of centrifugal clarification and microfiltration, or microfiltration alone, is effective (Hanemaaijer 1985; Kuo and Cheryan 1983; Rinn et al. 1990). These physical treatments will probably have less detrimental effects on the product and, indeed, may even improve the functional properties of the whey protein concentrate (WPC) by removing the fat.

Today, whey protein concentrates produced by ultrafiltration are well established in the food and dairy industries. A typical process train for the production of whey protein concentrates in dried form is shown in Figure 8.18. The first commercial scale UF/RO plant for fractionating cottage cheese whey was probably Crowley's Milk Company at LaFargeville, New York, in 1972 (Goldsmith and Horton 1972), which utilized Abcor tubular cellulose acetate membrane modules. Average UF fluxes of 15–40 LMH are common, depending on the

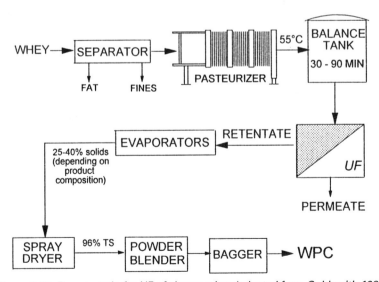

Figure 8.18. Process train for UF of cheese whey (adapted from Goldsmith 1981).

Table 8.3. Some uses of whey protein concentrate.

Baked custard	Cream filling	Meat extenders
Beverages	Cream icings	Meat loaf
Acid	Cream sauces	Meringues
Neutral	Cream desserts	Noodles
Biscuits	Cultured beverages	Pasta
Breads	Doughnuts	Potato flakes
Cakes	Egg white replacer	Puddings
Cake fillings	Egg yolk replacer	Sausage
Candy	Gravies	Sherbert
Caramel	Hot dogs	Snack foods
Nougats	Ice cream	Tortillas
Milk chocolate	Imitation cream cheese	Whipped toppings
Canned refried beans	Imitation milk	Yogurt
Chocolate drinks	Macaroni	
Coffee whitener	Meat analogs	

equipment, type of cheese whey, and operating parameters. Figures 4.5, 4.8, 4.16, 4.23, 6.1, 6.16, and 6.17 show performance data for processing cheese whey.

Owing to the relatively mild process conditions of temperature and pH, the functionality of the whey proteins remains good, giving rise to a wide range of applications (Table 8.3). The initial protein content of 10–12% (dry basis) can be increased by UF to result in 35%, 50% or 80% protein products, with a concomitant decrease in lactose and some salts. Figure 7.2 shows the change in composition of the retentate during UF (see also calculations in Tables 7.1–7.4). Whey is concentrated to 10–30× in one step, followed by DF to purify the WPC above 50% protein (dry basis). In the final spray-dried form, the yield is about 1.5 kg of 35% protein [dry basis (d.b.)] WPC per 100 kg of whey.

WPC can be further fractionated into β-lactoglobulin and α-lactalbumin fractions, as shown in Figure 8.19, or be used for the manufacture of caseino-macropeptide (CMP), a compound that may have a pharmotherapeutic value (Figure 8.20). This peptide is released into the whey from the κ-casein fraction when acted on by the enzyme (chymosin/rennet) during cheese manufacture. It contains a carbohydrate moiety and is thus also referred to as the glyco-macropeptide. The process takes advantage of the change in conformation and size of the peptide at different pH (Figure 8.21). The theoretical molecular weight of the CMP is 7000, but at pH 7, the effective MW is much larger. Thus, the first UF stage uses 50 K MWCO membranes at a low pH of 3.5. After a VCR of 5 and diafiltration to V_D of 5, the CMP-containing permeate is neutralized and fed into a second UF system with a 20 K membrane. After diafiltration and concentration, the final yield was 63% and the purity of the CMP was 81% (Kawasaki et al. 1993).

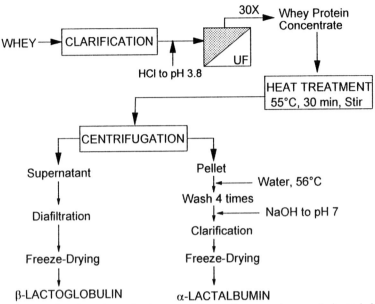

Figure 8.19. *Fractionation of whey proteins using membranes (adapted from Maubois et al. 1987; Maubois 1989).*

8.B.5.
MICROFILTRATION OF WHEY

The major application of MF is as a pretreatment for UF of whey. Whey usually contains small quantities of fat (in the form of small globules of 0.2–1 μm) and casein (as fine particulates of 5–100 μm). Centrifugal separation of whey does not completely remove the fat and casein fines. Thus, when

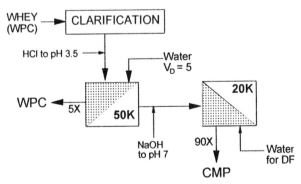

Figure 8.20. *Production of caseinomacropeptide (CMP) (adapted from Kawasaki et al. 1993).*

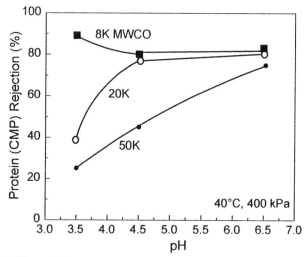

Figure 8.21. Effect of pH and membrane pore size on rejection of caseinomacropeptide (adapted from Kawasaki et al. 1993).

the whey is ultrafiltered, these components can prevent the attainment of high purity, as well as having detrimental effects on the functional properties of the WPC. MF (both conventional and CPF with tighter membranes of 0.1–0.2 μm) can effectively remove substantial quantities of these undesirable components. Fat/protein ratios of 0.07–0.25 in whey can be reduced to 0.001–0.003 by MF (Van der Horst and Hanemaaijer, 1990). In addition, some of the precipitated salts may be removed, and there is a considerable reduction in microbial load.

The key factor in MF of whey is the pretreatment. The most effective appears to be "thermocalcic" aggregation of the phospholipids and precipitation of calcium phosphate (Daufin et al. 1992; Fauquant et al. 1985; Gesan et al. 1995; Maubois et al. 1987). This involves the addition of $CaCl_2$ at 2–4°C to a final concentration of 1–2 g/L, after which NaOH is added to bring the whey to pH 7.5. The whey is heated at 55°C for 8–15 min, while maintaining the pH at 7.5. This treatment helps to form large (50–100 μm) aggregates of phospholipoproteins, which do not penetrate the pores of the MF membrane, thus resulting in higher flux during the MF treatment. Operating the MF system at low transmembrane pressure (TMP) during MF is also important to minimize fouling (Gesan et al. 1995). In addition, the high pH and temperature cause precipitation of calcium phosphate, which is also removed during MF. This reduces the fouling in the subsequent UF step, which is done at pH 7.5 and 55°C, resulting in higher flux (Figure 8.22).

If the MF system is part of a WPC operation, then it might be better to partially concentrate the whey proteins by UF before MF. This serves two purposes: it

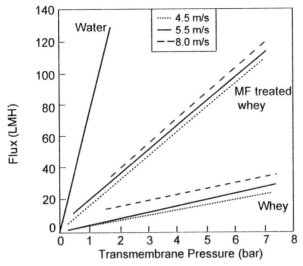

Figure 8.22. *Ultrafiltration of whey (untreated) and MF-treated whey using Carbosep M5 inorganic membrane (10 K MWCO). The effect of pressure and cross-flow velocity are shown. The MF-treated whey was prepared by thermocalcic aggregation described in the text and then microfiltered with Carbosep M14 (0.1 μm). The permeate from the MF step was ultrafiltered (adapted from Daufin et al. 1992).*

reduces the flow rate through the MF system and thus its cost (being ceramic, it tends to be much more expensive per unit membrane area), and it improves the efficiency of fat removal by increasing its concentration and enhancing coalescence during pumping through the UF system. For example, the raw whey is first clarified by centrifugation, pasteurized, and held for about an hour before UF at 50°C to 3–10×. The UF retentate is sent to the MF system, which is operated at 10×. A substantial removal of fat, fines, and bacteria takes place in this step. MF flux of 30–45 LMH have been reported by Alfa-Laval for a 4× UF feed with 0.1-μ membranes (with raw whey at 1×, flux is about 100 LMH). The MF permeate is led to a second UF plant for about 10×, bringing the retentate up to 22–24% solids, which contain 85% or more protein and less than 0.4% fat on a dry basis. Removal of the fat by MF improves the second UF flux by about 50–100%.

8.C.
WATER TREATMENT

There are two types of water of interest to membrane technologists: process water for the manufacturing industries and potable (drinking) water for human consumption.

8.C.1.
PROCESS WATER

The United States uses about 10 billion gallons (38 million m^3) per day
for its manufacturing industries (Parekh 1991), of which about 15% is "ultra-
pure water" used in critical applications. Examples include the semiconductor
and electronics industries, which need an assured supply of high-quality water
for washing integrated-circuit chips and other devices. Pharmaceuticals and
biotechnology need pure water for tissue culture media, bacteriological media,
buffer solutions, analytical solvents, formulation aids, drug and intravenous
solutions, standards and reagents, and rinsing of equipment. In the polymer
(plastics) industry, suspended matter interferes with polymerization reactions.
The electric and steam power industry is affected by seasonal variations in the
colloidal silica content of boiler feed water, which has a pronounced effect on
the blowdown rate that prevents excessive scale formation on heating surfaces.
While much of the process water needs of the manufacturing industry can be
met by natural surface and groundwaters with minimal pretreatment, ultra-pure
water (UPW) must be manufactured from raw water.

Raw water may contain the following contaminants to varying degrees: particulates and colloidal matter, organics, oxides of iron and manganese, disinfectants such as chlorine, acids or bases, microorganisms, inorganic salts, and
pyrogens. Surface waters are rich in particulates, suspended colloids, and organics, while groundwater tends to have relatively high levels of hardness (calcium
and magnesium salts) and alkalinity in the form of bicarbonates (Parekh 1991).
In agricultural areas, nitrates, phosphates, and psesticides may also be present.

Historically, the classic method of producing high-purity water is by distillation. However, as Table 8.4 shows, not all common contaminants in water can
be removed by any one method. The main problem with distillation is that it
will not remove volatile organics that have boiling points below that of water.
Distilled water must be stored, and high-quality water begins to deteriorate
quickly when stored.

Several trade and industry organizations have water quality standards. The
most common standard is its electrical resistance. Theoretically, pure water
has a resistance of 18.3 million ohms across a 1-cm path length. According to
ASTM, Type I water (the best quality available in practice) is required to have
a resistivity of 16.6 megaohm-cm, while the National Committee for Clinical
Laboratory Standards (NCCLS) considers 10 megaohm-cm as Type I water.
However, NCCLS also requires a 0.2-μm final filter and recommends activated
carbon to remove organics.

Figure 8.23 shows schematics for producing various grades of high-quality
water from various sources. Proper pretreatment (mechanical or chemical) before entering the membrane system is critical. Natural waters (rivers, brackish,
sea) may require clarification with sand filters to reduce turbidity and particulates. If excessive levels of bacteria are present, it may be necessary to chlorinate

Table 8.4. Water purification process comparison.

Purification Process	Dissolved Ionized Solids	Dissolved Ionized Gases	Dissolved Organics	Particulates	Bacteria	Pyrogens
Distillation	E	P	G	E	E	E
Deionization	E	E	P	P	P	P
Reverse osmosis	G	P	G	E	E	E
Carbon adsorption	P	P*	G**	P	P	P
Filtration	P	P	P	E	E	P
Ultrafiltration	P	P	G†	E	E	E
UV oxidation	P	P	G‡	P	G§	P

E= excellent, G = good, P = poor
*Activated carbon will remove chlorine by adsorption
**Special grades of carbon
†Depends on molecular weight of organic
‡Some types will remove organics
§Selected bactericidal capabilities, depending on intensity, contact time, and flow rate

the water, but then the chlorine must be removed if chlorine-sensitive RO or nanofiltration (NF) membranes are being used subsequently in the process train. This can be done with carbon columns or by adding bisulfite. Excessive hardness may require water softening, which can also be done by NF (Raman et al. 1994a). Another prefilter is usually required after all these pretreatments and just before entering the first membrane unit.

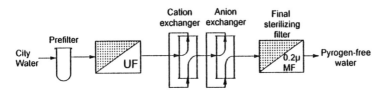

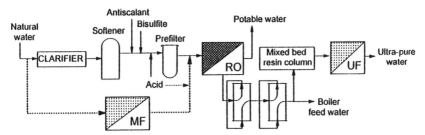

Figure 8.23. Process schematic for industrial production of water: top: high-purity, pyrogen-free water from city water; bottom: potable, boiler feed, or ultra-pure water from natural, brackish, or sea water.

Many of the above pretreatments can be replaced by microfiltration (Figure 8.23, bottom schematic). Microfilters can remove bacteria, suspended solids, and large organic molecules without the need for chemicals. In addition, ceramic and hollow fibers are backwashable, which allows them to be operated under low or no cross-flow and with high yields and flux. Membrane filtration generally provides a more consistent quality product even if raw water quality changes. They lose operational performance (e.g., flux becomes lower, pressures may have to be increased or more frequent cleanings needed) rather than comprising water quality. Figure 2.30 showed a ceramic installation for pretreating river water ahead of an RO installation.

If RO is done after MF, chemicals may be needed to reduce water hardness. Lime [$Ca(OH)_2$] precipitates calcium carbonate and magnesium hydroxide in a fine particulate form that can be successfully removed by MF down to less than 1 ppm. Silica can be reduced by up to 90% with ceramic membranes after addition of magnesium hydroxide and lime. Acid dosing may be required to bring the pH down appropriately for the membrane. Carbon treatment serves to remove organics, while the subsequent mixed-bed, ion-exchange treatment reduces inorganics. At this point the water should approach 18 megaohm-cm resistivity. In some installations, UF is done before the ion exchange in order to increase the lifetime of the resins; in this case, demineralization is followed by a 0.2-μm final filter at the point of use.

Pyrogens in the water supply also pose potential problems; e.g., tissue culture cells require pyrogen-free water for optimum growth. A pyrogen is defined as any substance that causes a fever (temperature rise) when injected into the bloodstream. Most pyrogens are lipopolysaccharides from bacterial cell walls, with molecular sizes ranging from 20,000 daltons to aggregates of 0.1 μm. Their threshold of detectability is 0.01–0.1 ng/ml. Pyrogens cannot be eliminated by autoclaving but have been successfully removed by UF. Hollow fiber and spiral modules with 10,000 MWCO membranes are used. The water plant located at the French National Blood Center in Orsay followed the top schematic shown in Figure 8.23 to produce 12,000 L/h of pyrogen-free water. A total of 40 Romicon hollow fiber type HF53-20-PM10 cartridges with an installed membrane area of 196 m^2 were used, requiring an installed power consumption of only 15 kW (*FiltratioNews* 1984). This is primarily because this water was a relatively clean feed, and it could be processed under low cross-flow, high-recovery (80–95%) conditions. The MembraStill system of USFilter uses 200 Å zirconia coated ceramic monoliths to effect a 4 log-cycle reduction in endotoxins (Figure 8.24). This is accomplished in a single pass with 96–99% yields when fed with USP purified water. The particular installation in Figure 8.24 is operated at 5.7 bar and gets a flux of 400 LMH/bar.

It should be pointed out that there is a strict distinction between quality standards for "water for injection" and purified water. In many countries, distillation and/or RO are the only methods permitted for producing the former. The

Figure 8.24. *MembraStill plant for producing 38 gpm (200 m³ per day) pyrogen-free water. The plant uses Membralox 200 Å zirconia ceramic elements and is equipped with special instrumentation for validation and data acquisition (Source: US Filter, with permission).*

disadvantages of distillation include its operating and capital costs, while RO systems are more difficult to sanitize and clean. UF-based systems are usually much cheaper. Operating and capital costs are quite sensitive to the flux, which in turn is dependent on the quality of the feed water.

8.C.2.
DRINKING WATER

Using membranes for cleaning potable water is potentially the largest single application of membrane technology, especially UF and MF. Almost all water supplies for residential and potable use are treated in some manner. Conventional water treatment may include chemical addition (aluminum sulfate, polymers, lime), coagulation, flocculation, sedimentation, filtration, and disinfection, usually with chlorine. Additional regulations may require the removal of trihalomethanes (THM) and synthetic organic chemicals. MF and UF are especially beneficial in removing micoorganisms that may constitute a health hazard. The U.S. 1989 Surface Water Treatment Rule requires a 3-log cycle (99.9%) reduction in *Giardia muris* cysts and a 4-log reduction of *Cryptosporidium parvum* oocysts and enteric viruses. The latter are resistant to traditional disinfectants such as chlorine and ozone. MF or UF can meet these standards while avoiding the formation of disinfection by-products (DBP), yet meeting the <0.5 nephelometric turbidity units (NTU) rule (Jacangelo et al. 1991, 1995; Madaeni et al. 1995).

Several membrane water treatment plants have operated successfully for several years in France (Baudin 1993; Bersillon et al. 1991). All were hollow fiber UF plants with capacities of 22–220 gallons per minute (gpm) (5–50m³/h), operating at pressures of 40–150 kPa (6–22 psi). The product water quality exceeded European standards for turbidity and microbial standards. Energy requirements for recirculation, backwash, pump losses, and purges were 1.6–1.75 kWh/m³. Membrane lifetimes of 5 years could be expected. Variations in raw water properties had much less of an effect on output water quality than a conventional plant. However, water cost to the consumer would have doubled with the addition of the UF plant at the time the study was done (1989–90). Cost is very sensitive to plant capacity: for example, typical treatment costs in 1995 were $0.5/1000 gal for MF water plants of >1 million gallons per day (gpd), but $1.12/1000 gal for a 0.1 million gpd plant (Adham et al. 1996). With increasing competition, lower manufacturing costs and larger plant sizes, UF water treatment should soon become economical and the preferred method. For small water plants in remote areas serving small communities, membrane technology may be the best method.

A demonstration of this technology was also conducted in 24 water purification plants in Japan in 1992–94 (Kunikane et al. 1995). Pressures were less than 300 kPa; some plants operated at less than 100 kPa. Average flux of MF plants were 11–80 LMH and of UF plants 16–75 LMH. Pretreament with powdered activated carbon was necessary for some MF plants to maintain a high flux. Particulate matter and coliforms were removed by both UF and MF membranes. Energy consumption was projected to be 0.5 kWh/m³ of produced water.

One potential problem is biofilm growth on the permeate side of the membrane. This can be treated with strong doses of disinfectant (chlorine) in backflushable membrane systems (hollow fibers, ceramics). UF membranes may be better than MF membranes in the long run since they can remove viruses more effectively. A hybrid process of powdered activation carbon (PAC) and UF could be effective as a total solution, i.e., for removing bacteria, turbidity and organic compounds, including DBP precursors, to meet anticipated drinking water standards.

Viruses can also be removed from biological and pharmaceutical fluids by UF, e.g., Filtron's Omega 200 K and 300 K PES membranes, Amicon's YM-100, and Millipore's Viresolve 70 and Viresolve 180 made of hydrophilic PVDF

(Michaels et al. 1995). Both cross-flow and dead-end modes are used. Pall's Ultipore VF DF50 is a hydrophilic PVDF membrane in a pleated sheet cartridge form operated in the dead-end mode. It can recover proteins such as bovine serum albumin (BSA) and immunoglobulin G (IgG) almost quantitatively while providing a titer reduction of $> 10^6$ for viruses that are larger than 50 nm in size (Pall 1996).

8.D.
WASTEWATERS

Almost every manufacturing industry (e.g., automobiles, food, steel, textiles, animal handling and processing, etc.) and service establishment (hotels, transportation, etc.) generates large quantities of wastewater daily (Table 8.5). The need for stringent pollution control (and legislation) provides tremendous opportunities for membrane technology in all aspects of pollution control, from end-of-pipe treatment to prevention and reduction of waste. It should be emphasized that, in this role, membranes merely serve to separate or fractionate

Table 8.5. Sources of wastewater.

Industry	Sources and Composition of Wastes
Metalworking	Surface cleaning (spent acid or alkaline cleaning solutions), cutting and grinding oils, lubricants used in machining operations, paint stripping, degreasing, water-soluble coolants, spent process baths (pickling, electroplating, anodizing, etching), discharges from parts washer tanks, floor washings, wastewater treatment sludges
Food processing	Bakery, vegetable oils, animal processing (rendering and slaughterhouse wastes), milk and cheese (floor washings, evaporator condensate, brine, caustic solutions), beverages, alcohol distilleries (stillage), steep or soak water in grain processing
Transportation	Discharges from tank car cleaning operations (sweeteners, oily wastes, latex, chemicals)
Textiles	Dyes, adhesives, sizing chemicals, oils from wool scouring, fabric finishing oils
Plating industries	Metal hydroxides
Laundries	Oils, grease, detergents, suspended solids
Printing	Flexographic ink, starch wastewater
Leather and tanning	Recovering sulfides from dehairing baths, desalting vegetable tanning baths, recovery of chromium
Pulp and paper	White water, color removal
Chemicals	Emulsions, latex, pigments, paints, chemical reaction by-products
Cities and municipalities	Sewage, industrial waste

wastewater components, hopefully into more useful and/or less polluting streams, and cannot break down or chemically alter the pollutants. At the very least, it can slash hauling and disposal costs and could render a permeate stream ready to discharge into the sewer. Compared to other treatment methods, membrane technology can result in fairly quick payback periods, operation is simple, effluent quality is consistently good, results are usually immediate, and it can substantially reduce chemical usage, although perhaps not eliminating them entirely, depending on the type of pretreatment required to minimize membrane fouling.

There are two approaches to wastewater treatment, depending on (1) if the permeate is to be reused, e.g., alkaline/acid cleaning baths, electrocoat paint, water, or (2) if the permeate is to be disposed of and the objective is to reduce the volume of solids, e.g., machining operations, food wastes, metal plating. However, the physicochemical properties of wastes vary widely, even within the same industry and sometimes within the same plant at different times of the year. Membrane suppliers sometimes suggest the same solution for all wastewater problems without proper analysis of the feedstream or inadequate on-site testing. Feed variations and proper selection of membrane and module should be considered during testing, especially if plant management is ignorant of or reluctant to discuss with suppliers the nature and magnitude of the waste problem. Wastewater treatment requires more extensive testing than most industrial membrane applications to account for possible feedstream variations, pretreatment options, cleaning problems, and issues related to recycling or disposal of permeate and retentate. This section illustrates typical examples of UF and MF in waste treatment, except for membrane reactors in anaerobic digestion, which are discussed in Section 8.K.

8.D.1.
OILY WASTEWATER

Oily wastes are generated in a wide variety of industries (Table 8.5). They can be grouped into three broad categories: free-floating oil, unstable oil/water emulsions, and highly stable oil/water emulsions. Free oil can be readily removed by mechanical separation devices that use gravitational forces as the driving force. Unstable oil/water emulsions can be mechanically or chemically broken and then gravity separated. However, stable emulsions, particularly water-soluble oily wastes, require more sophisticated treatment to meet today's effluent standards. Prior to the development of UF, the standard chemical treatment resulted in a sludge in which the dirt, floc and trapped water remained with the oil phase. This sludge always required further stabilization (by cracking or filtration) before disposal. The water phase from chemical treatment needed additional treatment to meet environmental standards for discharge into a sewer.

The UF method of treatment, on the other hand, produces a water phase that is usually clean enough to be discharged to a sewer with no post-treatment and an oil phase that can be incinerated if concentrated enough. Even if it cannot be burned, only a small fraction (typically 3–5%) of the original volume of oily wastewater will have to be hauled away. There are no chemicals to mix or store, no additional sludge is created, and the system operation can be made automatic, even during the cleaning cycle (which may be as little as once a week). However, careful selection of the membrane must be made, depending on the type of oily wastes to be processed. Waste cutting and grinding oil–water emulsions generally have a pH less than 10, but it could contain various foulants that will require both acidic and basic cleaning agents. Waste oils from steel rolling mills and alkaline metal degreasing baths require membranes that are stable up to 70°C and can tolerate alkaline conditions.

Figure 8.25 is a general schematic for a UF-based system for treatment of oily wastes. As with every membrane process, some pretreatment is usually required. Large particles and free oil must be removed, the former to prevent damage to the membrane and the latter to minimize membrane fouling. If large-diameter tubular modules are used, the pretreatment can be simply done with an equalization tank from which the free-floating oil and the denser solids are removed. Equalization time can be 1–2 days. Otherwise, particle removal can be done with a simple in-line screen filter, a rotating brush strainer, and/or a centrifugal separator, depending on feed channel dimensions and particle size and concentration. The membrane unit is usually operated in a semibatch cycle (Figure 7.10), whereby the clean permeate is drawn off continuously, while the retentate, containing the oils, is recycled to the process tank. Feed is added to the process tank to maintain a constant level. At a predetermined time, depending on the final concentration of the retentate and the maximum allowable oil

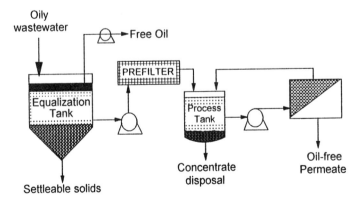

Figure 8.25. *General schematic for treatment of oily wastewaters by UF.*

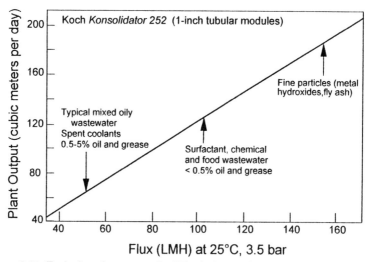

Figure 8.26. *Typical performance of a UF tubular module system for treatment of different types of wastewaters. The Konsolidator 252 consists of 252 tubes, each of 1" (25 mm) diameter and 10 ft (3 m) long, for a total effective area of about 60 m². Actual performance will depend on the feed and operating conditions (adapted from Koch 1996).*

concentration in the permeate (since permeate concentration usually increases with retenate concentration), the feed is stopped and the retentate allowed to concentrate. This will result in a final concentrate volume that is only 3–5% of the initial volume of oily wastewater fed to the process tank. The system may then be cleaned.

Cutting oil wastes typically contain 0.1–10% oils and grease. Concentrates up to 40–70% oil and solids can be obtained by conventional cross-flow UF. Typical flux that can be obtained with Koch's tubular modules are shown in Figure 8.26 for a variety of wastewaters. With emulsified oil at 1–2% feed concentration and 60% retentate concentration, flux is typically 50 LMH at 25°C and 50 psi. Synthetic oils tend to foul less and have high flux. Natural fats and oils exhibit low flux and foul the membrane more (Porter 1990). As mentioned in Section 6.D.2., free oils can coat hydrophobic membranes resulting in poor flux (emulsified oil is usually not as much of a problem, unless it is concentrated to such a high level that the emulsion breaks, releasing free oils). Hydrophilic membranes preferentially attract water rather than the oil, resulting in much higher flux. If a relatively hydrophobic membrane is used, a high cross-flow velocity should be maintained to prevent oil wetting of the membrane, which occurs at stagnant points (Porter 1990). This may favor tubular modules (e.g., Figure 8.27) over spiral modules. C_G values of 75–80% oil have been observed.

Figure 8.27. *A Konsolidator 252 tubular UF system (Source: Koch 1992, with permission).*

Most membrane companies recommend using membranes with 50,000–200,000 MWCO. This can typically result in a permeate with less than 10–100 ppm of oil, unless high concentrations of a soluble surfactant or polar solvent are present. In some cases, MF membranes with pore sizes of 0.1 μ have also been used, especially if it is necessary to recover surfactants in the permeate. If the salt content of the oily wastewater is too high for direct reuse of the permeate in the plant, it can be treated by RO or NF. The time required for a complete concentration and cleaning cycle depends on the waste characteristics and the plant size. Most installations are designed around a 1-week cycle. Depending on system capacity, equipment costs are $5 to $20 per gpd, while operating costs are from 0.3–1.5 cents per gallon of wastewater treated. Oily wastes is a successful application of membrane technology with more than 3000 polymeric UF installations and over 75 inorganic/ceramic units worldwide. Even polymeric membranes are reported to last 3–7 years, depending on the severity of the application, in part because of the low frequency of cleaning.

One example is an automobile transmission plant in Sweden that generated more than 1 million liters of spent coolants per year in its machining operations.

Hollow fiber modules (Romicon HF26.5-45-XM50) were used, requiring the feed to be prefiltered to remove particles larger than about 400 μm, i.e., to about one-third the channel diameter using a rotating brush strainer and a centrifuge. The pretreated oily wastewater was then sent to a process tank, from where it was pumped to ten UF modules. The total surface area was 25 m^2 for a design feed capacity of 1500 m^3/year. Two pumps were used: 7.5 kW for recirculation and 0.55 kW for the feed (Cheryan 1986).

In the metalworking industry, aqueous cleaning and rinsing is the predominant method of removing oils and greases from metal parts before they can be painted or further processed. The cleaning solution soon becomes exhausted due to accumulation of oils and solids. This results in high cleaner replacement costs and expensive oily waste disposal. UF can remove the particulates, oils, and greases from the rinse water tank, allowing reuse of the water, and can increase life of a parts washing tank fourfold. UF permeate quality in metalworking waste treatment is usually less than 100 ppm fats, oils, and grease (FOG), 10–50 ppm hydrocarbons, 25 ppm suspended solids, and less than 0.5 ppm of several metals such as chromium, lead, nickel, and zinc. Rolchigo (1995) describes the case of an automobile parts manufacturer with a 2000-gal wash tank and a 1000-gal rinse station. A Membrex ESP spiral-wound system kept the FOG levels below 150 ppm and 50 ppm in the wash and rinse tanks, respectively. This not only improved the cleaning quality of the parts, but reduced annual labor costs by $20,000, waste hauling by $65,000, and cleaner chemical consumption by $30,000. This total annual savings of $115,000 resulted in a payback of less than 1 year. Similar successes have been achieved with alkaline cleaners by Karrs and McMonagle (1993), who used ceramic membranes to decrease effluent generation by 98% and reduce cleaner consumption by 54%, and by Schwering et al. (1993) who used MF to accomplish 90% and 86% reduction, respectively. Table 8.6 shows typical costs for a tubular polymeric module operating on metalworking oily wastes.

The treatment of wastestreams from W. R. Grace Company's Dewey and Almy Chemical Division plant in Chicago has a long history. At the time of installation in the late 1970s, the principal wastestreams at this plant consisted of concrete additives (saponified tall oils and oleic acid, calcium lignin sulfonate, triethanolamine phenol), plastisols (dioctyl phthalate, zinc resinate, paraffin wax, PVC resin, mineral oil), a latex water base and centrifuge washes containing styrene–butadiene resin latex and mineral oils (Kirjassoff et al. 1980). The combined streams flowed at 18.6 m^3/day through two sumps and a settling tank, which removed latex particles. Before entering the main process tank, a nonionic surfactant was injected into the stream to stabilize the emulsion and minimize fouling of the membrane. The plant used Koch tubular modules containing 72 m^2 of membrane area. Operating pressures above 1.5 atm produced only slight increases in flux, and so the pressures were set according to the required recirculation flow rate. An inlet pressure of 3.5 bar and an outlet pressure

Table 8.6. Economic summary of ultrafiltration of metalworking oily waste.

Basis: 20,000 gallons (76 m^3) per day oily wastewater
UF equipment: 1″ tubular module with HFP 276 membranes, 248 tubes,
50 m^2 area, 40-HP circulation pump. Cost $138,000
Operating mode: Semibatch recycle, 24 hours per day, 250 days/year,
4 hours/week for cleaning

Item	Annual Cost ($)
Membrane replacement, 3-year life @$300/ tube	24,800
Electric power @$0.05/kWh, 50% efficiency	18,000
Labor 8 hours per week @$20/hour	8,000
Maintenance @1% of capital cost	1,380
Cleaning cost (0.5% Koch liquid detergent@ $5/kg, 4.72 kg/week)	1,185
Total annual operating cost	53,365
Cost per unit wastewater treated: ($/m^3)	2.8
($/1000 gallons)	10.6

Source: Data from Koch (1996)

of 1.4 bar were optimum, resulting in a flow rate of about 120–130 L/min per pass of membranes. Average flux was 41 LMH, equivalent to about 70,000 L of permeate per day. Typical UF process cycles were 700–1000 h. FOG was reduced from 3530 ppm in the feed to 35 ppm in the permeate, suspended solids reduced from 1640 ppm to 63 ppm, and COD from 21,200 ppm to 1333 ppm. Permeate quality remained well within city standards. Cost was less than $3/m^3 of permeate over a 3- to 4-year period in the early 1980s. The major contributors to the cost were surfactant (30% of total cost), labor (19%), and disposal of the sludge (17%). Membrane replacement was only about 10% of the total cost.

Another Membrex ESP spiral module system is processing 40,000 gal per month of spent coolant from a machine tool plant that contains 5–6% of oil by volume. The volume is reduced to 4000–6000 gal of concentrate, containing 30–50% oil. The FOG and total petroleum hydrocarbon (TPH) of the permeate is less than 75 ppm and 10 ppm. The in-house processing cost is less than $12 per 1000 gal ($3.2/m^3) of wastestream, including membrane replacement, prefiltration, membrane cleaning chemicals, electricity, and labor. Payback for the unit was 3 months. Prior to installing this UF system, the plant was paying $1/gal to haul away this waste ($480,000 per year). Now the only disposal cost is for the retentate, which is now only $0.25 per gallon since the oil is more concentrated. The net savings is over $450,000 per year (Rolchigo 1995).

A more challenging application required a hybrid UF system to treat over 30,000 gal per day of oily waste in a General Motors automobile manufacturing plant in Toluca, Mexico. The first stage is a spiral module system and the second stage is a rotating disc module, the Discover system from Membrex (Figure 5.46). Both membrane units were installed with the hydrophilic polyacrylonitrile membrane developed by Membrex (its structure was shown in Figure 6.18). The feed consists of all the wastewaters generated in the plant, including metalworking fluids, compressor condensate, rainwater run-off, alkaline and acid cleaners, and process rinse waters. Feed FOG is 1000 ppm and suspended solids is 500 ppm. The spiral system concentrates the feed 20× to under 1500 gal of concentrate, with permeate levels of 10 ppm FOG and 10 ppm TPH. The disposal of the 1500 gal of retentate would have cost over $250,000 per year. Instead, it is fed into the Discover rotary module, where it is concentrated another 10-fold to 150 gal containing 50% oil and 10% suspended solids. Flux data for these two membrane systems was shown in Figure 5.47. The payback period of the Discover system was under 1 year, based on the disposal costs of the 1500 gal from the primary UF system (Rolchigo 1995).

"Drumtop" units—self-contained systems that can be mounted on top of a standard 55-gal drum—are also available. These units are typically sized to process one 55-gal drum of oily wastes per day to a 90% volume reduction. It costs about $5000.

Petroleum oil refinery wastewaters, even after conventional biological treatment, can contain 20 ppm total hydrocarbons and 30 ppm suspended solids. Ultrafiltration with the M9 Carbosep membrane produced an effluent free of these compounds with a flux of 150–200 LMH at a pressure of 0.5 bar (Elmaleh and Ghaffor 1996).

8.D.2.
STILLAGE FROM BIOETHANOL PLANTS

An ethanol plant uses yeast to ferment sugars to produce 8–12% by volume of ethanol. The remaining 88–92% of the fermentation broth is waste. As shown in Figure 8.28, the ethanol is stripped away by high-pressure steam in the "beer still," leaving behind the stillage (also called *vinasse*). The stillage is hot (85–95°C) and contains a total solids of 5–8% w/w if starch or sugars is the fermentation substrate (TS is much higher if whole grain is used, as in corn dry milling ethanol plants). The solids consist of suspended matter in the form of dead yeast cells and cell debris (2–3%), traces of residual sugars (glucose, maltose, isomaltose), nitrogen compounds, and by-products of the fermentation (glycerol, succinic acid, acetic acid, lactic acid). The stillage is usually evaporated and dried for sale as animal feed or sent to waste treatment plants.

Large volumes of water are used in an ethanol process. Consequently, large volumes of wastewater are generated in the form of stillage. A plant producing 100 million gal (378,000 m³) per year of ethanol may have 900 million gal (3.4 million m³) of stillage to handle. Some ethanol plants may partially clarify the stillage (today, it is done mostly with centrifuges) and the "thin stillage"

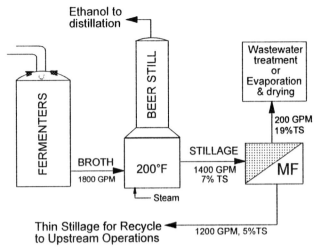

Figure 8.28. *Flow sheet in a bioethanol plant. The data is for a corn wet milling plant with a capacity of 100 million gal of fuel ethanol per year, assuming ceramic membranes are used for stillage clarification (adapted from Cheryan and Bogush 1955).*

recycled to upstream operations. This recycle operation saves wastewater treatment and evaporation costs. MF can be used quite effectively to clarify the stillage (Chang et al. 1994; Cheryan and Bogush 1995; Rane and Cheryan 1996), as shown in Figure 8.28. With 0.2-μ ceramic membranes from Cera-Mem (model PMA), flux in the feed-and-bleed mode at VCR 5 dropped from an initial 250 LMH to 140 LMH at the end of 12 h at a process temperature of 80°C. The TMP was initially set at 15 psi and allowed to increase gradually to about 25 psi, while the pressure drop was maintained at 15 psi, equivalent to a cross-flow velocity of about 3.5 m/sec. Flux was improved when a backpressure was maintained. The standard mode (without backpressure) gave a flux of 149 LMH over a 24-h period at a TMP of 15 psi. MF with 10-psi backpressure resulted in an overall flux of 210 LMH for a 24-h period. Higher or lower backpressures (keeping the TMP at 15 psi) were not as effective (Figure 6.26).

Typical concentration of total solids in the retentates and permeates at VCR 7 is shown in Figure 8.28. The quality of the permeate was excellent, with no suspended matter. In contrast, thin stillage from a centrifuge overflow still contains some suspended solids. The clearer permeate from MF is expected to be beneficial in those upstream operations where it is reused.

A suggested plant design and costs for this application are described in Section 7.G.2. As shown in Table 7.16, operating cost comes to $1.70 per 1000 gal ($0.45/m³) of stillage. If no stillage treatment is being done now, the savings in evaporation and wastewater treatment would justify the capital and operating cost. In addition, NF, electrodialysis (ED), or RO could be done to the MF permeate to recover some valuable by-products (Cheryan and Parekh 1995).

8.D.3.
CAUSTIC AND ACID RECOVERY

Caustic (alkalis such as sodium hydroxide) and acids (e.g., nitric, phosphoric, citric) are used in many industries as cleaning agents. When the cleaning solution becomes loaded with soil and other residues, it loses its effectiveness and is usually discarded. However, membrane technology offers the opportunity to recover these chemicals by removing suspended matter and dissolved components, depending on the pore size of the membrane. Due to the aggressive nature and high temperatures of the solution, inorganic membranes have been favored in this application. Up to 90–96% recovery of the feed volume can be accomplished with >99% rejection of colloidal and insoluble impurities (Niro 1994). Soluble impurities do pass into the permeate, but its level can be controlled by adding fresh chemical to make up the concentrate volume. Cleaning of these membrane systems is usually accomplished by backflushing.

A Membralox ceramic system is processing 23,000 gpd of spent caustic at Sunkist Growers, Tipton, California. The caustic solution was used to clean citrus juice processing lines. The MF system has resulted in a 50% reduction in caustic requirements (saving more than $120,000 annually), as well as a 30% reduction in energy for waste treatment, a major reduction in environmental discharges, and elimination of frequent shutdowns of the evaporator that was previously used for concentrating used caustic (USFilter 1996).

Stainless steel membranes are incorporated in the "Micro-Steel" caustic recovery system for dairy plants (MSS 1995). It is designed to remove suspended solids from clean-in-place (CIP) solutions used for evaporators and other processing equipment. Typical fluxes are 190 LMH for 80% recoveries and power requirements are 20–25 hp. A dairy plant using only 100 gal per day of 50% caustic could see a payback period of less than a year.

In the textile industry, caustic is used for scouring and mercerizing cotton and cotton blends prior to bleaching and dyeing. The wastewater contains about 5–8% NaOH. The "Scepter" stainless steel membrane is used by several textile mills to recover and reuse 90–95% of the caustic solution, resulting in significant savings in chemicals, energy, and waste treatment costs (Graver Separations 1996). Some pretreatment is required to remove lint and large particles. At a caustic price of $400/ton, a 14,000 gal per day system can save $300,000 per year.

8.D.4.
BRINE RECOVERY

The recovery of brine (NaCl solutions) is another interesting application of MF and UF. When used for salting cheese, the brine picks up cheese components (particulates, protein, fat, salts, lactose) and bacteria and becomes so cloudy that it must be discarded, usually by land spreading, an option that is rapidly disappearing. Cheese brine typically contains 17–25% NaCl and is at a pH of 4.8–5. Early trials by Merin et al. (1983) showed that 0.2-μ MF membranes were effective in reducing bacteria count to less than 10 per milliliter in the permeate. Today, spiral UF and ceramic MF membranes can efficiently reduce bacteria from 10^5 per milliliter to less than 100 per milliliter, as well as remove many other contaminants such as turbidity, foam, scum (fats, proteins, etc.), making the brine like new again. Flux with 0.8-μ Membralox ceramics was 400–600 LMH over an 8-h period at concentration ratios of 35–100× (Pedersen 1992). In all cases, bacteria retention was over 99%, with less than 10% retention of calcium and nitrogen.

In one mozzarella cheese plant, four spiral modules with 30-mil spacers, for a total area of 32 m^2, process about 9000 gal per day at normal brine temperatures of 10–20°C (MSS 1995). The yield of clean brine is 99.5%. The 200× retentate is sent to the waste treatment facility. Sometimes a warm water rinse is sufficient for occasional cleaning. Membrane lifetimes should be long in this application, about 3 years.

Going one step further is a ceramic Membralox system installed by Niro Hudson at Foremost (Morning Glory) Farms, Appleton, Wisconsin. The cooker/stretcher water from the mozzarella cheese plant contains some salt, but mostly fat (4–7%) and milk solids. Previously, centrifuges were used to separate fat and protein, but this required dilution with two to three times its volume of water to maximize separation, which sometimes was as low as 60% efficient. Adding water also increased waste volume, and the chloride could cause corrosion problems with metal surfaces. With the ceramic system, fat recoveries were 99.5%, resulting in a retentate with a fat content of 40% and total solids about 55%. This retentate has a market value, e.g., for use in whey butter manufacture. Waste treatment volume was reduced by 50%, and BOD was reduced (but not eliminated with the MF system; further BOD reductions required an NF/RO system). With disposal cost at $22/100 lb for BOD, $18 per 100 lb for suspended solids, and $0.80 per 1000 gal volume, the savings are about $20,000 per month. The ceramic system with four P37 modules and a 25-hp pump cost about $150,000 in 1991, which resulted in a payback of less than 1 year. The unit is equipped with a backpulse system using compressed air at 110 psi, which operates for a few seconds every 5 min. Chemical cleaning is done every day for 3–4 h.

8.D.5.
PRINTING INK

Flexographic ink is a water-based ink that is widely used. It is almost completely retained by an ultrafiltration membrane, which makes it ideal for treating the wash water from the printing process that can otherwise create a disposal problem. Tubular modules have been found to reliably process these low-volume (usually 5000–10,000 L/day), dilute wastewater streams. Average fluxes are 50–60 LMH, and volume reductions of 40- to 60-fold are common. The system operating costs compare favorably with the typical disposal alternative, which is usually hauling (Hayward 1982).

8.D.6.
LAUNDRY WASTEWATER

Wastewater from all laundry sources make up 10% of municipal sewer discharges (Tran 1985). It contains more than 1000 ppm suspended solids, 5000 ppm chemical oxygen demand (COD), 1100 ppm FOG, and 1300 ppm BOD, in addition to metals and organic solvents such as toluene, benzene, and perchlorethylene. Conventional treatment involves lime coagulation, flocculation with polymers, clarification by dissolved air flotation (DAF), and final polishing of the underflow by filtration using sand or diatomaceous earth. In one approach, chemical conditioning and physical separation is still done, and membranes are used only for the final polishing of the discharge water. A 50-gpm (11.3 m^3/h) Memtek system at an Aratex plant in Texas has a flux of 425–850 LMH, reducing BOD to less than 30 ppm, COD <100 ppm, FOG <10 ppm, and zero suspended solids in the permeate (about 30% of the permeate is recycled). The other approach seeks to replace many conventional steps such as DAF and flocculation polymers (a protective screen will still be needed). A Membralox ceramic membrane system in France with 45 m^2 of area has been producing 6.8 m^3/h (30 gpm) of treated laundry water at 70°C since 1993. The permeate is recycled. Flux averages 155 LMH and the COD of the laundry water has been reduced by 90%. This $400,000 system has shown a payback of about 1 year (Short 1994).

8.D.7.
MICELLAR-ENHANCED ULTRAFILTRATION

Many low molecular weight compounds that are normally permeable to UF or MF membranes can be removed if they can be attached to or complexed with larger impermeable molecules. The binding of multivalent cations to proteins is one example. However, surfactants can also form complexes with organic molecules and metals under certain conditions. Surfactants are amphiphiles; i.e., they contain a hydrophobic tail, which is usually a hydrocarbon chain, while the hydrophilic head consists of anionic, cationic, or nonionic groups. This characteristic allows them to be effective cleaning agents (Chapter 6), since the hydrophobic interior of the micelle serves as a solvent for organic dirt and soils, while the polar hydrophilic group orients itself outwards towards the water. In micellar-enhanced ultrafiltration (MEUF), we take advantage of a surfactant's ability to associate with itself and form "micelles" above a certain concentration known as the critical micelle concentration (CMC). Micelles can be 20–200 molecules in size; if solute–micelle interactions take place, it

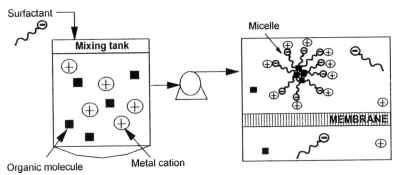

Figure 8.29. *Micellar-enhanced ultrafiltration with an anionic surfactant that interacts with organic molecules and metal cations.*

is possible to separate the micelles with UF membranes and, with them, the complexed low MW solutes.

As shown in Figure 8.29, the appropriate surfactant is added to the wastewater or other aqueous feed. The surfactant concentration is adjusted such that micelles are formed, but there should only be a small amount of free monomeric surfactant and solutes. The permeate should contain very little, if any, of the feed components or surfactant, while the retentate contains the surfactant and the solutes. The retentate could be treated to recover the solutes by conventional techniques, such as distillation (for organic solvents), crystallization, precipitation (for heavy metals), or extraction. Excellent removal of alcohols, benzene, cadmium, chromate, copper, cresols, lead, nickel, phenols, and zinc has been reported (Christian et al. 1988; Dunn et al. 1985, 1989; Geckeler and Volchek 1996; Huang and Koseoglu 1993; Klepac et al. 1991).

Some of the factors important in MEUF are solubilization capacity of the solute, type and concentration of surfactant, pore size of the membrane, and phase changes that may occur at high surfactant concentrations (Dunn et al. 1985). Nonionic surfactants (e.g., polyethoxylated alkylphenols or alcohols, Triton X-100) have low CMC values and large micelle sizes, thus allowing large-pore UF membranes to be used, and presumably result in higher flux. However, better solubilization can be obtained with ionic surfactants. Anionic surfactants [e.g., sodium dodecylbenzene sulfonate (SDS)] have negatively charged hydrophilic groups, form small micelles, and thus need UF membranes as tight as 1000 MWCO. They have a high CMC and relatively low solubilization capacity, thus usually not making them desirable either. However, anionics have to be used if the objective is to remove both nonionic organic compounds and divalent metal cations (Dunn et al. 1989; Klepac et al. 1991). Cationic surfactants (e.g., hexadecylpyridinium chloride, CPC) have low CMC values, form large micelles at room temperature, and have a relatively large solubilization capacity (Dunn et al. 1985).

The hydrophilicity of the membrane used in MEUF is important. Membranes that are opposite in charge to the surfactant or are hydrophobic gave poor flux. The closer the surfactant concentration was to the CMC, the greater the flux decline with hydrophobic membranes (Jonsson and Jonsson 1991). Rejection increases with surfactant concentration, especially above the CMC value (Akay and Wakeman 1993). Temperature is also important: rejection of surfactant increases sharply above the cloud point of the surfactant (the temperature at which the surfactant's appearance changes from clear to cloudy). Flux and permeation of surfactant increases with temperature, probably because of a reduction in viscosity and increase in adsorption of the surfactant by the membrane. Above the cloud point, both flux and permeate concentration decrease, probably because of phase inversion and subsequent pore blockage.

Heavy metals can also be separated from wastewaters if macromolecules (e.g., sodium polystyrene sulfonate, polyacrylic acid, polyethylenimine, polyvinyl alcohol) are added to bind the metals, thus forming large complexes that can be removed by UF membranes. This concept has been used to reduce hardness in water (Tabatabia et al. 1995) and to remove copper, zinc, cadmium, silver, mercury, nickel, and cobalt from a variety of wastewaters (Chaufer and Deratani 1988; Geckeler and Volchek 1996; Huang and Koseoglu 1993; Juang and Liang 1993; Nguyen et al. 1980; Strathman 1980).

8.E.
TEXTILE INDUSTRY

The textile industry uses synthetic warp sizing agents such as polyvinyl alcohol (PVA), polyacrylate and carboxymethyl cellulose (CMC) in cotton blends, in place of starch and natural gums. After weaving, the size agents must be washed out, which requires large volumes of wash water. These sizing agents are, however, expensive and nonbiodegradable; thus, they pose challenging waste treatment and/or recovery problems. Figure 8.30 shows an ultrafiltration-based system for the recovery and reuse of PVA and CMC. The feed is from the desize washer or the washing stage of the desize range. After prefiltration to remove lint, it enters UF modules. The feed is usually about 0.5–2.0% solids, and the retentate is usually about 8–16% solids (depending on the sizing agent), which is slasher strength, ready for reuse. The permeate from each stage is recycled directly to the washer or range. A small amount of desizer waste is purged from the system to prevent buildup of low molecular weight solutes (Porter 1990).

Flux in spiral wound modules is low (about 10–15 LMH), although stable. At steady state, the systems need only weekly cleaning since these sizing agents are relatively nonfouling. Weekly detergent and peroxide cleaning have enabled Koch spiral elements to achieve over 2 years of useful life, despite continuous operation at 70–85°C (high temperature is required because of retentate

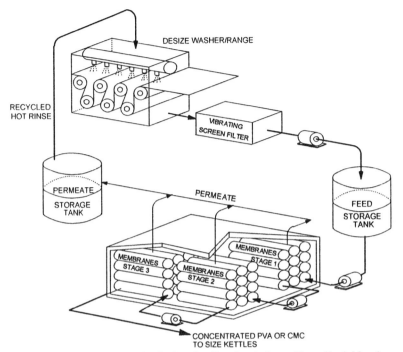

Figure 8.30. *Recovery of textile size by ultrafiltration (adapted from Koch Membrane Systems literature).*

viscosity). PVA yields of up to 97% are possible, with only the small degraded PVA molecules permeating the membrane. This actually improves the quality of the recycled sizing agent. Scepter stainless steel membranes are also in use for this application (Graver Separations 1996), utilizing a formed-in-place UF membrane over the titania surface of the porous stainless steel tubes. Depending on plant capacity, even the stainless steel systems have shown payback periods of less than 1 year.

Wool scouring effluents have also been successfully treated by UF (Bilstad et al. 1994; Turpie et al. 1992). Raw wool has to be treated (cleaned or scoured) with hot water containing sodium carbonate and nonionic detergents before it can be spun into yarn. Large volumes of scouring solution are used, about 30 kg per kilogram of wool. The spent scouring effluent contains particulates, oil and grease, salts, and detergent. It has a pH of 8 and a BOD of 20,000–150,000 ppm. Traditional treatment methods of screening, flotation, flocculation, and biofiltration are not as effective as membrane technology.

A PCI tubular UF plant in the UK with 182 m^2 of PES membranes (25 K MWCO) processes 8000 L per hour of wool scouring effluent at 50–60°C and 1 MPa pressure, reducing COD to 25,000–125,000 ppm in the permeate while the retentate COD is 350,000 ppm. The UF permeate (containing some of the detergent) is of good enough quality to recycle back to the scouring tank, at least early in the cycle when the COD is low. Daily cleaning with 0.5% NaOH is supplemented with occasional addition of hypochlorite. The retentate, combined with sludge from the pretreatment settling tank, can be evaporated and disposed as solid waste. Pretreatment can be as simple as a settling tank ahead of the UF modules.

Another tubular UF plant in Norway has been operating since 1989. Today it has 31 m^2 of membrane area and processes 40 m^3/h of effluent for a 10× volume reduction. It operates for about 15 h per day at a pressure of 8 bar, velocity of 3.8 m/sec, temperature above 40°C to prevent solidification of the fat, and has a 10-kW power consumption. It has consistently reduced COD, fat, and solids by more than 80%, allowing the permeate to be discharged to the sewers directly (Bilstad et al. 1994). Membrane replacement is once every year. Fluxes are 120–140 LMH with new membranes and reduce to 20–40 LMH near the end of the life of the membrane.

Dyes can be recovered by membrane technology. UF can be used with polymeric dyes and indigo, which is in great demand for the manufacture of denim material that goes into blue jeans. Indigo has an MW of 262, but it is insoluble in its oxidized state and can be recovered with a 50 K MWCO membrane. Wafilin developed a process for the recovery of indigo dye out of textile wastewaters. The dye in the retentate stream is then reduced to the soluble form and recycled. The indigo wastestream is strongly fouling, which may necessitate the use of tubular systems (Wafilin 1983). One Wafilin plant in operation since 1978 uses a two-stage UF system with 214 m^2 of tubular non-cellulosic, type WFA membranes. The plant's feed capacity is 144,000 L/day and the retentate volume is only 3000 L/day. Energy consumption is about 8 kWh/m^3 permeate. Interestingly, Direct Red 2, an anionic dye of 1–1.5 nm in size, was 82–99% rejected by 0.2-μ inorganic microfilters (Porter and Zhuang 1996).

Figure 8.31 shows a flow sheet for the processing of dyes, where the major objective is to remove salts. In one case, UF and DF of reactive Remazole Turquoise Blue G21 with Permionics spiral modules increased the strength of the dye 2.25×. Typical plant capacity is 1 ton of salt-free dye per 80 m^2 of membrane area. With another dye (Black B), the feed dye concentration is 8% and salt concentration 2%. Ultrafiltration to 3.3× followed by diafiltration results in salt reduction to 0.02% and dye concentration of 24% in the retentate. Dye in the permeate is less than 0.3%. The spray-dried dye powder had the advantage of higher bulk density (Permionics 1995).

The Ciba Textile Products Division plant in Toms River, New Jersey, has been using two ceramic UF systems and an RO plant to recover dyes from its wastewater (Short 1993). About 19,200 gpd is processed through the UF systems to produce a retentate with 15% solids and a permeate with less than 1%

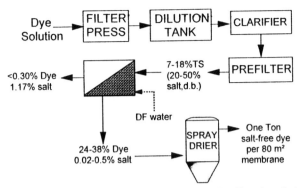

Figure 8.31. *Desalting and concentrating dyes by ultrafiltration (adapted from Permionics literature).*

solids. The UF retentate is blended in with the next batch of the same color. The permeate goes through an RO system to remove solubles and residual color. About 500 lb/day of dye is recovered for an annual savings of over $900,000. Wastewater treatment costs have been reduced by $228,000. The UF + RO system costs $6 million. Annual savings are $1.18 million and operating costs are $0.41 million (cleaning chemicals are $3000–5000 and pump power is $5300 for both UF systems). Payback period for the UF + RO system based on recovered dye and avoided waste treatment costs is 7–8 years. Long payback periods are typical in most dye recovery operations: a payback period of 3.8 years was estimated for the indigo dye recovery (Cheryan 1986). If only waste minimization is considered, with no credit for the recovered dye, a combination of UF and NF can cost less than $4 per 1000 gal of wastewater from a stonewashed jeans manufacturing plant (Short 1995a).

8.F.
LATEX EMULSIONS

Latex emulsions consist of particles of about 0.05–0.3 μm in size, stabilized by ionic and nonionic surfactants. It was one of the very early suggested uses of UF (Zahka and Mir 1977). Some end-of-pipe latex streams, such as styrene butadiene (SBR) and PVA polymerization kettle wash waters, are dilute, and the principal UF application is in concentrating these "whitewater" streams from 0.5% to about 25%, generally as a pollution control measure. Conventional treatment is complicated and labor-intensive, involving pH adjustment, chemical coagulants, and settlers. It generates a sludge that must be disposed of by landfilling and a supernatant that may require further treatment prior to discharge (Breslau and Buckley 1995). When the latexes are stable, high flux can be obtained with large-diameter tubular modules, which are quite concentration-

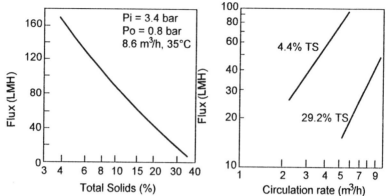

Figure 8.32. Ultrafiltration of styrene-butadiene latex in a (Koch) Abcor tubular module: left: effect of feed concentration on flux; right: effect of recirculation rate and latex concentration on flux (Zahka and Mir 1977).

dependent and flow rate–dependent (Figures 8.32 and 8.33). High recirculation rates are generally recommended since flux-velocity exponents are typically 0.7 for laminar flow modules and 1.8 for turbulent flow modules (much higher than theory, as explained in Chapter 4).

Addition of surfactant can greatly enhance the performance. For example, adding 5% Triton X-100, based on polymer weight, to a 0.5% solids SBR feed resulted in a fairly stable flux of 340 LMH at 50°C, 2.7 atm pressure, and recirculation rate of 150 L/min per tube. Without the Triton, the flux dropped from 100 LMH to 17 LMH in 2 h. Economic analyses for the concentration

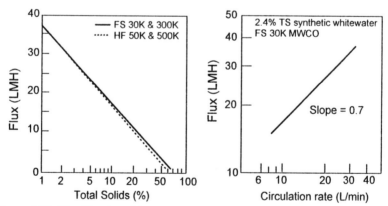

Figure 8.33. Ultrafiltration of latex whitewater with polyethersulfone flat-sheet (FS) and polysulfone hollow fiber (HF) membranes. Two different MWCO membranes were evaluated in each module. Both modules operated under laminar flow conditions (adapted from Breslau and Buckley 1995).

and recovery of dilute, SBR-like latex streams indicate a payback period of 1–2 years in the United States.

With in-process latex streams such as polyvinyl chloride (PVC) where the latex concentration is much higher, UF is an attractive processing alternative to evaporation, despite some inherent occasional latex instability problems (Hayward 1982). The latex surface tension should be maintained below 35 dynes/cm, and by ensuring a high degree of turbulence in the UF module (e.g., over 150 L/min flow rates per pass in tubular Abcor modules), a reasonably stable flux can be obtained. PVC latex can be concentrated from about 35% solids to about 55% solids prior to spray drying, thus reducing or eliminating the need for thermal evaporation. Flux typically averages 50–75 LMH. The permeate stream contains surfactant in addition to water, and this may be recycled back to the reactor vessels. This feature allows UF to aid in closing the loop around the PVC production process. The total operating cost of UF in this application is usually about 1.5–3% of the salable value of the recovered product (Beaton 1976).

Cleaning of the membrane is performed after most batches with a water flush or detergent washing and by spongeballing in the large-diameter tubular modules. Sometimes the membrane may get severely fouled by a coherent polymer film, formed by unstable latex particles, which surfactant cleaning alone will not remove. This will require cleaning with solvents, such as pure methyl ethyl ketone and methyl isobutyl ketone for PVC and SBR, and lower molecular weight alcohols, such as propanol for polyvinyl acetate, for an hour on a monthly basis. Thus, membranes resistant to solvents are required.

8.G.
PULP AND PAPER INDUSTRY

The pulp and paper industry produces enormous amounts of highly polluted water, e.g., up to 100,000 L of bleach liquor per ton of pulp. Effluents are at extremes of pH, highly colored, and nonbiodegradable for the most part. It is quite difficult to meet stringent environmental regulations by conventional treatment techniques such as coagulation and activated sludge processes. Most of the BOD comes from low MW carbohydrates that may require RO to treat. However, UF can be used to concentrate and recycle some of the effluents prior to discharge. Some applications include color removal from kraft mill bleaching effluents, concentration of dilute spent sulfite liquor (SSL) from the sulfite process, recovery of lignin from Kraft black liquor (from the sulfate process), and recovery of paper coatings. Membranes with MWCOs of 3000–5000 are most promising in these applications. SSL typically contains 12–16% solids, of which half is lignin sulfonate. Black liquor contains 17–22% solids, of which 40% is alkali lignins.

In the color removal application, treating the first alkaline stage alone can recover 70–95% of the color bodies and a major part (60–70%) of the BOD,

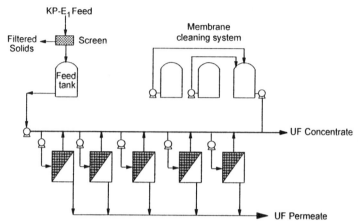

Figure 8.34. *Flow diagram of the Taio UF plant for processing KPE effluents from a Kraft Paper Mill (adapted from Nitto-Denko 1983).*

COD, and organic chlorinated compounds. The total solids are usually around 0.7%, and volume reductions of $25\times$–$99\times$ can be achieved, depending on the module. It is better to segregate the highly colored streams from the mill effluent. Proper pretreatment is necessary: a side-hill screen and prefiltration down to 100 ppm suspended solids can result in spiral-wound fluxes of 40–50 LMH, although the performance is somewhat site-specific.

Nitto-Denko (1983) has described the treatment of the KP-E$_1$ stream in the Kraft pulping process of the Taio Kraft mill in Japan (Figure 8.34). This stream typically has COD of 1200–2000 ppm. It is first screened to remove wood fibers and large particulate matter before recirculating through the UF modules. Figure 8.35 shows gel permeation chromatograms of the feed and the two UF streams after processing with Nitto's NTU-3008 polysulfone membranes.

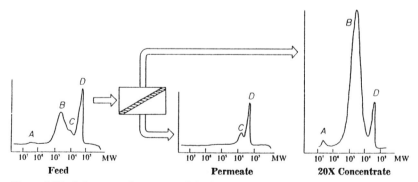

Figure 8.35. *Gel permeation traces of the feed, permeate, and 20× concentrate from the Taio plant shown in Figure 8.34 (adapted from Nitto-Denko 1983).*

The constituents of peak B in the feed are high in aromatic content and are particularly difficult for conventional activated sludge treatment. The permeate is essentially free of peaks A and B, which now enables it to be treated by conventional activated sludge processes. A portion of the permeate is recycled to a holding tank where it is mixed with NaOH and detergents to be used for periodic cleaning of the membranes. The concentrate, high in peak B and aromatics, can be incinerated. The Nitto UF plant shown in Figure 8.34 is a six-stage feed-and-bleed design, processing more than 1 million gallons (about 4 million liters) per day of KPE_1 effluent with a membrane area of 1480 m^2.

As far as SSL is concerned, a large part of the solids is lignosulfonates. Some fractions in the 10 K–50 K molecular weight range have potential uses as a raw material for glue, dispersants, vanillin, and adhesives. Yields of over 90% and purity of over 97% lignin can be achieved with appropriate DF. The retentate stream is usually 25–30% total solids, which should reduce the load on the evaporators significantly. The permeate contains mostly sugars that can be used as a fermentation substrate. The lignosulfonate stream contains vanillin, detergents, binders, and adhesives that could be valuable enough to result in a short payback period for the UF plant.

The PCI tubular UF plant at Borregaard Industries in Norway processes 50,000 L/h of liquor with 12% solids (5% high MW lignins) into a retentate stream of 22% solids at VCR 3.1. The feed is prefiltered through $500-\mu$ screens; process temperature is 60–65°C and pH 4.2–4.5. Membranes are polysulfone 20 K MWCO, total area 1120 m^2, and operating pressures 10–15 bar, and a total of 260 kW of electric power are consumed by the pumps. Cleaning is done once a day, and membrane lifetime is about 15 months since the plant started up in 1981. Another system using PCI membranes is operating at Nymolla, Sweden (Figure 8.36). It processes 300 m^3/h (1320 gpm) of effluent from bleaching softwood and hardwood. The UF plant achieves a 50% reduction of COD by retaining the higher MW fractions, which are then sent to the chemical recovery boiler. With a $50\times$ volume reduction, most of the sodium compounds are removed. Interestingly, while the ES404 membrane (Table 5.2) and low velocities of 2.4 m/sec worked well for bleaching effluent from softwood, a special membrane had to be developed for treating hardwood effluent, and higher velocities had to be used (Merry 1995). Cleaning is done with a caustic detergent every 2–4 days. Spiral modules have also been used to fractionate SSL into a high-purity lignin stream and low-lignin sugar stream (Hayward 1982).

Kraft black liquor contains alkali lignins and can be treated by UF in a similar manner. The retentate solids is 24% solids, of which 90% is alkali lignin. About 50–60% of the lignin can be recovered by UF and used in phenol formaldehyde resins. The permeate is evaporated and combusted.

"Cross-rotating" filters, originally manufactured by ABB Flootek, Sweden, are being used in Finland and Sweden for UF of machine white water to remove

Figure 8.36. An effluent treatment system using PCI tubular membranes at Stora Nymolla AB mill, Sweden. The plant has 1784 B1 modules (3.6 m long) with 4650 m^2 in 13 stages (Courtesy of PCI Membrane Systems).

suspended and dissolved solids. These are similar to the rotary module shown in Figure 5.44, with solid rotating plates between stationary membrane plates. The permeate is used as shower water. The retentate is either evaporated and burned or sent to a treatment plant (Nuortila-Jokinen and Nystrom 1996).

It should be noted, however, that despite these successful examples, membranes are not yet widely used in the pulp and paper industry, except in Scandinavia and Japan where there are specific site-related benefits.

Paper coating is a mixture of pigments, binders, and additives and is applied to paper to improve its quality. Pigments may be clays, calcium carbonate, titania, and polymeric pigments. Binders may be latex, starch, soy, and milk proteins and make up 5–15% of the solids. Additives (1–2% of the solids) may be dispersants, optical brighteners, and cross-linking and lubricating agents. Coating effluents arise from waste coating from machine breaks, washdown, and change of production. These can be concentrated with Koch-Abcor 1″ diameter tubular modules from ~1% coating solids to 40–50% solids, at which level the viscosity may be as high as 300–500 cp. The system is operated in the modified batch mode (Figure 7.10). When a predetermined solids or viscosity is reached, feed is stopped and the concentrated coating diverted for reuse. Typical process runs are 24–48 h, with a cleaning cycle of 2–4 hours using caustic and surfactants. The permeate is clear and can be reused in the plant or disposed in the drain, since it is reduced in BOD and COD by 70–90%. The retentate contains all coating solids, pigments, and binders. No flocculants or chemicals are needed, and thus coating properties do not change, which allows easy reuse of the retentate back in the coating kitchen. For a large mill of 1000 tons per day of low-weight cardboard (LWC) paper with 12 tons per day of coating being lost, the savings may be as much as $5000 per day (Koch 1995).

CR filters ("Cross-rotating") are being used in Frovifors, Sweden, for UF of dilute coating color effluents. Two of these units are designed to operate at constant 20% solids in the retentate. With polyamide membranes of 50 K MWCO, flux with standard coating formulations was 120 LMH at 10–15% solids in the retentate and 50 LMH at 40% solids. These are apparently double the value obtained with Koch-Abcor tubular modules (Jonsson et al. 1996). The CR filters are operated continuously for 10 days twice a month and cleaned only between runs. Membrane life is more than a year. Payback is 1–2 years, depending on the value of the pigment recovered. Other benefits include a reduction in fresh water demand and lower loading on waste treatment plants.

8.H.
TANNING AND LEATHER INDUSTRIES

Effluents from leather industry tanneries are usually quite difficult to handle. The initial step is to soak animal hides in hot (90°C) saturated brine to remove FOG, hair, etc. Table 8.7 shows the composition of dehairing bath effluents,

**Table 8.7. Composition of
dehairing bath in the
leather industry.**

pH	12–14
Sulfide (as Na_2S)	4,700 mg/L
COD	40,000 mg/L
Alkalinity (as NaOH)	16,800 mg/L
Suspended solids	6,441 mg/L
Dissolved solids	74,193 mg/L
Protein nitrogen	9,000 mg/L
BOD_5	16,250 mg/L

Source: Data from Drioli and Cortese (1980)

although in some cases, the spent brine can have much higher concentrations (see later). In this industry, UF or MF can be used to recover sulfides from spent dehairing baths, recover and desalt vegetable tannin baths, and recycle or at least remove dissolved chromium from spent chrome tannin. Drioli and Cortese (1980) found that both Abcor HFM-180 (MWCO of 18,000) and the Berghof BM-500 membrane (MWCO of 50,000) showed zero rejection of electrolytes, about 2% rejection of the sulfides, and 85% rejection of the proteins and colloidal matter. With proper pretreatment, average fluxes for the Abcor HFM-180 tubular membrane are about 70 LMH, and a 2-year life time can be expected (Hayward 1982). Recovery of sulfides in the permeate is usually 65%, and there is a significant pollution reduction and savings in traditional chemical treatment costs. The retentate can be used for animal feed or fertilizer.

Some spent brines can have much higher concentrations than shown in Table 8.7, e.g., a BOD >200,000 ppm, total dissolved solids (TDS) >160,000 ppm, and suspended solids >12% (Seprotech 1995). In this case, a decanter and rotary drum screw filter are used ahead of Membralox ceramic membranes to remove particles larger than 120 μm. The membranes are operated at 80°C and 50 psi. Backpulsing has proven to be useful in extending the operating cycle between cleanings and maintaining high flux. The resulting brine solution (permeate) has a TDS of 180,000 ppm NaCl, BOD <1000 ppm, suspended solids <25 ppm, and FOG <10 ppm. The MF retentate is further heated, and the oil is separated to be sold as tallow, while the nonfat retentate solids are sold as a natural fertilizer. Membrane cleaning is done with caustic and acid. This MF system has reduced salt purchases by 95% and—because of value received from the sale of retentate by-products—has resulted in a short payback period.

In contrast to dehairing baths, it is the retentate that is valuable during UF of vegetable tannin baths. Tannin recoveries of 85% can be expected. As for the dissolved chromium, UF can retain essentially 100% of the suspended chromium, if the precipitation step is carried out properly. Permeate from the chromium recovery plant is usually less than 1 ppm of dissolved chromium. The retentate can be reused several times after reacidification.

8.I.
SUGAR REFINING

Sugar processing is one of the most energy-intensive processes in the food and chemical industries; thus, it would seem that membranes would find several applications in this industry. Modern sugar refineries typically process 67% sugar solutions at temperatures of 70–80°C. The high osmotic pressure, coupled with the high viscosities, limits the role of membrane processes to dilute streams, and thus the major attention has been given to clarification and purification of the juices at the extraction stage where lower solids concentration and lower temperatures are used.

Figure 8.37 shows schematics of typical cane sugar and beet sugar processes. Beet and cane sugar extracts contain, in addition to sucrose, objectionable amounts of polysaccharides, lignins, proteins, starches, gums, fat, waxes, and other colloidal impurities that contribute color or taste to the crystalline product and reduce the product yield. This raw juice is subjected to several steps such as heating, addition of lime, and clarification to remove proteins and colloidal matter. Some mills add flocculants and enzymes to help eliminate macromolecules. The clarified thin juice is then evaporated and crystallized. With cane sugar, additional steps are needed to refine the sugar (which may be

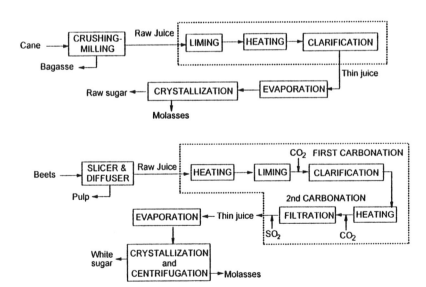

Figure 8.37. *Possible applications of MF and UF in sugar processing. The operations that could be replaced by UF or MF are shown in the enclosed areas.*

done in a different plant). This is achieved by washing, remelting the raw sugar, and subjecting it to ion exchange, electrodialysis, or activated carbon treatment and crystallization to achieve the required 99% + purity.

Considering the nature of the impurities that have to be removed, UF or MF could be used at the mill to remove the colloidal and macromolecular impurities, with little or no need for addition of lime, carbon dioxide, or sulfite, resulting in a clear decolorized thin juice. If ion exchange is done immediately after UF, lime could be eliminated completely. In addition, removing macromolecules and reducing lime levels will reduce fouling and scaling of the evaporator, which in turn reduces downtime and energy costs for cleaning. Higher yield of sugar and better crystallization are also possible. Another place for UF or MF is to clarify thick juice (after evaporation), which reduces bacterial counts and storage losses. Treating thick juice has the advantage of handling lower volumes, but this is partially compensated for by higher viscosity and lower flux.

Early experiments by Zanto et al. (1970) and Madsen (1973a) indicated that low molecular weight nonsugar impurities could be separated from sugars using a fairly tight UF membrane. Several other studies have essentially concluded that UF membranes produce a superior juice compared to the traditional process: much lower viscosity and 60–90% color removal (Kishihara et al. 1989; Mak 1991).

Hydrophilic membranes perform much better than hydrophobic membranes (Figure 8.38). With raw cane juice, even the tight YM-5 membrane of Amicon's gave better flux with little or no flux decline compared to the polysulfone PM-10 and PM-30 membranes, even though the water flux of the PM membranes was much higher. Adding lime to increase pH to 7 improved flux of the YM-5 membrane by 30%, but reduced the flux with the other membranes (Tako and Nakamura 1986). In general, increasing pH (without necessarily adding lime) has beneficial effects (Verma et al. 1996). Flux increases by 20–50% at pH 7–8.5 compared to pH 5.5–6.5, perhaps caused by a change in the conformation of the protein impurities, similar to what has been observed with cheese whey and other proteins as discussed in Chapter 6.

Batstone (1990) reported on semiworks scale trials conducted in the Pak Sap sugar mill, which has a capacity of 200 tons of cane per day. Cane juice is prefiltered to 100 μ and then heated and pH adjusted to prevent inversion of the sucrose. Sixty spiral-wound modules are used, each 6″ diameter × 37″ long, with 1.4 mm nonmesh spacers. They are placed two in series and installed vertically, rather than horizontally as is the norm for spirals. The system produces 10,000 L/h of sparkling clear juice with lower viscosity and lower color compared to conventional clarified juice. Near white sugar could be made in a single crystallization step.

In cane sugar refining, the main objective is removal of color compounds that affect the quality of the final product. This is done by affination, carbonation or

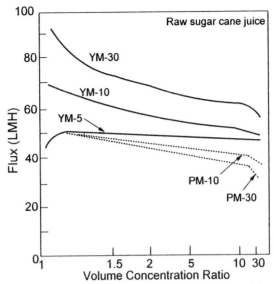

Figure 8.38. *Effect of membrane type on flux of raw sugar cane juice (18–20 °Brix). The YM membranes are regenerated cellulose; PM membranes are polysulfone. Numbers are MWCO in thousands (adapted from Tako and Nakamura 1986).*

phosphatation, and decolorization by resins or carbon prior to crystallization. Attempts have been made to replace the color removal steps by one MF operation (Dornier et al. 1995; Herve et al. 1995). Raw sugar from the mill would be remelted and microfiltered before being sent directly for ion exchange. The retentate would be used for molasses manufacture.

With beet sugar, UF is an alternative to the liming and carbonation purification steps (Hanssens et al. 1984). Large-diameter tubular modules could purify raw beet juice without any pretreatment. The juice is a strong foulant, and thus high velocities (4 m/sec) were needed to minimize fouling and to extend run times to 8–20 h before cleaning. With Wafilin WFA or WFS membranes, a concentration factor of two could be obtained with a flux of 40–50 LMH at 50°C. In order to diminish sugar losses, the final concentrate was subjected to diafiltration (Figure 8.39). A thin juice of the same quality as conventional processes was obtained.

With hollow fiber and ceramic MF membranes operated at 90°C, higher flux of 170–200 LMH at 3–5× has been obtained with raw beet juice containing 0.3–0.4% suspended solids (Kochergin 1996). However, since MF membranes have relatively large pores, more nonsugar molecules go through the membrane, but significant color reduction was obtained. The MF or UF pretreatment is well suited for subsequent ion-exchange softening and chromatographic purification (Herve et al. 1995). Another advantage is that a high-quality soft cane molasses

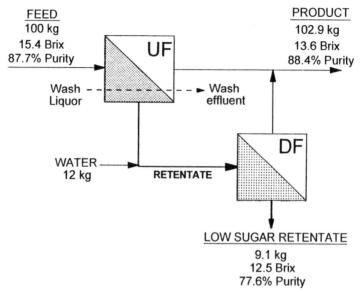

Figure 8.39. UF and diafiltration of sugar solutions (adapted from Wafilin 1983).

is obtained, which can go directly to chromatographic separation to recover sucrose and fructose or to NF to remove some of the monovalent salts.

8.J.
SOYBEAN AND OTHER VEGETABLE PROTEINS

Vegetable protein products have been consumed for centuries in Asia and Middle Eastern countries. It was not until this century that the western world recognized their food value, especially that of soybeans. With the growing interest in reducing the consumption of animal products for health and economic reasons, vegetable proteins are making up a higher proportion of the human diet in recent years. Soybeans (18–20% oil) are the major source of edible oil in the United States and by far the most important source of vegetable protein ingredients. Once the oil is removed, however, the major portion is used mostly as animal feed, with perhaps only about 3–5% being used directly as human food. The problem with soybeans is that they also contain some undesirable components that must be removed or reduced to increase the usefulness and functionality of the soybean protein (almost all sources of plant protein share this problem). For example, soybeans contain oligosaccharides that have been implicated with gastrointestinal stress. Lipid–lipoxygenase interactions must be avoided to prevent painty off-flavors from developing. Phytic acid forms insoluble chelates with minerals and can form complexes with proteins that reduce bioavailability of the minerals and proteins. Trypsin inhibitors are proteinaceous compounds that affect the efficiency of protein digestion.

Traditional processing techniques for producing soy protein concentrates and isolates partially overcome these problems. These methods involve extraction, heat treatment, and centrifugation to separate the protein and fat from the other components. These conventional methods are time-consuming, they sometimes result in products with poor functional properties, and they can generate a whey-like wastestream that contains a significant portion of the proteinaceous compounds of the starting material.

Since the undesirable oligosaccharides, phytic acid, and some of the trypsin inhibitors are smaller in molecular size than proteins and fat components, it should be possible, by careful selection of the membrane and operating parameters, to remove these undesirable components selectively and produce a purified protein isolate or lipid–protein concentrate (depending on the starting material) with superior functional properties. The processes developed for producing full-fat soy protein concentrates and soy isolates are shown in Figures 8.40 and 8.41. The blanching step shown in Figure 8.40 was necessary in order to prevent lipoxygenase-induced off-flavors during grinding. The first separation step (filtration or centrifugation) serves to remove insoluble carbohydrate and fiber and to ensure the particle size is appropriate to the membrane being used. Ultrafiltration of full-fat soy extracts ("soymilk") with 20,000–500,000 MWCO membranes have been reported (Ang et al. 1986; Berry and Nguyen 1988; Omosaiye et al. 1978; Omosaiye and Cheryan 1979a). The composition of the full-fat products obtained with a 50,000-MWCO membrane is shown in Table 8.8. Using higher MWCO membranes did not change the final product composition much but resulted in much higher flux (Berry and Nguyen, 1988). Reducing the pH of the extracts to pH 2 improved flux but resulted in off-flavors in the product due to the hydrolysis of the oligosaccharides.

To produce soy protein isolates or concentrates, the raw material is defatted soy flour (Figure 8.41). After extracting under optimum conditions of

Table 8.8. Composition of soybean products (% dry basis) produced by ultrafiltration of water extracts of soybeans.

Product	Protein	Fat	Ash	Oligo-saccharides	Phytic Acid	Other*	Trypsin Inhibitor**
Soybeans	44.3	24.0	4.7	13.2	1.3	13.5	81.2
Water extract (VCR 1)	48.3	26.4	5.3	16.3	1.7	2.0	42.6
UF retentate (VCR 5)	56.7	32.5	3.4	3.5	0.8	3.1	39.7
Diafiltered retentate	59.7	34.2	2.9	0.6	0.1	2.6	33.1

*By difference. Includes fiber, other carbohydrate
**Trypsin units inhibited/mg solids
Source: Data from Omosaiye and Cheryan (1979a)

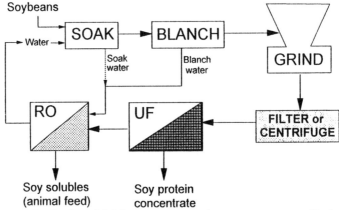

Figure 8.40. *Production of full-fat soybean protein concentrate or purified soymilk from whole soybeans by ultrafiltration (adapted from Omosaiye et al. 1978).*

temperature, meal-to-water ratio, and pH, the extract can be directly ultra-filtered with large diameter tubular membranes to remove the oligosaccharides. The retentate's final composition will approximate a soy protein concentrate (70% protein, dry basis). To produce isolates (90% protein), the fiber and insoluble carbohydrate are removed by centrifugation or filtration prior to UF. The underflow from the centrifuge (or the filter cake) can be re-extracted if necessary. Ultrafiltration at pH 9 increases flux 40% compared to pH 5.2 (Liu et al. 1989). A sequence of UF, continuous DF, and UF appears to be optimum (Nichols and Cheryan 1981).

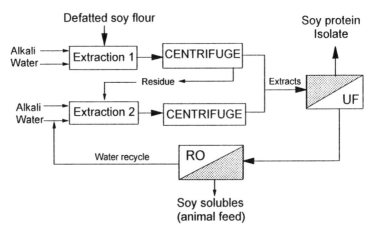

Figure 8.41. *Production of soy isolates by ultrafiltration (adapted from Nichols and Cheryan 1981).*

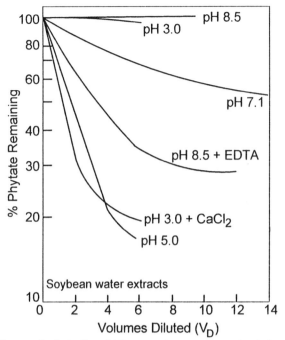

Figure 8.42. *Removal of phytic acid from soybean water extracts by continuous diafiltration: effect of pH, ionic strength, and salts (adapted from Okubo et al. 1975).*

Removal of oligosaccharides follows theoretical predictions and is relatively unaffected by operating parameters. However, removal of phytic acid is affected by pH. Phytic acid has a molecular weight of 880 and should be easily removed by UF. However, it is a negatively charged compound and can bind directly with soy protein by charge interactions at low pH or through a divalent cation bridge at high pH (Cheryan 1980). Thus, the pH, nature, and concentration of other charged compounds present during ultrafiltration has a profound effect on the removal of phytic acid (Ang et al. 1986; Okubo et al. 1975; Omosaiye and Cheryan 1979b). As shown in Figure 8.42, UF at pH 8.5 or pH 3 results in strong binding of phytate to the protein, and thus no removal will occur. Charge interactions are minimized near the isoelectric point of the protein (pH 5). Divalent cations become more soluble, and binding of phytate to the protein decreases, thus allowing it to be removed. Adding EDTA to the extract at high pH or excess calcium at low pH also effectively reduces the binding.

The manufacture of soy products by UF usually results in higher yields because of the inclusion of the whey proteins that are normally lost in conventional manufacturing methods. These whey proteins could also be contributing to the superior functional properties of the UF soy products, in addition to the benefits of the nonthermal and nonchemical nature of the UF process (Lah and Cheryan 1980a, 1980b; Lah et al. 1980).

There have been several attempts to fractionate and concentrate proteins from potato processing wastewaters (Ericksson and Sivik 1976; Oosten 1976; Wafilin 1983; Wajnawska et al. 1979). Potato juice (or fruitwater) is another notoriously bad fouling stream; furthermore, since it contains fibrous material remaining from the potatoes, large-diameter tubular membranes operating under high Reynolds numbers are to be preferred. Not only are the energy consumption and total cost for water removal comparatively low, but the heat coagulation of the protein is easier, cheaper, and more efficient after ultrafiltration. Furthermore, downstream equipment can be reduced in size by some 65%, and there is a net increase in yield of the protein by about 10%. A Wafilin tubular system installed in 1975 used an 18-stage recirculation system with a total membrane area of over 1900 m^2. The feed rate was 100,000 L/h of 4.5% w/v suspension. The retentate flow was 35,000 L/h of a 6.9% solids suspension, while the permeate flow was 65,000 L/h of a 3.2% solids solution. The energy consumption of the plant was about 10 kWh/m^3 of permeate. The plant operated about 125 h per week, with a daily cleaning with chlorine solutions. The COD of the permeate was about half that of the feed.

The application of UF in several other vegetable protein systems have been studied, among them alfalfa (Eakin et al. 1978), chickpeas (Ulloa et al. 1988) coconut (Chakraborty 1985), corn (Section 8.L.), cottonseed (Lawhon et al. 1977), faba beans (Gueguen et al. 1980; Olsen 1978), jojoba protein (Nabetani et al. 1995), leaf protein (Tragardh 1974), *lupinus albus* (Pompei and Lucisano 1978), navy beans (Lawhon and Lusas 1987), pasture herbage (Ostrowski 1979), peas (Buechbjerg and Madsen 1987; Gueguen et al. 1980), rapeseed (Diosady et al. 1984; Finnigan and Lewis 1989; Kroll et al. 1991; Tzeng et al. 1988), and sunflower seeds (Cheryan and Saeed 1989; Culioli et al. 1975; O'Connor 1971). In all cases, some manipulation of the micro-environment and diafiltration was necessary to maximize the removal of undesirable components.

8.K.
VEGETABLE OILS

Figure 8.43 (left) shows the basic unit operations in vegetable oil processing. The crude oil that is extracted from plant materials is predominantly triglycerides. It also contains small amounts of free fatty acids (FFAs), mono- and diglycerides, phosphatides (lecithins/gums), pigments (chloropohyll, carotenoids), sterols, and tocopherols (e.g., vitamin E). Trace amounts of metals, flavanoids, tannins, and carbohydrate (glycolipids) may also be present. The refining process removes as much of these nontriglyceride fractions as possible to improve the quality and stability of the edible oil. Some of these fractions may have value by themselves, such as the fatty acids, lecithins, and vitamin E. Both chemical and physical refining methods are used.

There are several drawbacks to today's technology (Raman et al. 1994b):

- High energy usage: After extracting the oil with a solvent (usually hexane), the oil–solvent miscella is then evaporated to separate the oil and hexane. This requires a considerable amount of energy (about 530 kJ per kg of oil). In addition, the explosive vapors in the vegetable oil plant also create a safety problem.
- Oil losses: In the refining step, the addition of caustic leads to saponification of triglycerides. The resulting soap can trap some oil, amounting to 50% of the FFA content, with total refining loss usually equal to the amount of FFA in the oil.
- Large amount of water and chemicals to be used
- Heavily contaminated effluents

Although the earliest publication in this field is probably the U.S. patent by Sen Gupta (1978) for a degumming process, membrane technology has not had a significant impact as of this writing. Conceptually, membranes can be used in almost all stages of oil production and purification as shown in Figure 8.43 (right). Many have been evaluated at the laboratory or pilot plant scale and,

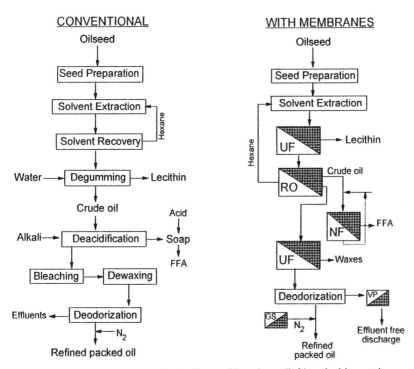

Figure 8.43. *Processing vegetable oils the traditional way (left) and with membranes (right). VP is vapor permeation and GS is gas separation.*

except for dewaxing, degumming, and gas separation for the production of nitrogen, there are few commercial membrane installations in the vegetable oil industry. The following examples focus only on UF and MF applications: other membrane technologies have been reviewed by Koseoglu and Engelgau (1990), Raman et al. (1994b), and Snape and Nakajima (1996).

8.K.1.
DEGUMMING

This step usually occurs after the oil is separated from the miscella. Gums are phospholipids that can be removed by adding water to the crude vegetable oil, then hydrating and settling or centrifuging them out. Sometimes (e.g., with soybean oil), dilute phosphoric acid is added instead of water to hydrolyze the nonhydratable phosphatides along with other gums. In contrast, membrane degumming can be done with the miscella itself. Being amphiphilic, the phosphatides form reverse micelles in the miscella with molecular weights exceeding 20,000 daltons and molecular sizes of 20–200 nm. Thus, UF could be used to separate them from the oil–hexane mixture. The permeate that flows through the membrane consists of hexane, triglycerides, free fatty acids, and other small molecules, while almost all the phospholipids are retained by the membrane. The membrane for this application should be hexane resistant, e.g., polyamide (PA), PS, PVDF, polyimide (PI), polyacrylonitrile (PAN), or inorganic.

Iwama (1989) has done an exhaustive study of this application using tubular Nitto-Denko membranes (NTU-4200) made of polyimide with an MWCO of 20,000 daltons. Their data showed very high rejections of phospholipids and over 97% recovery of the oil up to VCR 40. The remaining oil could be recovered by DF with hexane, resulting in essentially oil-free lecithins, which have a marked added value compared to conventional lecithins. With 25% miscella at 40°C, flux increased linearly with pressure up to 4 bar (60 psi) and became asymptotic above this pressure at 95 LMH. Flux increased with temperature, but since the boiling point of hexane is less than 69°C, it is best to operate the system at 40–50°C to minimize the danger of explosion while maximizing the flux. High cross-flow velocities may be required to maximize flux. A membrane plant for degumming 250 tons per day of soybean oil would require five stages with a total of 1610 m^2 of membrane area, producing 1029 tons/day of degummed miscella with a phospholipid level of 10–50 mg/L oil and 25 tons/day of concentrated phospholipids. Suzuki et al. (1992) used inorganic UF membranes for degumming.

8.K.2.
DEACIDIFICATION

The deacidification step has a major economic impact on vegetable oil production. Chemical (i.e., alkali) refining is the most common method of FFAs removal, but it has several drawbacks: (1) there is loss of oil due to hydrolysis by the alkali and by occlusion in soapstock, (2) soapstock itself has low commercial value; however, since FFAs in their native state find many uses, the soapstock is usually split with concentrated H_2SO_4, but this results in heavily polluting streams; (3) the water used to wash the oil after alkali treatment needs to be treated to meet effluent disposal regulations.

In theory, membranes should overcome many of these problems. The ideal process would use a hydrophobic membrane with pores so precise that they could effectively separate the FFAs from the triglycerides (Raman et al. 1994b). However, fatty acids have about one-third the molecular weight of triglycerides, and the difference in their molecular size is too small if membranes alone are to be used in their separation. The separation is more likely in a solvent such as hexane; in fact, a partial separation of the fatty acids with the appropriate NF membrane has been observed (Raman et al. 1996a). Another method is to modify the properties of FFAs, e.g., chemically associating them into large micelles; it is then theoretically possible to separate the FFAs using UF membranes (Sen Gupta, 1985). For oils having low FFAs and high phospholipid concentrations (e.g., soybean and rapeseed oils), the miscella is directly neutralized with ammonia and then ultrafiltered. But for oils containing low amounts of phosphatides (e.g., fish oil), lecithin is added to the miscella, the FFAs are neutralized by ammonia and then the oil–hexane solution is passed over the membrane to effect the deacidification. Less than 6% of the FFAs and almost none of the phospholipids permeated through the Amicon PM-10 membrane with rapeseed and soybean oils; however, due to the high FFA content of fish oil, comparatively larger amounts of fatty acids are found in the permeate. Nevertheless, the membrane rejected >90% of the FFAs.

A major benefit of this process is the possibility of obtaining FFAs from the soapstock in a simple manner. Heating the retentate will vaporize the ammonia, leaving the FFAs and lecithin behind. The ideal membrane for this application should have a 5000 MWCO and be solvent resistant. The main drawback of this process is the necessity of ammonia and associated environmental restrictions on its use. Methanol could instead be used as the solvent to extract the FFAs from the crude oil. This will form two immiscible phases that can be separated by centrifugation. NF is used to separate the FFAs from the methanol that is recycled to the extractor (Raman et al. 1996b).

Figure 8.44 shows a two-membrane system for separating FFAs from crude oil. For convective flow to occur through the pores of an MF or UF membrane, the membrane surface should be wetted by the permeating liquid. If an oil–water emulsion comes in contact with a hydrophilic membrane, water wets the pore wall preferentially and is transported across the membrane. The oil, on the other hand, has to overcome the Laplacian (ingress) pressure barrier in the pore wall to move across the membrane. If the membrane system is operated below the Laplacian pressure of oil, water would preferentially permeate through the

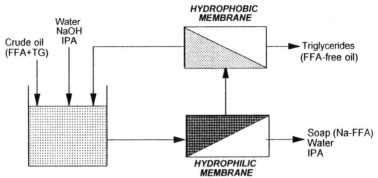

Figure 8.44. Two-membrane system for removal of fatty acids from oil. IPA is iso-propyl alcohol, FFA is free fatty acids, and TG is triglycerides (adapted from Keuren-tjes 1991).

membrane. In contrast, a hydrophobic membrane would allow the oil, instead of water, to pass through preferentially.

In actual operation, FFAs in the oil are neutralized by aqueous NaOH. The soap that is formed is solubilized with isopropyl alcohol (IPA). This results in a two-phase system: the aqueous phase consists of water, IPA, and soap, and the oil phase contains primarily triglycerides, with traces of IPA and soaps. The emulsion is first passed over a hydrophilic membrane operating at pressures below the ingress pressure of the nonwetting phase. Water, IPA, and soap pass through the membrane while almost all the oil is retained. The retentate still contains significant amounts of the aqueous phase components. It is now directed over a hydrophobic membrane, operating below the Laplacian pressure of water. This membrane rejects almost all the aqueous substances while allowing the free passage of oil. This oil is substantially deacidified. The retentate from this stage is recycled to the reactor and the process is continued.

Of the different membranes tried for the hydrophilic first stage, PAN, with an MWCO of 30,000 manufactured by Rhône-Poulenc, was identified as the most appropriate. Any suitable hydrophobic membrane could be used in the second stage. One of the major drawbacks of the two-membrane system is its complexity. The relative amounts of different components in the aqueous phase affect membrane performance. Higher water content severely hampers the soap solubilization process. An increase in the IPA content, on the other hand, aids the solubilization process. The optimum composition of the non-oil phase was found to be 1:6.5:3 of fatty acid:water:IPA. The water phase to oil phase ratio determines the nature of the two-phase system. At levels of up to 20% of aqueous phase in the system, a dispersion of water-in-oil is formed; between 20–65%, both oil and the aqueous solution are present as a continuous phase, and above 65% water, the system behaves as an oil-in-water emulsion. Thus, 20% is approximately the lowest water phase content the hydrophilic membrane can attain. In addition, flux falls sharply below 20% water content. In fact, no flux is obtained at 18% aqueous phase concentration (Keurentjes, 1991).

8.K.3.
BLEACHING

Color compounds in vegetable oil, such as carotenoids and chlorophyll, are removed by absorption on activated clay or carbon. The current bleaching process has several disadvantages, related primarily to the retention of some oil (equivalent to 30–70% of the weight of the activated earth). Due to oxidation, the entrapped oil is of little commercial value, even if it is recovered successfully. Further, the adsorbent has to be periodically disposed of, which results in high costs. Residual impurities from the degumming process tend to overload the activated earth, preventing the adsorption of compounds like chlorophyll. Consequently, it causes the earth to become "sticky," retaining excessive oil, resulting in higher dosage of adsorbents and longer processing times. An economic analysis of membrane bleaching suggests that $730,000 could be saved each year for a 250-ton/day plant (Raman et al. 1994b). If membranes are being used for degumming in the plant, then a separate bleaching step may be unnecessary since it also reduces color compounds. Despite considerable research, however, there appears to be no satisfactory membrane that will selectively remove color compounds from oils, in part because of the similarity in molecular sizes. A combination of techniques will have to be used (Diosady et al. 1992).

8.K.4.
REMOVAL OF METALS

Edible oils contain traces of heavy metals such as copper and manganese at concentrations of 0.1–0.7 ppm. They can originate from the oilseed itself and from storage tanks and pipes. The trace metals in oils are undesirable because they catalyze the oxidation reactions. Most of them are removed by the "acid washing" process, where phosphoric, citric, or tartaric acid is used to chelate and separate the metals. Another source of metals is the hydrogenation process, which converts unsaturated triglycerides to saturated fats at 120–200°C in the presence of a nickel catalyst. Unlike the molecular solutions of trace elements as in the previous case, the catalyst exists as finely dispersed particles of nickel. They are removed by filter presses and reused. The process is labor-intensive since the filters are manually opened to remove the catalyst. In addition, some oil is lost with the filter material each time the filter is opened. Ceramic membranes with pore sizes of $0.01–0.2\,\mu$ have proven effective in recovering the nickel catalyst and reducing nickel levels in the finished oil to less than 1–10 mg/L (Vavra and Koseoglu 1994).

8.K.5.
DEWAXING

Chemically, waxes are esters of long-chain aliphatic alcohols or fatty acids. The main sources of these materials are the seeds and seed coat of the oil-bearing material and are undesirable in the finished oil. Since waxes have higher melting points than triglycerides, they can be removed by cooling the oil, crystallizing, and then filtering. Clogging of the filter media, entrapment of neutral oil in the filter earth, and disposal costs associated with the process are some of the unfavorable factors associated with this method of processing.

Some of these problems can be overcome by MF using pore diameters of 0.05–1 μ. Depending on the type of the oil, the oil temperature is adjusted to −10 to 20°C to crystallize the wax before MF. Most of the waxy substances are retained on the upstream side of the membrane, with a very low amount of wax in the permeate. For example, sunflower oil containing 2600 ppm wax was precooled to 5°C and processed through a polyethylene hollow fiber with 0.12-μ pores. At a pressure of 2 bar and filtration temperature of 10°C, the average flux was 10 LMH. The permeate's wax content was 30 ppm and showed no cloud during a cold test. Membrane fouling by the wax was compensated for by periodic backwashing with high pressure nitrogen (Mutoh et al. 1985). In another case, Microza TP-313 hollow fibers (0.2 μ) was fed through the shell side with oil containing 700 ppm of wax. The permeate that came through the tube side contained less than 20 ppm wax with an average flux of 5-27 LMH at 3 bar (Snape and Nakajima 1996).

Membralox ceramic filters with 0.5-μ pores were also effective in decreasing wax of sunflower oil from 900 ppm to 10 ppm (Muralidhara et al. 1996). The flux averaged 42 LMH at 18°C, and backflushing was done with permeate every 3 min for 5 sec. Membranes could be cleaned using clean warm oil. The rate of cooling the oil affects crystal size, which could have an effect on the efficiency of MF.

Phospholipids and FFAs can also be removed along with the wax (Mutoh et al. 1985; LaMonica 1994). Dilute phosphoric acid is added to hydrate the phospholipids, and caustic soda is added to neutralize the FFAs before filtration. The waxes and other impurities are retained by the membrane. A UF membrane will probably have to be used instead of an MF membrane, e.g., a poly(ether)imide membrane with 20 K MWCO (LaMonica 1994; Tanahashi et al. 1988). The membranes need not be placed in a cold environment like filter presses. In addition, it is also possible to carry out this separation in the miscella (i.e., in the presence of an organic solvent) rather than directly with the oil. This is the preferred method of dewaxing hydrocarbon (petroleum) oils (Gould and Nitsch 1996).

8.K.6.
CLARIFYING FRYING OILS

Large amounts of oil are used as a frying medium at fast-food restaurants and institutional kitchens. During the frying process, particles from the food that is fried enter the oil, catalyzing and/or causing undesirable reactions. The oil itself undergoes several physicochemical changes, such as formation of dimers of oil molecules, lowering its frying quality. Hence, food regulations do not permit the repeated use of frying oils. Most restaurants use some kind of treatment, like adsorption, to extend the life of these oils. Membranes may offer an alternative way for rejuvenating the frying oils. The technology is in a developmental stage, and preliminary results are very encouraging. The treatment removes polar compounds and dimers of triglycerides in significant amounts, thereby improving frying quality (Koseoglu, 1991).

8.L.
CORN AND OTHER GRAINS

There are two types of corn processors: those using "dry" milling and those using "wet" milling technology. Wet milling accounts for more than 75% of the corn processed into food, feed, and industrial products in the United States. This includes 6.5 million tons of corn sweeteners and 1.3 million tons of corn starch annually. Corn is also the feedstock for more than 5.3 million m^3 (1.5 $\times$ 10^9 gal) of ethanol per year. As shown in Figure 8.45, corn wet milling today is a fairly sophisticated and energy- and capital-intensive process. Raw corn contains about 3.8% oil, 10% protein, 16% moisture, 61% starch, and the rest fiber and minerals. The first step is the "steeping" of corn, where the raw corn is soaked in water containing sulfur dioxide at 45–55°C for about 24–40 h. This helps to loosen the starch–protein matrix for easier separation during the subsequent unit operations. The amount of water used is fairly large, about 1.0–2.2 L/kg of corn. This water comes mainly from the downstream starch washing operation.

After steeping, the steep water is evaporated and sold as corn steep liquor for use as a fermentation medium or mixed with corn fiber and germ meal, dried, and sold as corn gluten feed (21% protein, dry basis). The steeped corn is lightly milled to release the oil-containing germ. The oil is extracted from the germ by expelling and/or solvent extraction. The meal from the germ is used to produce corn gluten feed as described above.

The de-germed corn is finely ground in attrition mills and passed through screens to recover the fiber (for corn gluten feed, as described earlier). The remaining protein–starch fraction is processed through a series of centrifuges and cyclones, with the purpose being to separate the starch from the protein. Large quantities of fresh clean water are introduced at this point to wash off the impurities in the starch. The protein-rich fraction is sent to a belt filter and

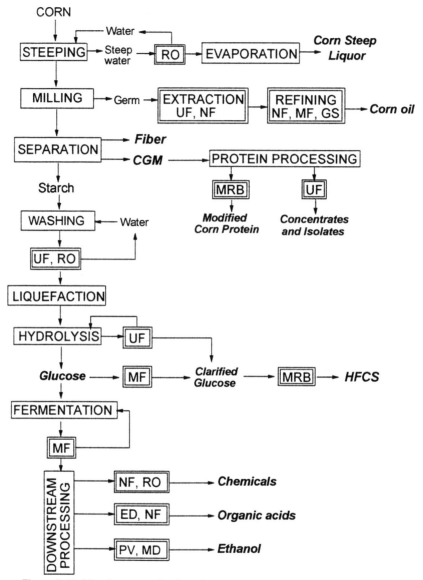

Figure 8.45. *Membrane applications in corn wet milling (Cheryan 1992, 1994).*

drier for the production of corn gluten meal (CGM) containing 60% protein (dry basis).

The starch fraction can be sold as such or is modified using chemical or biological to produce a variety of high-value products. It can be hydrolyzed by enzymes to glucose (dextrose). The dextrose stream, usually at 30–34%

solids, is clarified, usually by pressure/vacuum filtration with a filter aid before evaporation and crystallization (Singh and Cheryan 1997b). For high-fructose syrups, the clarified glucose is further purified by ion exchange and reacted with another enzyme (glucose isomerase). It can also be fermented to a variety of products such as ethanol, organic acids, lysine, etc., all of which can benefit from membrane technology in downstream operations for recovery, isolation, and purification.

The numerous applications of membrane technology in corn refining are shown in Figure 8.45 in the double boxes. Only MF and UF applications are discussed here. A complete review of potential applications is available elsewhere (Cheryan 1992, 1994; Singh and Cheryan 1997a).

8.L.1.
DEXTROSE CLARIFICATION

Dextrose clarification is perhaps the most developed and widely used application in the industry. Its primary benefit is reducing or eliminating filter aid. Glucose syrups are produced from starch using acid or enzymes or both as catalysts. Also known as "dextrose," it is produced using a thermostable bacterial α-amylase for liquefying (thinning) starch to 10–20 dextrose equivalent (DE), followed by saccharification with glucoamylase to 95 DE, i.e., containing 93–96% dextrose on a dry basis, with disaccharides and higher sugars making up the rest. Most starch hydrolysates such as this dextrose syrup are clarified to remove fat and protein impurities, which may form 0.3–1% of the dry weight of the starch. The clarified hydrolysate is treated with powdered or granular carbon to remove off-flavor and color compounds. This may be followed by ion exchange to remove the last traces of flavor and color bodies, as well as ash (soluble salts) and other inorganic impurities. The refined and purified syrup is evaporated to 50–55% solids and processed further as needed (Singh and Cheryan 1997b).

The traditional method of clarifying dextrose syrups or other starch hydrolysates is rotary vacuum precoat filters (RVPF) using diatomite (diatomaceous earth) as filter aid. In comparison to a membrane process, RVPF is more complicated (Figure 8.46). The major problems with RVPF include

1. Cost: RVPF process requires 4–5 kg of filter aid per ton of dry dextrose.
2. Disposal of the rejected filter cake and spent filter aid: In the past, diatomaceous earth was spread on land or added to animal feed. These options are no longer viable in many places, which makes membrane filtration even more attractive for new plants.
3. Plant wear and tear: Since the diatomaceous earth media is abrasive, plants annually replace pipelines, valves, and pumps through which the slurry is pumped. Furthermore, upsets or breakthroughs means that the filter drum has to be recoated with diatomaceous earth media, resulting in extra labor and diatomaceous earth consumption.

RVPF system

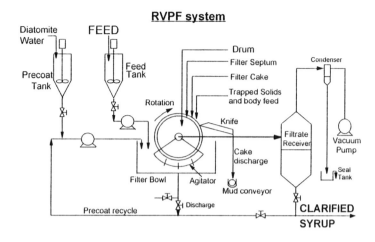

Microfiltration

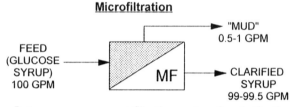

Figure 8.46. Top: Rotary vacuum precoat filtration system for clarification of dextrose syrups. Bottom: Microfiltration method of clarification. The flow rates are given as gallons per minute (gpm) and are dependent on the properties of the syrup and the membrane–module combination (adapted from Singh and Cheryan 1997b).

Microfiltration eliminates filter aid, thus simultaneously eliminating all problem of purchasing, handling, and disposing spent diatomaceous earth. MF reduces overall wear and tear of the plant and decreases downtimes in the plant, which also significantly reduces operating cost. MF requires little operator attention and tremendously reduces the labor costs. The quality of the membrane-clarified syrup is superior in terms of turbidity (much less than 1 NTU with no suspended solids), color, and microbiological cleanliness compared to conventional centrifugation or filtration using filter aid. The positive membrane barrier results in a consistent clarification of the syrup day in and day out. In contrast, RVPF syrup has variable clarity because of possible rupture of the filter cake, which will allow suspended solids to go through with the clarified syrup.

Centrifugation can also be used for mud removal, but the clarity of the syrup is not as good, maintenance costs are high, and it requires careful control of pH and discharge pressures to minimize syrup losses.

A typical dextrose stream from a corn starch saccharification reactor is 95 DE, 30–35% TS, containing about 0.2% suspended matter (termed mud in

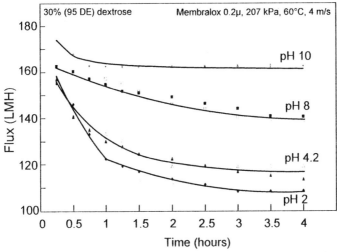

Figure 8.47. *Effect of pH on MF of dextrose syrup from a corn refinery. The syrup contains 30% dextrose and "mud" components. The membrane is Membralox 0.2 μ (adapted from Singh 1997).*

the industry). The MF retentate contains protein and fat and can be used for animal feed. The usual fouling patterns are observed with 95 DE syrup (Figures 8.47) and 43 DE syrup (Figure 8.48). Despite the low level of rejected material, the feed shows significant polarization effects with Membralox MF membranes. With 43 DE syrup containing 20% total solids, flux becomes pressure-independent above 3 bar at 80°C, attaining a maximum value of 140 LMH (Amar-Rekik et al. 1994). With 95 DE corn syrup of 30% total solids, maximum flux was 175 LMH at 2 bar, 5 m/sec and 60°C (Singh 1997). Too high a pressure is detrimental [Figure 8.48(B)]. Temperature had a significant effect on flux [Figure 8.48(C)]. With both syrups, the flux–velocity exponent was close to 1, as expected for tubular modules operating in turbulent flow. Pore sizes larger than 0.1 μ generally fared worse in terms of flux [Figure 8.48(A)], and periodic backwashing for 2–5 sec every 2 min had a significant beneficial effect [Figure 8.48(D)]. The pH of the syrup has a profound effect on flux (Figure 8.47) but can also affect the color of the final product (Singh and Cheryan 1997b).

Early studies indicated annual savings of $328,000 using cross-flow MF instead of RVPF for a plant processing 300 gpm of 95 DE syrup (Short and Skelton 1991). However, the total impact of MF is much more, if the economics of the unit operations that follow clarification are considered (ion exchange, decolorization, and isomerase reactions). Microfiltered syrup can cut carbon requirements in downstream processing by as much as 60–70% and ion exchange by 20–30%. These reductions, along with the elimination of diatomaceous earth

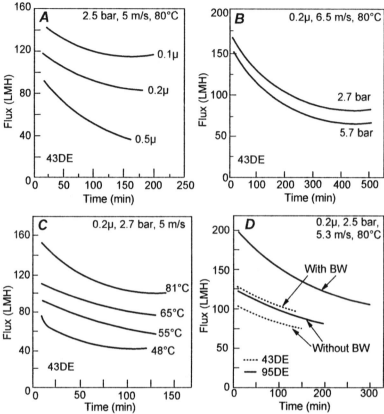

Figure 8.48. *Microfiltration of 43 DE dextrose syrup from wheat starch hydrolysis with Membralox ceramic membranes: (A) effect of membrane pore size; (B) effect of transmembrane pressure; (C) effect of temperature; (D) effect of periodic backwashing (BW) on 43DE syrup (full line), and 95 DE syrup (broken line) (adapted from Amar-Rekik et al. 1994).*

filter aid, result in annual savings of $2–3 million per year for a 500-gpm line (Graver Separations 1996).

8.L.2.
PROTEIN PROCESSING

Corn proteins have little commercial value apart from animal feed. This is partly because of their poor functional properties (as compared to dairy and soy proteins). They are composed of zein, which is a water-insoluble, alcohol-soluble storage protein (40% of the protein in corn); glutelin, which is soluble in alkali (40% of the kernel protein); and albumins and globulins (20%, found

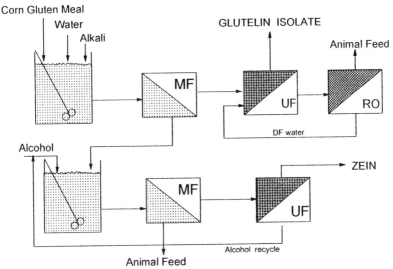

Figure 8.49. *Separation of corn proteins by solvent extraction and membrane technology (Cheryan 1992, 1994).*

mostly in the germ). A small quantity (500 tons/year) of zein is produced in the United States for specialized applications. Corn protein concentrates and isolates could be produced from raw ground corn or from corn gluten meal, a co-product of the wet milling industry which contains 60% protein and 40% starch. The general concept is similar to vegetable protein processing (Section 8.J), except that this process takes advantage of the different solubilities of the glutelin and zein in various solvents. As shown in Figure 8.49, extracting corn gluten meal with an alkaline solution will solubilize the glutelins, which can be separated from the rest of the meal components by centrifugation or microfiltration. Subsequently, the zein can be extracted with aqueous ethanol or propanol solutions. The suspension is microfiltered to allow permeation of the zein–alcohol solution after which the zein is recovered by UF and the alcohol recycled for extraction. The sequence of operations may be reversed if necessary, i.e., the zein–alcohol treatment followed by alkaline solubilization of the glutelin.

Methods have also been developed to enzymatically modify the zein in the alcohol and use another UF membrane to recycle the enzyme and separate the zein hydrolysate. The resulting products have substantially better functional properties—such as water solubility and water binding—than the native corn proteins (Mannheim and Cheryan 1993). There have also been attempts to extract the protein from stillage of dry milling ethanol plants by UF (Wu et al. 1985).

8.M.
ANIMAL PRODUCTS

Animals provide some of the most desirable, high-quality proteins known to man. This section covers membrane applications in red meat, eggs, gelatin, fish, and poultry processing.

8.M.1.
RED MEAT

Abbatoir by-products such as blood, offals, and gut tissues, which are all proteinaceous materials, are generally wasted. Typical wastewater loadings for a meat packing operation are 12 kg solids, 14.6 kg BOD, and 1.7 kg of nitrogen per 1000 kg of live-weight kill (Hansen 1983). Based on nitrogen in the wastewater, about 10% of the protein of live-weight kill is lost to the sewer. Even if half of this wasted protein were recovered, more than 200 million kg of additional protein would be available per year in the United States alone, representing a value of about $200 million per year.

Blood is probably the abattoir waste product of the greatest volume. Blood is typically about 70% plasma/serum and about 30% red blood cells (RBCs)/ heme fraction. The plasma contains 92% water and 7% protein while RBCs are 70% water and 28% protein. Several studies on RO and UF of these fractions have been done by Porter and Michaels (1971) and Fernando (1981) on whole blood; by Delaney (1975), Delaney et al. (1975, 1979), Ericksson and von Bockelmann (1975), and Goldberg and Chevrier (1979) on the plasma fraction; and by Delaney (1977) on the RBC fraction.

Figure 8.50 shows the effect of some operating parameters on the flux of blood plasma and red blood cells. The system is polarization-controlled, and thus a low-pressure, high velocity operation is probably best. Flux is linearly related to temperature, with slopes of 0.2–0.5 LMH per °C (Delaney 1977; Ericksson and von Bockelmann 1975). The optimum temperature is around 43°C, since blood coagulates above this temperature (Fernando 1981). In general, the flux of RBCs is higher than that of the plasma/serum fraction because of the colloidal nature of the former, which could enhance mass transfer by the "tubular pinch" effect (Chapter 4). Continuous DF resulted in much faster rates than UF, primarily because of a sharp decrease in viscosity during DF, which results in an increase in flux during DF (Delaney 1977). Membrane fouling is severe with abattoir effluents. Cleaning could be done with a combination of enzymes and detergents (Maartens et al. 1996).

Wafilin had an industrial multistage installation in the Netherlands, processing about 1600 L/h of blood plasma to 3.2× concentration. At 35°C, the flux drops from about 30 LMH at 13% TS to 5 LMH at 25% TS. The operating costs were estimated to be Dfl. (Dutch guilders) 25/m^3 (Wafilin 1983). Fernando (1981) estimated the cost of concentrating homogenized blood using

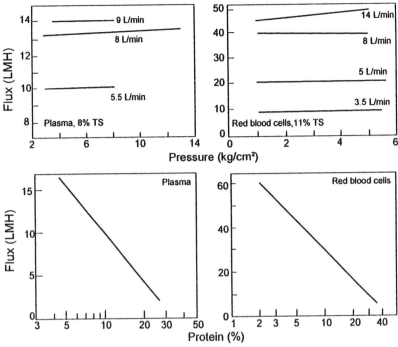

Figure 8.50. *Ultrafiltration of porcine blood fractions: left: plasma; right: red blood cells; top: effect of pressure and cross-flow rate on flux; bottom: effect of protein concentration on flux. The UF unit was a DDS laboratory module with 20 cm type 600 membranes (adapted from Delaney et al. 1975; Delaney 1977).*

a Dorr-Oliver Iopor ultrafiltration unit as $NZ 5.9/m^3$ permeate, whereas concentration by thermal (vacuum) evaporation costs $NZ 8.8/m^3$.

Wastewater from meat processing plants, just like any food industry effluent, has a high amount of biologically degradable matter. Proteins and fat that come from the carcass debris and blood are the major pollutants. With Kalle tubular modules, average flux was 50 LMH with PA-100 membrane and 30 LMH with PA-40 membrane. A reduction in COD of 72–75% was obtained (Table 8.9).

8.M.2.
GELATIN

Gelatin is another animal product that has benefited greatly by UF. Gelatin is not exactly a "waste" product, although it is obtained from animal by-products that are not ordinarily consumed. Gelatin is widely used as a glue, in pharmaceutical preparations, photographic products, and in its edible form, as a popular dessert. It is extracted in aqueous acid or alkali at high temperatures from the

Table 8.9. Ultrafiltration of animal processing industry waste streams using Kalle membranes.

Membrane	Component	Feed (mg/L)	Retentate (mg/L)	Permeate (mg/L)	Percent Reduction
PA-100	COD	15,156	76,000	4,202	72
	Fat	2,800	11,900	33	99
	Protein	6,440	16,580	3,200	50
PA-40	COD	14,514	76,000	3,677	75
	Fat	3,700	14,600	84	98
	Protein	3,100	15,500	2,000	36

Source: Data from Beer (1979)

skins, hides, and bones of animals. The extract is typically dilute, about 2–5% protein, and contains high quantities of ash as a result of the acid or caustic. This gelatin extract is de-ashed (usually by ion exchange) and concentrated (by evaporation) to yield a product of about 90% protein, less than 0.3% ash, and no more than 10% moisture.

UF has a number of specific applications in the gelatin industry: (1) preconcentration of the dilute extract solution prior to evaporation; (2) simultaneous reduction in ash components, thus upgrading the product; and (3) reduction in lower molecular weight components to improve the gelling properties of the product. In addition, overall yields can be increased by increasing the number of extraction stages. The extracts from the additional stages will have a lower solids content, which would normally be uneconomical to process. Figure 8.51 shows a typical process for gelatin. MF could be used ahead of the UF to clarify gelatin and thus avoid the use of filter aids and filter presses. An economic analysis of the process by Koch, based on a 20,000 kg/h plant feed capacity, indicates a potential savings of over $329,000 per year if UF was used in place of a triple-effect evaporator (Table 8.10). Similar benefits have been observed by Chakraborty and Singh (1990).

UF flux is strongly dependent on protein concentration (Figure 8.52). The C_G values of gelatin are 20–30% protein, which limits practical UF to about 18–20% protein in the retentate. This usually implies a 65–80% reduction in volume. Flux with De Danske Sukkerfabrikker (DDS) plate units averaged 4–18 LMH for the GR8P membrane and 10–60 LMH for the GR6P UF membrane over the range of 3–20% gelatin concentration (Akred et al. 1980). Flux with Koch spiral-wound modules is 50–80 LMH at 2–3% gelatin concentration, TMP of 50 psi, a pressure drop of 20 psi per module, and a temperature of 50°C. It drops to about 3–4 LMH at 18% gelatin concentration. Ceramic MF modules have also been evaluated for this application (Freund and Rios 1992).

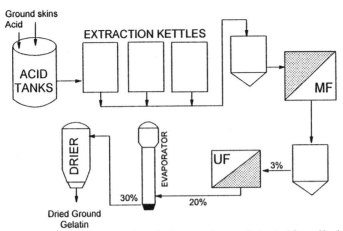

Figure 8.51. *Membrane process in gelatin manufacture (adapted from Koch 1984).*

Donnan effects play an important role in the UF of gelatin. Negative rejections of calcium occurred below pH 4, and sometimes calcium rejections of up to −380% were observed (Figure 8.53). Most negative rejections were obtained at low pH and high concentrations of gelatin. The Donnan-enhanced transport was increasingly counterbalanced as the membrane became more gel-polarized (e.g., at higher TMP and/or higher gelatin concentrations). Calcium was rejected at pH values above 4, possibly because of to its binding to the protein, which has a net negative charge above its isoelectric point of 4.5. Faster rates of desalting were observed by applying dialysis UF instead of conventional UF (Section 7.B.3.).

Table 8.10. Operating costs (US$) for concentrating 20,000 kg/h of gelatin from 2% to 18% for 6600 hours per year.

	Evaporation	Ultrafiltration
Steam (triple-effect evaporator, $13.2/1000 kg)	516,120	—
Electric power ($0.05/kWh)	—	41,250
Membrane replacement	—	117,000
Cleaning cost	—	9,000
Maintenance	—	19,500
Total	516,120	186,750
Net savings by using UF = $329,370		

Source: Data from Koch (1984)

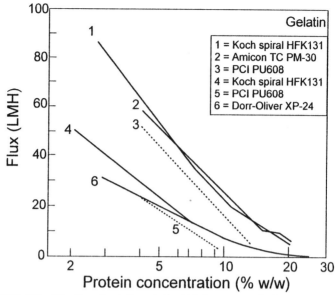

Figure 8.52. *Ultrafiltration of gelatin: relationship between gelatin concentration and flux for several UF membranes. Data sets 1 and 3 were obtained with low strength gelatin, 4 and 5 with high strength gelatin. All data obtained at 49–53°C except 3, which was at 70°C (adapted from several sources listed in Dutre and Tragardh 1995).*

The first UF unit went into Atlantic Gelatin (a division of Kraft General Foods) in 1984. It was a seven-stage, spiral-wound system with HFK-131 membranes (5000 MWCO) processing an average of 140 gpm of gelatin solution containing 2.5% protein, 0.2% nonprotein nitrogen, and 0.8% ash. A second UF unit was added 3 years later, handling 81 gpm. Each removes 90% of the water and rejects 98% of the gelatin. Inlet pressures are 100–130 psi, and temperatures are 50–55°C. The UF retentate is 18–20% gelatin, 1% nonprotein nitrogen, and 0.8% ash. The UF permeate is typically 0.01% protein, 0.07% nonprotein nitrogen, and 0.8% ash. It is used upstream to wash animal hides prior to cooking. Cleaning is done with caustic every 2 days (EPRI 1992).

Compared to the six evaporators that the UF system replaced, maintenance and labor were less, and there is better control of the concentration process. Because of short residence times (20 min) in the UF system, the quality of the product is superior. The first UF system at Atlantic Gelatin (32,000 kg/h) cost $1.2 million, and the second unit (18,000 kg/h) cost $0.7 million. Power consumptions were 165 kW and 100 kW, respectively, giving a total annual power bill of $145,000. Membranes are replaced every 3 years at a cost of $90,000 per set. Payback was 1.7 years for the first unit and 2.7 years for the second unit. The company has installed additional units to recover grease from

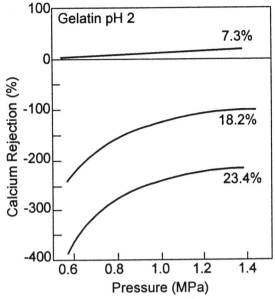

Figure 8.53. *Effect of gelatin concentration and transmembrane pressure on calcium rejection. The membrane was DDS type GR6P, and the feed solution was at pH2 (adapted from Akred et al. 1980).*

the animal hides and could eliminate all evaporators. RO could polish the UF permeate and reclaim additional water (EPRI 1992).

8.M.3.
EGG PROCESSING

A large number of food products contain eggs as an ingredient, either in the form of fresh, frozen, or dried egg solids. Egg white, in particular, is in great demand in the baking, confectionery, and cake mix industries for its whipping and foaming properties. About 200 million kg of egg white are produced each year, of which about 65–70% is in the form of dried egg white solids. Liquid egg white is about 12.0% total solids (10.7% protein, 0.5% salts, and 0.5% carbohydrate in the form of free glucose). Whole eggs contain an additional 11% fat and slightly higher quantities of the other compounds, for a TS of 25%. The protein is extremely sensitive to shear and heat damage. Eggs become spoiled by microorganisms very easily. The presence of the glucose is especially undesirable since it causes a detrimental effect on storage stability and quality of the product, by forming brown pigments during processing. Traditional methods for removing glucose include fermentation and enzyme hydrolysis. Egg products are rarely concentrated before spray drying because of the undesirable effects of heat and shear during conventional thermal evaporation.

RO was first suggested for concentrating egg white by Morgan et al. (1965) and Lowe et al. (1969). Ultrafiltration would be more useful since the glucose could be removed and the egg white concentrated simultaneously. Several studies on the UF of egg white have been conducted on both laboratory- and industrial-scale equipment (Madsen 1973b; Payne et al. 1973; Plotka et al. 1977; Tsai et al. 1977; Wafilin 1983). In general, it was observed that

- Although a fairly strong temperature dependence of UF was observed, with an average activation energy of 5 kcal/mole, microbial problems and potential heat and shear damage control the operating temperatures.
- The system is strongly velocity-dependent, with a flux-velocity slope close to 1.0 in tubular units (Wafilin 1983). This is not unexpected for feed material of such a high protein content.
- UF fluxes are fairly low, initially averaging about 10–15 LMH at 50°C for whole egg and egg white in the Wafilin tubular system (at 300 kPa TMP), in the Abcor tubular module with HFA-180 membrane, and in DDS plate units (Payne et al. 1973; Plotka et al. 1977; Tsai et al. 1977). The flux drops to about 4 LMH at 25% solids with egg white and to about 3 LMH at 37% total solids with whole egg. Typical compositions of the product stream are shown in Table 8.11 and plant data in Table 8.12.

In contrast to spray drying, which costs about 0.3 Dfl. per kilogram egg powder, the cost of UF varies from Dfl. 0.071–0.210 per kilogram egg powder, depending on the final solids content of the retentate (Wafilin 1983). Tsai

Table 8.11. Typical compositions (%) of feed and product streams in egg processing.

Component	Feed	Retentate	Permeate
Whole egg*			
Total solids	24.8	40.7	14.7
Ash	0.6	0.5	0.8
Protein	22.4	38.8	2.4
Egg white**			
Total solids	11.7	23.2	
Ash	0.7	0.5	
Protein	10.4	22.5	
Glucose	0.50	0.42	
Calcium	0.008	0.015	
Potassium	0.160	0.173	
Sodium	0.128	0.130	
Phosphorus	0.011	0.020	

*Data from Wafilin (1983), with noncellulosic UF membrane at 55°C
**Data from Froning et al. (1987)

Table 8.12. Plant specifications for egg processing.

Plant Configuration	Stage 1: Standard Flow Spacers, 60 m^2 Stage 2: High Flow Spacers, 28 m^2 Positive Displacement Pumps		3 Stages, 60 m^2 Each with Standard Flow Spacers. Centrifugal Pumps
Egg product	Whole eggs	Egg white	Egg white
Feed capacity (L/h)	1700	2200	2950
Concentrate TS (%)	38.3	22.9	23.0
Concentration factor	1.6	2.1	2.1
Cleaning	Caustic pH 10.5, chlorine 250 ppm, Ultraclean II (0.5%)		

Source: Data from Koch Membrane Systems

et al. (1977) have concluded that a savings as much as $0.09 per kilogram of concentrate can be realized by a twofold concentration of egg white by UF. In addition, the ultrafiltered product has superior functional properties because of the removal of free glucose and salts, which should enable it to command a premium price compared to conventionally produced egg solids. A combination of UF and affinity chromatography can be used to isolate and purify egg white lysozyme (Chiang et al. 1993). Several UF plants for processing whole egg up to 40% solids and egg white up to 24% solids have been installed around the world.

8.M.4.
FISH PROCESSING

Most applications of membranes have been in waste treatment applications or for recovering enzymes used for hydrolysis of fish proteins (Almas 1985). Several studies with hollow fibers of 50,000 MWCO have been done on effluents from processing blue crab, scallops, and mince fish (Chao et al. 1984). Reductions of 50–70% in BOD and COD were obtained with fluxes of 43 LMH at VCR 1 to 23 LMH at VCR 10. Mameri et al. (1996) studied UF modules (Membralox and PCI) and obtained 70–80% rejection of proteins. Optimum conditions were 220–380 kPa and velocities of 6 and 0.47 m/sec for Ceraver and PCI membranes, respectively. The BOD was reduced by 80%.

In Scandanivia, stockfish is treated with lye to produce Lutfish. About half the protein in the fish is in the effluent, which contains 0.5–1.5% TS. Protein recovery by isoelectric precipitation increased by 20% when the effluent was preconcentrated by UF (Ericksson et al. 1977). Hollow fiber UF with a 10,000 MWCO membrane at 20°C resulted in a flux of 15 LMH at 1.2 bar for a fivefold concentration.

Surimi production wastewaters have received some attention (Miyata 1984; Ninomiya et al. 1985). After removal of suspended matter and fat by

Table 8.13. Ultrafiltration of fish processing refrigeration brines. Feed was 9% salt brine, pH 6.2, 1173 ppm nitrogen, and 8720 ppm COD. Data obtained with Amicon PM-30 membrane.

Temperature (°C)	Flux (LMH)	Rejection (%)		
		Nitrogen	NaCl	COD
4	22	52	—	69
27	87	47	30	59
38	105	9	40	60
49	111	47	—	57

Source: Data from Welsh and Zall (1984)

centrifugation, UF could recover up to 90% of the protein with MW of 4000–15,000 with a 10-fold concentration. Brine solutions used to refrigerate or thaw frozen fish have been recycled after UF with PS membranes of 20,000 MWCO (Paulson et al. 1984). Payback periods of 6–9 months are reported, based on the savings in salt and sugar alone (Rajagopalan 1996). Table 8.13 shows typical flux data obtained in this application.

Watanabe et al. (1984) used "self-rejecting" membranes to recovery essentially 100% of proteins of >10,000 MW from jelly fish processing. The membrane was formed on a ceramic support, which had a pore size of 0.05 μ. Best conditions were at 15°C, 1.4 m/sec and 500 kPa.

8.M.5.
POULTRY INDUSTRY

Water is used in poultry operations in two main areas: scalding and chilling. Total water use can be as high as 18–30 m^3 per ton of finished product in the slaughtering operation and 15–100 m^3/ton for processing (Zhang et al. 1997). The main contaminants are blood, FOG, and microorganisms. Suspended solids are 600–1000 ppm. Shih and Koznick (1980) reported on the use of an Abcor tubular membrane with 50,000 MWCO for processing poultry wastewaters. They obtained an average flux of 17 LMH at a pressure of 2.8 atm. Flux obtained with a ceramic MF membrane with different chicken process waters is shown in Figure 8.54. The UF permeate is lower in solids by 85% and lower in COD by 95% (Table 8.14) and completely free of microorgansims, color, and suspended solids (Zhang et al. 1997). However, total organic carbon (TOC) could be unacceptably high (260 ppm was observed even with the tightest 4.5 K UF membrane), thus necessitating the use of NF or RO to meet today's discharge limits. The dried concentrates contained 30–35% protein and 25% fat. In most cases the permeate could be discharged to the sewer with no additional treatment or surcharges. The retentate could be sold as "brown grease" or incorporated into animal feed, fertilizer, or as a soil conditioner.

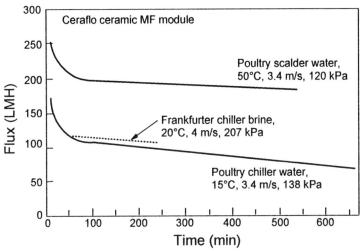

Figure 8.54. Flux of poultry processing wastewaters (adapted from Hart et al. 1990).

8.N.
BIOTECHNOLOGY APPLICATIONS

Biotechnology is the glamorous term used to describe a wide range of biological processes that can be used for the production of a variety of products, from foods, flavors, and pharmaceuticals to fuels and agricultural chemicals. A biocatalyst is usually involved, such as an enzyme, microorganism, plant cell, or animal cell. Sometimes the biocatalysts themselves are the end products. Biotechnology also refers to the genetic manipulation of plants, animals, and microorganisms such as yeasts, bacteria, and fungi. The euphoria associated with the widespread rush into biotechnology by the corporate world has abated somewhat with the realization that successful commercial exploitation of biotechnology will be limited, not by the microbiology or genetic engineering per se, but by the downstream processing (the separation, isolation, purification,

Table 8.14. Ultrafiltration of poultry wastewater.

Component	Feed Wastewater (mg/L)	Concentrate (mg/L)	Permeate (mg/L)	Reduction (%)
Total solids	1276	5386	240	85
COD	1968	9116	131	95
Ash	104	276	48	63
Total nitrogen	82	372	14	86
Protein	492	2013	37	94

Source: Data from Shih and Koznick (1980)

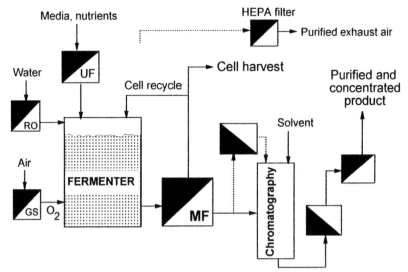

Figure 8.55. *Examples of the use of membrane technology in bioprocessing. Some, like gas separations and water purification, may be used upstream of the bioreactor or fermenter, while others, such as MF, UF, and NF, are used downstream for harvesting, purification, and concentration.*

and recovery of the product). Fermentation broths tend to be very dilute and contain complex mixtures of compounds. In addition, they are highly impure and sensitive to almost any extremes of operating parameters normally used in conventional separation processes.

As shown in Figure 8.55, membrane technology is well suited to take advantage of the rigorous demands of biotechnology. The major applications of membranes are

1. Separation and harvesting of enzymes and microorganisms
2. High-performance bioreactors for enzymatic and microbial conversion processes
3. Tissue culture reactor systems
4. Production of high-purity water (Section 8.C.)
5. Production of enriched nitrogen or oxygen (gas separation membranes)
6. Filtration of exhaust air

Depending on the product, downstream processing will involve clarification, isolation, and/or purification operations. Figure 8.56 shows the selection criteria for deciding which membrane technique is appropriate. If the objective is clarification of a fermentation broth, then the physical properties of the broth determine the type of module to be used, as discussed in

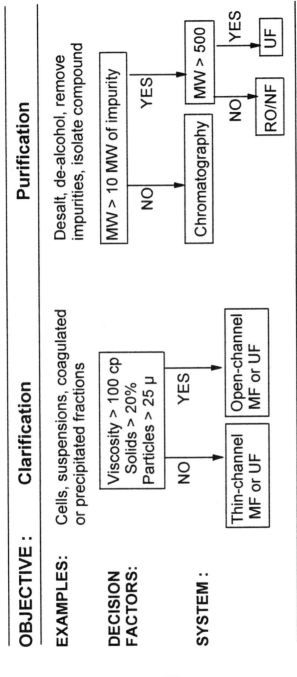

Figure 8.56. Selection criteria of separation methods in bioprocessing (adapted from Raaska and Kelly 1987).

Table 8.15. Concentration of components of typical fermentation broths.

Component	Concentration (% w/w)
Bacteria or yeast (e.g., ethanol, lactic acid broths)	1–5
Fungi (e.g., citric acid, penicillin production)	1–3
Animal cells (e.g., mammalian tissue culture)	0.1–5
Acetic acid in fermentation broth	0.2–5
Lactic acid in fermentation broth	8–10
Citric acid in fementation broth	5–10
Extracellular enzymes	0.5–1
Vitamins	0.005–0.1
Antibiotics	1–5
Ethanol	7–10

Section 8.N.1. Since membranes will not be able to separate components with an MW difference of less than five to ten times, isolation and purification of the compound of interest will have to be done with a column technique such as chromatography. Membranes could be used prior to the column to purify or preconcentrate the component and after elution from the column to purify, desalt, and concentrate the component.

8.N.1.
SEPARATION AND HARVESTING OF MICROBIAL CELLS

The first downstream step in a biological process is usually the removal of the suspended cell mass and other particulate/colloidal debris from the suspending medium. Table 8.15 shows the concentrations of solids in some typical biotechnological processes, and Table 8.16 shows some characteristics of fermentation broths. In the past, the preferred methods of cell harvesting were filtration or centrifugation. To be economical on an industrial scale, the method must be a continuous process. Continuous centrifuges, where the cells

Table 8.16. Characteristics of fermentation broths.

Component	Size (μm)	Viscosity of a 2% (w/w) Suspension (cP) at 25°C	Resistance to Mechanical Shear
Cell debris	0.4 × 0.4	1.5	—
Bacterial cells	1 × 3	1.5	Good
Yeast cells	2 × 10	1.5	Good
Mammalian cells	40 × 40	3	Poor
Plant cells	100 × 100	3	Poor
Fungal hyphae	1–10 (matted)	8000	Fair

are removed by the action of high G forces, are expensive to buy, operate, and maintain. The production rate of a centrifuge is a function of many variables, among them the square of the particle diameter, the difference in density between the particle and the suspending medium, the G force, and the viscosity of the suspending medium. The production rate and the cell recovery (yield) are strongly dependent on particle size and inversely proportional to each other. In addition, there may be substantial heat generation requiring complicated cooling systems; there may be production of aerosols, and the capacity of most tubular centrifuges is limited and may require frequent teardowns and cleaning.

Filtration, on the other hand, is not limited in throughput by the particle size, and the separation rate does not go down as smaller microorganisms are harvested. The most common method is rotary drum filters, where the broth is drawn to the drum surface by a vacuum (Figure 8.46). The microbial cells and other insoluble solids accumulate on the filter cloth and are scraped off with a knife. The major problem with this and other dead-end modes of filtration is the rapid drop-off in filtration rate caused by the accumulation of the solids, and the cost of the filter aid (Table 8.17).

Many traditional cell harvesting processes are being replaced by cross-flow membrane filtration. With proper selection of the membrane, the module, and operating parameters, 100% cell recoveries can be obtained, and capacity can be increased by merely adding more membrane area. However, as shown in Table 8.16, some products of biotechnology are not suited to harvesting by cross-flow membrane filtration. Plant and animal cells are extremely sensitive to shear damage and thus would not survive the shear stresses caused by the pumping and flow through the retentate channel. Some fermentation broths (e.g., those containing mycelial fungi) exhibit extremely high viscosities at low concentrations and produce sticky or gelatinous by-products that would require extremely high cross-flow velocities to keep concentration polarization

Table 8.17. Comparison of operating cost ($/m³ filtrate) between ultrafiltration and rotary vacuum filtration.

Cost Item	Precoat Vacuum Filtration	Ultrafiltration
Filter aid	6.34	—
Membrane replacement	—	2.22
Energy	0.08	0.25
Labor	0.32	0.08
Maintenance	0.14	0.04
Cleaning chemicals	—	0.43
Depreciation	0.28	1.20
Total operating cost	7.16	4.20

Source: Adapted from Bemberis and Neely (1986)

Table 8.18. Typical flux obtained in various biotechnology applications (Garretson 1983). All were Amicon hollow fiber membranes with 100,000 MWCO, except as noted.

Application	Example	Flux (LMH)
Cell harvesting	E. coli to 100 g/L	147
	E. coli to 90% v/v	88
	Streptomyces sp.	110
	S. cerevisiae to 100 g/L	176
	Leptospira sp.	88
	Pseudomonas sp. to 10^{12}/mL	74
	Algae	96
Cell washing	S. aureus (protein in filtrate)	162
	B. thuringiensis (40 g/L)	81
Virus concentration	Rubella in 2% serum	110
	Swine influenza	55
Cell debris removal	Interferon	110
Broth purification	Soy hydrolysate*	176
Product purification	Serum desalting (6% protein)**	33
	Interferon concentration*	66
	Peptide concentration[†]	11

*10,000 MWCO membrane
**30,000 MWCO
[†]1000 MWCO

effects low and permeate flux high. Of course, high viscosities also decrease the performance of centrifugal separators.

Tables 8.18 and 8.19 show typical performance data for biotechnology applications. There are several factors that are important when considering membrane technology for cell harvesting:

- Module design: Almost any type of module can be used for clarification of a fermentation broth if the solids concentration in the feed/retentate is low enough and if the permeate is the required product. However, at high solids levels, or if the microbial cells are the desired product, modules that use spacers or screens in the retentate flow path (such as the spiral-wound, pleated-sheet, and some plate designs) are not recommended. The spacer is an impediment to the smooth flow of feed through the channel. It creates dead spots behind the spacer and causes blockage of the channel caused by accumulation, or "hang-up," of solids within the mesh. This not only worsens the already severe "parasitic drag" in the feed channel, but it makes it difficult to recover all of the retained solids and to clean these systems properly. This was shown schematically in Figure 5.40.

Table 8.19. Typical performance data in various biotechnology applications with Millipore membranes.

Application	Initial Concentration of Cells	Membrane	Flux (LMH)
Actinomyces sp.	3 O.D.*	HVLP (0.45 μ)	240
Brevibacterium thiogenitalis	12 g/L	GVLP (0.22 μ)	105
E. coli	10^7/mL	HVLP	731
	20 O.D.	HVLP	155
Haemophilis sp.	1.5 O.D.	HVLP	358
Lactobacillus sp.	6 g/L	PTHK (100 K)	67
Streptomyces sp.	—	HVLP	128
Diafiltration (cell washing)			
Brevibacterium thiogenitalis	128 g/L	GVLP	36
Streptomyces sp.	40 O.D.	HVLP	240
Streptomyces sp.	4.5 O.D.	HVLP	720
UF Applications			
Antibiotic clarification	—	PTGC	63
Enzyme concentration	—	PTGC	315
Glycan concentration	—	PTGC	300
Guar gum concentration	—	PTGC	390
Human γ-interferon	—	PTGC	210

*Optical density
Source: Data from Millipore Corp. (1983)

The module design also affects the velocity in the feed channel, which in turn affects the shear forces and efficiency of removal of the cake or concentration polarization/gel layer. High Reynolds numbers are possible with tubular units, while high shear is possible with thin-channel designs. Patel et al. (1987) observed that yeast cells could be concentrated as high as 250–290 g/L in hollow fiber UF modules with average flux of 40–50 LMH, provided adequate velocities (>1 m/sec) were maintained in the system, corresponding to shear rates of >5000 sec^{-1}. Cell recoveries were also better than 99%. In contrast, when a pleated filter cross-flow module was used, velocities were restricted to less than 0.1 m/sec, corresponding to shear rates of less than 500 sec^{-1}, owing to the spacers in the feed channel and the general design of the module. This resulted in much greater fouling rates and lower cell recoveries. High shear in clear unobstructed channels assumes even greater importance if one considers microbial cells to be colloidal particles, where the "tubular pinch" and shear-induced enhanced back-diffusion effects discussed in Chapter 4 can be used to advantage. These factors tend to favor tubular ceramic and hollow fiber membranes for cell harvesting applications.

- Pore size: Although it appears logical to use microfilters for harvesting microorganisms, the short-term advantage of higher initial flux with the larger pores will be outweighed by the long-term problems associated with higher fouling rates, as discussed earlier in Section 6.D.1. and shown in Figures 6.9–6.11. If the size of the particle (microorganism) to be separated is of the same order of magnitude as the general range of pore sizes being considered, it is possible that some of the smaller microbial cells in a culture population will lodge themselves in the pores (without necessarily going through them), causing a physical blockage of the pores. In this scenario, of course, the larger pores of the membrane will plug up first, probably the very ones that have a significant portion of the water flux going through them. Thus, these membranes will appear to "foul" very rapidly. High pressures in the system will aggravate the problem by causing a compression of the adsorbed "cake" of microorganisms and other by-products of the fermentation (Figure 6.21). In contrast, if the pores are much smaller than the particles to be separated, the particles will not be able to "sit" on the pores but will simply roll off under the shear forces generated by the flow. Experimental evidence of the inverse effect of pore size and long-term flux was shown in Figure 6.10.

- Chemical nature of the membrane: Frequently, it is the media components that are responsible for fouling and not the microbial cells themselves. As indicated in Section 6.D.1., hydrophilic membranes tend to foul less with fermentation broths and biotechnology media. In addition, since most microorganisms have a net negative charge on their cell surfaces, negatively charged membranes would be able to better resist charge-induced fouling.

- Ultrastructure of the membrane: As discussed in Chapter 2, filtration can be done with either "depth" filters, or "screen/surface" filters (Figure 2.1). Surface filters, in turn, can be further classified as either "isotropic/anisotropic" or "microporous/asymmetric," depending on whether a "skin" is present on the membrane surface. These distinctions are very important when selecting membranes for cell harvesting or biotechnology work. Since separation of particles occurs by retention on its surface, as well as by entrapment within its depths, a depth filter has a high particle loading capacity. This makes it suitable if the permeate is the desired fraction and if the retained solids content of the feed is low. But by the same logic, depth filters will foul rapidly and are unsuitable for applications where the retained solids is the desired fraction. Also, depth filters are unsuitable for the cross-flow mode of operation.

Screen or surface filters are a better choice if quantitative microbial cell recoveries are needed. However, although microporous screen filters are absolute filters, retaining all particles larger than the rated pore diameter, many particles with sizes approximately those of the pores may become trapped in the matrix. If enough of these pores get plugged by particles being trapped in the pores, the membrane will become irreversibly fouled. Asymmetric, or "skinned," membranes, on the other hand, which include almost all UF membranes, are less susceptible to fouling. Furthermore, since fouling tends to occur in the "depths" of the pores of a microporous membrane, high flow rates will not be as beneficial as with asymmetric membranes.

- Operating parameters: These are very important, especially the applied pressure and cross-flow velocity, as discussed in Section 6.D. Figure 8.57 shows relationships between flux and Reynolds numbers for several microorgansims and module types. It is interesting that the hollow fibers could get almost as high fluxes as tubular units even though hollow fibers operate at much lower Reynolds numbers. No doubt the high shear in hollow fibers had a beneficial effect. Values of α and C_G for several microorganisms with different modules are given in Tables 4.2 and 4.3.

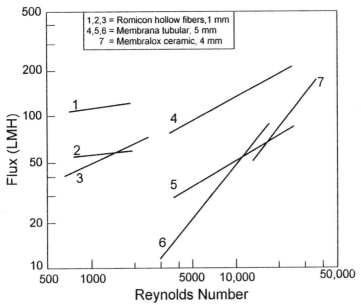

Figure 8.57. *Relationship between flux and Reynolds number for separation of microorganisms. Data sets 1, 4, 7 are yeast at 10 g/L; data sets 2 and 5 are yeast at 50 g/L; data set 3 is* L. bulgaricus *at 10 g/L; data set 6 is* A. niger *at 7 g/L (adapted from Patel et al. 1987, Saglam 1995, Sims and Cheryan 1986, Tejayadi and Cheryan 1988).*

Taniguchi et al. (1987) were able to obtain a cell concentration of 54.4 g/L with *Bifidobacterium longum*, which is more than seven times that normally found in batch culture. Using a ceramic membrane, *S. cremoris* and *L. casei* were also concentrated to 81.5 g/L and 49.0 g/L, giving cell productivities 19- and 9-fold higher, respectively, compared to conventional batch cultivation.

Periodic backwashing has been found to be extremely effective in maintaining high fluxes with microorganisms. For example, Kroner et al. (1984) observed an almost 50% increase in flux with Membrana/Enka polypropylene hollow fibers (0.3-μ pore size) while processing an *E. coli* fermentation broth (Figure 6.24). Backflushing was done with 1.5 atm air, 5 sec every 5 min. Similarly, Matsumoto et al. (1987) obtained a flux of 300 LMH during MF of 8.5 g/L yeast suspended in 1% NaCl with periodic backwashing 5 sec every 5 min. In contrast, flux was 20 LMH without backwashing. Cell concentration reached 100 g/L with a Fuji 0.45-μ microfilter before the flux went to zero. The effectiveness of backpressure methods of MF are shown in Figures 6.24–6.26. Rotary membranes have also been studied for this purpose (Membrex 1989; Riesmeier et al. 1990).

The type of pump used in cell harvesting has a small effect on the viability of cells. Shimizu et al. (1992) observed that cell damage caused by shear was minimal with rotary pumps and caused about 5–8% cell breakage with single-screw and centrifugal pumps after 60 h of recirculation.

• Effect of broth components: Although microbial cells may be the largest single component in a fermentation broth, it is rarely the cells per se that is the main cause of fouling or low flux, but rather the media components. Prefiltration of the fermentation media prior to fermentation significantly improved the flux during cultivation of *Lactobacillus bulgaricus* (Tejayadi and Cheryan 1988). Flux with the cells in media was much lower than the cells suspended in water (Patel et al. 1987). It is only at extremely high cell concentrations that the cells become the limiting factor. Backflushing may be only partially successful if media components are causing severe fouling (Figure 6.24).

A complicating factor in processing fermentation broths with membranes is the antifoam compounds used during fermentation. Antifoams can cause severe fouling of membranes (Section 6.D.2.). Silicone-based antifoams are particularly notorious foulants, while vegetable oils are the most benign. Antifoams with inverted cloud points can cause dramatic reduction in flux, especially with hydrophobic membranes such as polysulfone, if operated above their cloud points. In this regard, the chemical nature of the membrane is also important. Hydrophilic membrane materials (cellulosics, ceramics) generally tend to foul less than hydrophobic membranes (polysulfone, polypropylene).

8.N.2.
ENZYME RECOVERY

There are numerous examples of the use of membranes for recovering enzymes from plant and microbial sources, many from the early days of UF and MF (Barbaric et al. 1980; Flaschel et al. 1983; Heinen and Lauwers 1975; Hummel et al. 1981; Porter and Michaels 1971). For this application, environmental factors such as pH, temperature, and ionic strength (including type and concentration of ions) are critical. A good example is the purification of the enzyme penicillinase (MW 20,000–22,500) from different fermentation broths using different Amicon membranes. As shown earlier in Figure 3.21 and also in Figure 8.58, enzyme recovery was affected by ionic strength and pH, in addition to membrane MWCO. This is probably because of changes in charge, hydration, and solute–solute interactions that, in turn, will affect the molecular shape and size of these proteins. Similar data is shown in Tables 8.20 and 8.21. It is not surprising that membrane MWCO affected throughput. Care should be taken in manipulating operating parameters such as temperature and shear rate, since

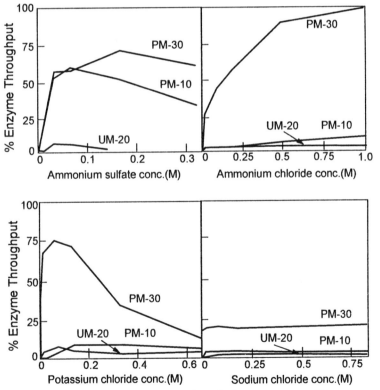

Figure 8.58. *Effect of type and concentration of salts on recovery of S. aureus penicillinase with various Amicon UF membranes (adapted from Melling 1974).*

Table 8.20. Effect of pH on enzyme throughput and specific activity during UF of E. coli penicillinase.

	% Enzyme Throughput		Change in Specific Activity*	
pH	Amicon PM-30	Amicon XM-100A	Amicon PM-30	Amicon XM-100A
5.1	42	66	5.9	3.1
5.6	46	82	7.2	3.6
6.1	50	83	7.4	3.0
6.7	34	82	4.4	2.1
7.1	28	81	4.4	2.1
7.8	20	75	3.2	1.7

*Expressed as specific enzyme activity in permeate/specific enzyme activity in retentate
Source: Data from Melling (1974)

it may affect enzyme activity; e.g., as shown in Table 8.20, pH affects specific activity of the enzyme favorably since there is a purification effect. However, sometimes it is harmful (Table 8.21).

Papain recovery and purification has been practiced since 1983 (Permionics 1995). The feed is typically 10% TS containing 3% ash. UF is done in 10,000-L batches every 8 h to 3.3×. This results in a concentrate of 22–24% TS and permeate with 0.5% nonprotein nitrogen and 3% ash. Cellulose acetate UF membranes are used in this application because of its low protein binding nature. Membrane life is 100 batches of enzyme solution if concentration is done to 22% TS, and only 20 batches if the concentrate is pushed up to 30% TS. Despite the relatively low membrane life, the quality of the enzyme and savings in evaporation cost resulted in rapid payback of less than 1 year (the equivalent evaporator costs three times more than the UF system and has twice the operating cost).

Table 8.21. Effect of pH on enzyme throughput and specific activity during UF of S. aureus penicillinase.

	% Enzyme Throughput		Change in Specific Activity*	
pH	Amicon PM-30	Amicon XM-100A	Amicon PM-30	Amicon XM-100A
4.0	77	100	1.2	1.1
5.0	43	100	0.8	0.99
6.0	29	100	0.6	0.95
6.5	11	100	1.0	1.2
7.0	6	100	1.0	1.1
8.0	0	100	1.1	0.98

*Expressed as specific enzyme activity in permeate/specific enzyme activity in retentate
Source: Data from Melling (1974)

8.N.3.
AFFINITY ULTRAFILTRATION

Affinity ultrafiltration is a relatively new technique that combines cross-flow membrane filtration with another well-established separation technique—affinity chromatography (Klein 1991). If the product compound of interest is present together with other undesirable compounds, all of which would normally pass through a membrane, then a high molecular weight binding agent (macroligand) can be added that will selectively bind to the compound of interest. Membrane filtration with the appropriate membrane is then performed, as shown in Figure 8.59. The product–macroligand complex will be rejected by the membrane while the other undesirable compounds pass through. The complex can then be broken by the appropriate eluant and membrane filtered again. The product will pass through the membrane in a pure form, while the macroligand can be recycled back into the process after suitable reconditioning if necessary. In this respect, the concept is similar to "micellar-enhanced UF" (Section 8.D.7.).

Table 8.22 lists some recent applications of affinity UF or cross-flow filtration (ACFF). Alcohol dehydrogenase was isolated from a yeast fermentation extract by adding starch-immobilized Cibacron Blue. The latter is a dye that can bind to NAD- and ATP-requiring enzymes. After UF with a 500,000 MWCO hollow fiber, the macroligand–enzyme complex was dissociated by adding potassium phosphate solution (0.55 M).

Similarly, Concanavalin A from Jack beans was purified using heat-shocked cells of *S. cerevisiae* as the macroligand. The affinity interaction was conducted in a mixing chamber, while the purification and washing were done in a membrane unit with 300,000–1,000,000 MWCO membranes. The eluant used in this case was 0.8M D-glucose.

Flaschel et al. (1983) describe the separation of pullulanase, which by itself is retained by the Amicon XM-300 membrane only to about 13%. However, when

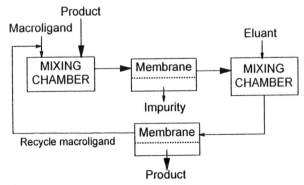

Figure 8.59. *Affinity ultrafiltration for purification.*

Table 8.22. Examples of affinity ultrafiltration.

Compound Purified	Raw Material	Macroligand	Reference
Alcohol dehydrogenase	*S. cerevisiae* extract	Cibacron Blue on starch	Mattiasson and Ling (1986)
β-galactosidase	*E. coli*	Modified agarose	Pungor et al. (1987)
Concanavalin A	*Conovalia ensiformis* extract	Heat-killed cells of *S. cerevisiae*	Mattiasson and Ramstrorp (1984)
Trypsin	Mixture with peroxidase	Cross-linked trypsin inhibitor	Bartling and Barker (1976)
Trypsin	Mixture with chymotrypsin	Soybean trypsin inhibitor on dextran	Choe et al. (1986)
Trypsin	Mixture with chymotrypsin	m-Aminobenzamidine on polyacrylamide	Luong et al. (1987)

1% DEAE-dextran was added and the solution kept at a low ionic strength, an interaction occurred between the enzyme and the DEAE-dextran, resulting in a 96% rejection by the XM-300 membrane. After separation from the enzyme reaction mixture, the DEAE-enzyme complex could be dissociated by increasing the ionic strength and then separating the enzyme using the same XM-300 membrane. Similarly, Bartling and Barker (1976) separated trypsin from horseradish peroxidase by adding cross-linked trypsin inhibitor to the solution. The mixture was then ultrafiltered through a 50 K MWCO membrane. The peroxidase could be eluted through the membrane at pH 8.3. When the pH was adjusted to 1.8, the inhibitor–enzyme complex dissociated and the trypsin could be separated. The cross-linked trypsin inhibitor was retained for further use. Luong et al. (1988) used a continuous affinity UF process for trypsin purification.

In all cases, recoveries were fairly high (55–90%) and purities ranged from 65–98%. This technique apparently has a wide scope for the purification of many biochemicals from fermentation broths. The choice of the macroligand is, of course, critical, as is the choice of the appropriate membrane. As a general rule, there should be a 10-fold difference in MW or molecular size between the macroligand–product complex and the undesirable components for most commercially available membranes to function properly.

In summary, membrane technology has had a significant impact on the biotechnology industry. For cell harvesting/separation applications, membranes are gradually replacing continuous centrifugation. The overall economics favor membranes for small-scale operations (regardless of particle size). With large-scale operations, centrifuges may still be more economical with larger particles such as yeast cells if a sharp and clean separation is not required (e.g., if all that is required is to concentrate the yeast). Membranes will be better with smaller particles such as bacteria. Membrane filtration will always be more economical than ordinary filtration with a filter aid, except perhaps for recovering fungi.

For enzyme recovery, MF will be useful to separate the cells and suspended matter, while UF will be used in concentrating and desalting by DF, perhaps combined with other more efficient means of fractionation.

8.N.4.
MEMBRANE BIOREACTORS

Chemical and biological conversions using enzymes and/or microorganisms as catalysts are commonly used in the production of organic chemicals, food products, pharmaceuticals, hormones, vitamins, and other biological products. The most common bioreactor is the batch reactor where the enzyme or microbial cells are used in their soluble, or "free," form. Since these are the currently practiced methods, they do have implicit advantages. However, they also have several limitations:

1. Inherent less efficiency than continuous processes, because of their startup and shutdown nature
2. High capital costs for equipment because of their low productivity
3. Batch-to-batch variation in the product
4. The need to inactivate or separate the enzyme or microbial cells at the end of the reaction, thus increasing processing costs: In enzymatic conversions, since the enzyme is not recovered and reused, it leads to extremely low productivity and high catalyst costs, since each fresh batch of substrate needs a fresh charge of enzyme. With fermentations, there is usually a lag period before cell numbers reach their maximum level, thus delaying the onset of maximum productivity.
5. Long times needed for completing the reaction, which is related to substrate depletion, product inhibition, and the generally low biocatalyst concentrations in the reactor, especially with fermentations

In addition, although enzymes have an untarnished safety record over the past 60 years, governmental legislation to protect the consumer from the presence of chemical carcinogens in the diet or the environment has tended to restrict the growth of new enzyme applications. Regulatory agencies could perceive added enzymes, not as food, in spite of their ubiquitous presence in all living things, but as food additives. Thus, food and pharmaceutical companies are expressing renewed interest in developing methods of using enzymes that will result in an enzyme-free product.

Immobilization of enzymes or microbial cells is one way of overcoming the disadvantages of using the soluble or free biocatalyst. Immobilization refers to the chemical or physical attachment of the biocatalyst (the enzyme or microbial cell) to solid surfaces or otherwise physically confining or localizing them in a defined region of space with retention of their catalytic properties (Cheryan and Mehaia 1985b; Mehaia and Cheryan 1990a). Several advantages for immobilization have been claimed and documented:

1. The opportunity to develop the more efficient continuous processes, since the immobilized enzyme or cell can be packed into a column and fed continuously pumped into the reactor
2. Lower enzyme costs for immobilized enzyme systems, since the enzyme is trapped within the bioreactor and is reused a number of times, thus increasing enzyme utilization and presumably lowering enzyme costs
3. Better process control and a more uniform and consistent product
4. Higher fermentation rates, since the concentration of cells per unit reactor volume will be higher in a packed-bed reactor

There are, however, several problems with immobilized biocatalysts. Depending on the method and the particular enzyme or microorganism, losses in activity of 10–90% have been reported. Steric hindrance, enzyme–substrate orientation, and diffusional restriction problems may occur, which will affect activity and specificity, especially with macromolecular or colloidal substrates. Difficulties that must be surmounted include possible contamination of the microbial culture during immobilization, the expense of the immobilization step, high pressure drops in packed-bed reactors, and for fermentations with immobilized whole cells, gas holdup and consequent pressure buildup. Thus, despite the early promise and the vast amount of research in this area, there are very few known industrial processes that have been commercialized to date (Mehaia and Cheryan 1990a).

An alternate approach to conventional immobilization methods and development of high-rate conversion processes is the membrane (bio)reactor, which uses synthetic semipermeable membranes of the appropriate chemical nature and physical configuration to localize the (bio)catalyst within the reaction vessel or separate it from the reaction mixture. By taking advantage of the size differences between the catalyst and the product(s) of the reaction, the membrane can be used in one of two ways:

1. To continuously separate the catalyst from the reaction mixture and recycle it back to the main reaction vessel for further reaction: This is the membrane recycle reactor described in the next section (Section 8.N.4.a.).
2. To trap and confine the biocatalyst on one side of the membrane (e.g., the shell side of a hollow fiber or ceramic module) and to pump the feed in and product out through the other side of the membrane (i.e., the tube side): These are most commonly operated in the plug-flow mode. A membrane "sandwich" configuration could also be used as a plug-flow reactor: the catalyst is trapped between two flat sheets of membrane, and the feedstream flows across the membrane (Section 8.N.4.b.).

The rest of this section focuses on biological reactions utilizing biocatalysts; however, extensive literature is also available on membrane reactors for chemical reactions (Govind and Itoh 1989).

8.N.4.a.
MEMBRANE RECYCLE BIOREACTORS

The basic concept behind the recycle bioreactor is shown in Figure 8.60. Typically, a reaction vessel, operated as a stirred tank reactor, is coupled in a semi–closed-loop configuration via a suitable pump to a membrane module containing the appropriate semipermeable membrane. In operation, the reaction vessel is first filled with the substrate solution or slurry at the appropriate reaction conditions. The biocatalyst is then added, and the contents of the reaction vessel are continuously pumped through the membrane module and recycled back to the reaction vessel. The membrane should be chosen to retain the biocatalyst while minimizing retention of the product molecules. Since most enzymes are of the order of 10,000–100,000 MW, UF membranes with these MWCO can be used. With microbial cells, MF membranes can be used, subject to the conditions laid out in Section 8.N.1.

At steady state, product molecules small enough to permeate through the pores of the membrane will be removed from the system while the biocatalyst will be recycled to the reaction vessel for further reaction. The total volume of the system is maintained constant by matching the incoming feed flow rate

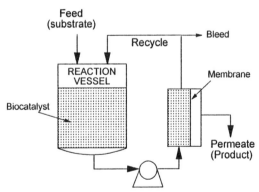

Figure 8.60. *Membrane recycle bioreactor. The biocatalyst (e.g., enzyme or microbial cell) is suspended in the main reaction vessel. Feed is pumped into the main reaction vessel at the same rate as the product (permeate) flow rate. The entire reaction mixture is pumped continuously through the membrane module, where the products of the reaction and any other compounds small enough to permeate through the membrane are removed, while the biocatalyst and other rejected compounds are recycled back to the main vessel for further reaction. Drawing not to scale: the membrane module will usually be much smaller in volume than the main reaction vessel.*

to the product outflow (the permeate flux). If the substrate is a macromolecule (e.g., starch or protein), unhydrolyzed or partially hydrolyzed substrates too large to go through the pores will also be recycled to the reaction vessel for further reaction. If the substrate is also a small molecule, such as glucose in a membrane fermenter, the unconverted substrate will also pass through into the permeate. Hence, although the recycle bioreactor can be used with substrates of any molecular size, it is the only configuration possible for the hydrolysis of macromolecules such as proteins, starch, or cellulose. On the other hand, plug-flow reactors (e.g., the hollow fiber or sandwich membrane reactor) requires both substrate *and* product to be of the same order of magnitude in size and much smaller than the biocatalyst. This is discussed later in Section 8.N.4.b.

The recycle bioreactor is usually operated under completely mixed conditions, depending on the relative volumes of the reaction vessel and the membrane module and the ratio of recycling flow rate to flux. Thus, it can be modeled as a simple CSTR (continuous, stirred-tank reactor), as shown by residence time distribution studies (Deeslie and Cheryan 1981a; Sims and Cheryan 1992a). In effect, the concentration of product in the outflow will be essentially the same as its concentration in the reaction vessel at any given time. In the case of low molecular weight substrates that are also freely permeable, the unhydrolyzed substrate concentration in the reaction vessel will be the same as in the product stream. Thus, the CSTR-type recycle bioreactor is more suited for substrate-inhibited reactions than product-inhibited reactions, when the conversion is high. Ideally, the bioreactor should then be operated such that the product concentration is below the level at which severe inhibition occurs. The common belief that membrane bioreactors operated in the recycle mode are more productive because "inhibitory end products can be removed as they are formed" is incorrect because of this equilibrium partitioning nature of the membrane process. The higher productivity is caused by the higher biocatalyst concentration and its continuous recycle and reuse.

The completely mixed conditions prevailing in the membrane recycle bioreactor (MRB) is a disadvantage for most common enzymatic and microbial conversion processes since they have a positive reaction order, i.e., the rate of reaction is proportional to the substrate concentration. These reactors will be operating at very low conversion rates since the concentration of substrate in the system is very low. Consequently, this system must be significantly larger than a plug-flow reactor used for the same conversion, especially if the system is product-inhibited, as are most fermentations. To overcome this limitation, several recycle bioreactors can be set up in series (Figure 8.61). The substrate with a flow rate of F and concentration S_o is fed into the first stage, which has a volume V_1 with biocatalyst concentration C_1. The permeate from the first stage, containing product at a concentration of P_1 is fed continuously into the second stage, which contains biocatalyst at a concentration of C_2 in a volume V_2. This continues until the nth stage where the required degree of conversion has been

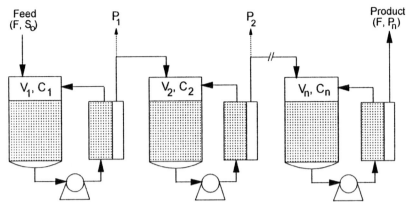

Figure 8.61. *Membrane recycle bioreactors in series. F is the feed flow rate, V is the volume, C is the concentration of biocatalyst, P is product concentration, and S_o is the inlet substrate concentration.*

obtained, and product leaves at a concentration of P_n at a flow rate of F. The total volume of the system should be less than if only one stage were used, since all stages except the last stage are operating at higher reaction rates. However, due to the complexity of the system and the cost for membranes, pumps, and associated hardware and control systems, it will rarely be economical to use more than three stages.

It may be more feasible to use just one membrane unit, attached to the last stage of a series of conventional CSTRs (Figure 8.62), especially if both substrate and product are freely permeable. In such a system, the overall performance will be a function of the recycle ratio (R/F in Figure 8.62). At high ratios, the catalyst will be distributed uniformly along the whole series of reactors, while at low ratios, most of the conversion will occur in the last stage. This

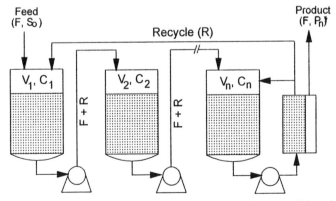

Figure 8.62. *Multistage stirred-tank reactors with the membrane module only in the last stage. R is the recycle flow rate.*

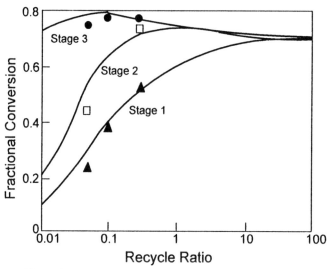

Figure 8.63. Effect of recycle ratio (R/F) on fractional conversion in multiple stage bioreactor as depicted in Figure 8.62. The reaction was the conversion of N-acetyl-D, L-methionine to L-methionine by acylase. Lines are drawn according to a theoretical model developed by the authors, points are experimental data (adapted from Wandrey and Flaschel 1979).

is shown in Figure 8.63 for the specific case of the conversion of N-acetyl-D,L-methionine to L-methionine by acylase. In this case, $V_1 = V_2$, and $V_3 = 0.1 V_1$. At low recycle ratios, most of the conversion occurs in the third stage, whereas at high recycle ratios, the contribution of the second and third stage is minor. For this particular case, R/F ratio of 0.1 resulted in approximately equal degrees of conversion in each stage.

A similar two-stage system was used for continuous fermentation of glucose to ethanol using Z. *mobilis*, with V_1 of 1.65 L and V_2 of 0.83 L. The biocatalyst (cell) concentration was more than 10 times higher in the second stage, which enabled a very high outlet ethanol concentration to be maintained with almost theoretical substrate utilization (Charley et al. 1983). For the fermentation of glucose to sodium acetate by *Clostridium thermoaceticum*, a two-stage MRB system of the type shown in Figure 8.62 resulted in 11% higher productivity and higher product concentration (Figure 8.64) than a single-stage MRB. The recycle ratio (R/F) was 0.25; feed and nutrients were supplied only to the first stage. Cell concentration was three times higher in the second stage. The MRB system was very stable for extended period of operating time (Shah and Cheryan 1995).

Using a membrane module in each stage, as shown in Figure 8.61, may be more expensive and complicated, but it could be useful in certain situations. For example, with multi-enzyme reactions (especially consecutive reactions), such as the production of high-fructose corn syrup from starch, stage 1 could

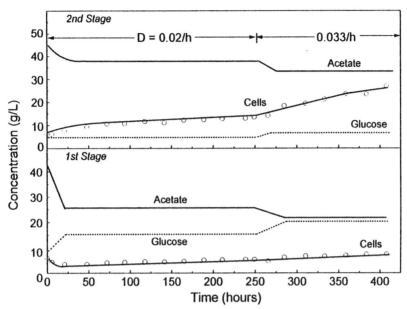

Figure 8.64. Conversion of glucose to acetic acid by Clostridium thermoaceticum in a two-stage membrane bioreactor with the membrane only in the second stage. Fermentation broth was maintained at pH 6.2 with NaOH. D is dilution rate (adapted from Shah and Cheryan 1995).

contain the starch-hydrolyzing enzyme(s) such as glucoamylase, and stage 2 could contain glucose isomerase. In this way, each stage could be optimized for its own particular reaction conditions rather than compromise and use suboptimum operating conditions if all enzymes were mixed together. This concept has been demonstrated for simultaneous saccharification and fermentation for the continuous production of ethanol from starch (Lee et al. 1982) and for the production of vinegar from date juice extracts, which is shown in Figure 8.65. The first stage MRB converts the sugars in date juice to ethanol using *S. cerevisiae*, and the second stage MRB contains *Acetobacter* to convert the ethanol to acetic acid. The MRB is sensitive to changes in dilution rate/residence time. If the dilution rate is too high in the first stage, the yeast stops metabolizing fructose. In the second stage, high dilution rates cause a reduction in ethanol utilization by the *Acetobacter* and a corresponding reduction in acetic acid levels (Figure 8.65, bottom). The yields of products in each stage are near theoretical (Mehaia and Cheryan 1991).

8.N.4.b.
PLUG FLOW BIOREACTORS

In this type of configuration, the membrane separation unit is used as the bioreactor itself (Figure 8.66). The biocatalyst is trapped on one side of the

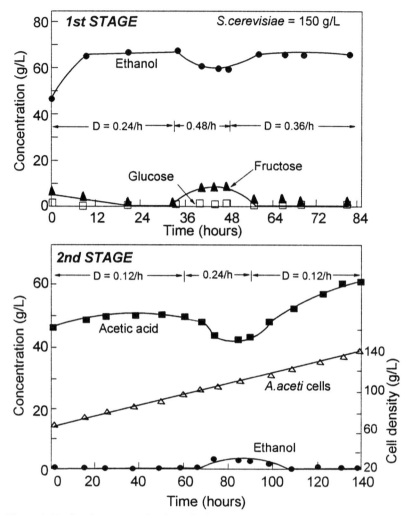

Figure 8.65. Continuous production of ethanol and acetic acid in a two-stage MRB with membranes in both stages, as shown in Figure 8.61. The first-stage yeast concentration was maintained at 150 g/L by bleeding. The second stage contained Acetobacter acetii cells, which were allowed to grow (no cell bleed). The feed to the first stage was date juice extract containing 138 g/L total sugars. The feed to the second stage was first-stage product containing ethanol (47 g/L for the first 90 h, 65 g/L from 90–140 h). The dilution rate (D) was changed as shown (adapted from Mehaia and Cheryan 1991).

membrane, usually the permeate side, while the substrate is pumped in from the retentate side. The substrate must diffuse across the membrane and react, and the product and unconverted substrate must then diffuse back into the flowing stream on the tube side and be removed from the reactor. Thus, in this bioreactor, the substrate and product must be of the same order of magnitude in size and much smaller than the biocatalyst. When the order of magnitude of product and substrate are vastly different or when the substrate is of the same order of magnitude in size as the biocatalyst, the only way to achieve substrate–biocatalyst contact is in the recycle bioreactor.

In theory, almost any module or configuration could be used as a plug-flow bioreactor. Hollow fiber cartridges and inorganic membrane modules are suitable because the membrane is self-supporting, with no need for a separate backing or support. This is important since the flow goes in both directions, i.e., from the retentate to the permeate side and vice versa. In addition, hollow fibers have high surface area to volume ratios, important in any catalytic reaction.

Hollow fiber devices are available in two basic configurations (Figure 8.66): the beaker type is suitable only for bench-top feasibility studies. The tubular type is more realistic for industrial applications. Several theoretical models for enzyme catalysis in hollow fiber enzyme reactors (HFERs) have appeared (Cheryan and Mehaia 1985b). With the enzyme in the tube side and feed pumped

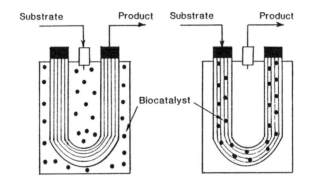

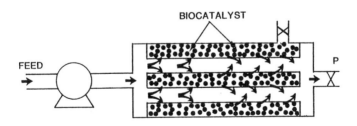

HOLLOW FIBER BIOREACTOR (PFR)

Figure 8.66. Schematic of plug flow membrane reactors: Top: hollow fiber beaker reactor; the biocatalyst can be trapped either in the shell side or the tubes; Bottom: tubular type hollow fiber bioreactor. The biocatalyst is loaded into the shell side through the permeate ports. Feed is pumped through the tube under pressure and the product stream removed through the retentate outlet. The reverse (i.e., tube-side loading of biocatalyst) is less common.

in through the shell side, the residence time distribution approximates a completely mixed model rather than the plug-flow model (Katoaka et al. 1980). With the biocatalyst in the shell side and feed pumped in through the tube side, the residence time distribution approximates plug flow (Kohlwey and Cheryan 1981), which is a better method of conducting product-inhibited reactions (Kawakami et al. 1980).

Early work on HFERs was done with the enzyme localized within the spongy matrix of the hollow fibers and with the shell-side space drained of any liquid. In this case, the transmembrane pressures had to be kept low to prevent the enzyme from washing off the fibers. Thus, this mode of operation relied principally on diffusion as the mode of mass transfer of substrate and product across the membrane. On the other hand, if the shell side is filled with the biocatalyst solution, the performance of the HFER will be affected by the pressure profiles within the hollow fiber cartridge. For example, if the cartridge is operated at an inlet pressure of 150 kPa and outlet pressure of 50 kPa, the shell side will equilibrate at about 100 kPa. Thus, there would be a net inflow of liquid from the tube to the shell side at the entrance region of the cartridge and a net flow of liquid from the shell side to the tube side towards the outlet of the cartridge. This will tend to cause a migration of the biocatalyst towards the outlet end of the cartridge, unless the enzyme is "fixed" onto the fibers by cross-linking (Breslau 1981). Membranes with MWCO at least 10 times smaller than the molecular weight of the enzyme should be used to prevent leakage of enzyme.

Since the plug-flow bioreactor is limited to reactions where both substrates and products must be of low molecular weight (both must freely permeate the membrane), it is not surprising that the most attention has been given to the hydrolysis of oligosaccharides and other sugars (Cheryan 1986; Mehaia and Cheryan 1990a), especially the hydrolysis of lactose or a synthetic substrate [o-nitrophenyl-β-D-galactopyranoside (ONPG)] by β-galactosidase (lactase). In some cases, the PS hollow fibers (Amicon's PM series) were inactivating enzymes by mere contact, e.g., β-galactosidase (Kohlwey and Cheryan 1981), α-galactosidase (Korus and Olson 1977a), and glucose isomerase (Korus and Olson 1977b). Preconditioning the fibers with a protein reduced the membrane poisoning, perhaps by reducing the hydrophobicity of the membrane. In contrast, these carbohydrase enzymes were unaffected by the Amicon's XM50 membranes. PS had little or no effect on proteases.

8.N.4.c.
DEAD-END VERSUS CROSS-FLOW SYSTEMS

Most laboratory studies on enzyme–membrane reactors have used dead-end stirred cells (DESCs) of the kind shown in Figure 5.3 as reactors. A typical system lay-out would be similar to that shown in Figure 5.4. The DESC is first charged with the biocatalyst. Substrate is then fed in under pressure while permeate is continuously withdrawn through the membrane (i.e., the membrane cell is operated under the continuous DF mode described in Section 7.B.2.).

Agitation, necessary to minimize concentration polarization as well as improve reaction rates in the bulk phase, is provided with a magnetic stirrer.

This type of dead-end operation differs considerably from the MRB system depicted in Figure 8.60. Although both are operated in the CSTR mode, the membrane is physically separated from the reaction zone in the MRB, while the dead-end cell serves both as a reactor and the separation device. The primary consideration in continuous bioreactors is obtaining high conversion and maintaining a steady-state operation. The former is controlled by the reaction kinetics of the system, while the latter is governed by the performance characteristics of the membrane module, the efficiency of controlling concentration polarization effects, membrane area-to-volume ratio, and fouling characteristics of the particular membrane–solute system. There will be fewer problems maintaining stable long-term operation as far as the membrane unit is concerned if the substrate is of low molecular weight and can easily permeate the membrane than if the substrate were an impermeable macromolecule.

In dead-end cells, owing to the pressure gradient and the convective flow of solutes to the membrane surface, concentration polarization of the biocatalyst, and substrate if it is a macromolecule, could be serious enough (even with adequate agitation) to inhibit the biocatalyst. This is especially a problem if the biocatalyst is an enzyme and a change in the conformation of the enzyme molecule or steric hindrance occurs due to the "close-packed" arrangement of molecules in the gel layer.

The recycle bioreactor, on the other hand, is much more flexible. Because the reaction zone and the separation zone are physically separated, the reaction can proceed in the absence of high pressures and concentration polarization effects. The degree of mixing provided in the reaction vessel can be adjusted for optimum reaction rates alone instead of being a compromise between the needs of the reaction and the need for minimizing concentration polarization, as it is with the DESC reactor. For these reasons, the DESC reactor may not be the best example of a membrane reactor, especially with macromolecular substrate. In some cases, the relatively poor polarization control of the dead-end cells has led researchers to conclude that a particular enzyme process is impractical in a membrane reactor due to fouling problems. However, had a recycle system been used instead, the conclusions may have been different. It is best to use membrane modules that have as high a surface area-to-volume ratio as possible, to minimize holdup in the membrane module to conduct as much of the reaction in the main reaction vessel itself. A listing of some DESC reactor studies is available elsewhere (Cheryan 1986; Mehaia and Cheryan 1990a).

8.N.4.d.
ENZYME PROCESSES IN MEMBRANE REACTORS

The major advantage of membrane bioreactors over conventional batch enzyme hydrolysis is that the enzyme is retained within the system, thus increasing

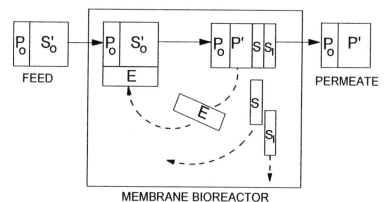

MEMBRANE BIOREACTOR

Figure 8.67. Sequence of reactions occurring in membrane recycle bioreactor (adapted from Deeslie and Cheryan 1981a; Sims and Cheryan 1992a). See text for explanation of symbols.

enzyme utilization, which is measured as weight of product per unit weight of enzyme. A few examples will be cited here to illustrate the methodology and show the potential of this technique.

Perhaps the earliest demonstration of the membrane bioreactor concept was by Blatt et al. (1968) who hydrolyzed milk proteins with α-chymotrypsin in a DESC reactor with an Amicon UM-10 membrane. Figure 8.67 shows the theoretical sequence of reactions occurring in the bioreactor. The feed at a concentration S_o is assumed to consist of the principal substrate for the enzyme (S_o') and a fraction that is already permeable through the membrane (P_o). For example, P_o may be nonprotein nitrogen with protein substrates and dextrose or oligosaccharides with liquefied starch. Upon entering the reaction vessel, it reacts with the enzyme (E), producing product (P'). There will be some unconverted substrate in the reactor at all times (S), the concentration of which depends on the activity of the enzyme and the sieving characteristics of the membrane. In some cases, a portion of the substrate (S_I) may be unhydrolyzable by a particular enzyme under the operating conditions used. In the case of soybean protein, it is a "hydrophobic core," which is 2–5% of the incoming protein substrate (Deeslie and Cheryan 1981a). This, along with any other impermeable components, will gradually build up in the reactor, eventually necessitating reactor shutdown or periodic purges.

The important variables that need to be optimized in any reactor are the enzyme concentration (E), substrate concentration (S_o'), volume of the reaction zone (V), and flow rate (F), which is flux (J) × area (A). Combining the simple Michaelis-Menten model for a single-substrate, uninhibited reaction with a mass balance for an ideal reactor gives the following relationships:

For recycle systems (e.g., MRB):

$$X + \frac{K_m X}{S'_o(1 - X)} = k_2 \frac{EV}{FS'_o} = k_2 \tau \tag{8.1}$$

For plug-flow systems (e.g., HFER):

$$X - \frac{K_m}{S'_o}\ln(1 - X) = k_2 \tau \tag{8.2}$$

where K_m and k_2 are constants, X is the fractional conversion (P/S'_o), and τ is a modified space time. The goodness of fit of these models is shown in Figures 8.68 and 8.69. For the soy protein–Pronase system described later, the model [Equation (8.1)] appears to overpredict at high conversions, which could be because it does not take into account product inhibition, which is expected to be significant at high conversions. The k_2 for the MRB was six times lower and the K_m value was 2.3 times lower than the values obtained for batch hydrolysis, indicating some loss of activity, probably because of enzyme leakage in the early stages of operation (Deeslie and Cheryan 1981a).

The HFER (Figure 8.69) was studied with *o*-nitrophenyl-β-D-galactopyranoside (ONPG) as the substrate and β-galactosidase as the enzyme. A good fit was obtained for the data using Equation (8.2) and the parameters shown in the figure legend. The K_m values for the HFER and batch reactor were similar, but k_2 for the HFER was 168 times lower; the loss of activity was probably caused by poisoning of the enzyme by the PS membrane (Kohlwey and Cheryan 1981).

A few general conclusions can be arrived at by reviewing the literature on this subject:

1. The enzyme must be carefully selected, since it should maintain the proper (high) activity and stability over long operation periods. Most of the scientific and trade literature on enzyme kinetics is based on short-term studies with free enzymes in batch reactors and cannot be simply extrapolated for continuous long-term use where enzyme stability is most important. For example, Figure 8.70 shows temperature stability for β-galactosidase. The optimum temperature suggested by the manufacturer for conventional batch hydrolysis is 40°C. However, at this temperature, the enzyme inactivates rapidly. The enzyme is much more stable at 15°C. Similarly, the best balance between activity and long-term stability of Pronase for soy protein hydrolysis in an MRB was 25°C even though 50°C is recommended for batch reactions (Deeslie and Cheryan 1982). For starch hydrolysis by glucoamylase in an MRB, the recommended temperature for conventional

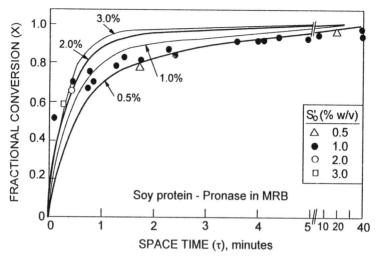

Figure 8.68. *Effect of space time (τ) on fractional conversion (X) of soy protein by Pronase in a membrane recycle bioreactor. Variable shown is substrate concentration (S_o', %w/w). Lines are drawn according to the model [equation (8.1)] with $K_m = 0.682$ mgN/mL and $k_2 = 2.56$ min^{-1}. Points are experimental data (adapted from Deeslie and Cheryan 1981a).*

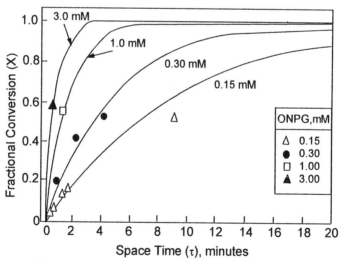

Figure 8.69. *Effect of space time (τ) on fractional conversion (X) in a plug-flow membrane bioreactor. The substrate was o-nitrophenyl-β-D-galactopyranoside (ONPG) and β-galactosidase was the enzyme. Variable shown is substrate concentration (S_o', mM). Lines are drawn according to the model [equation (8.2)] with $K_m = 2.31$ mM and $k_2 = 1.80$ min^{-1}. Points are experimental data (adapted from Kohlwey and Cheryan 1981).*

456

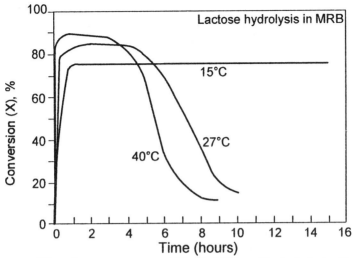

Figure 8.70. Effect of temperature on stability of a MRB used for hydrolysis of lactose by β-galactosidase. Initial lactose concentration = 5%, enzyme concentration = 20 mg/mL, residence time = 60 min (adapted from Mehaia et al. 1993).

batch operation is 60°C, with an enzyme half life of 1 day. On the other hand, at 45°C, the half life is 30 days though the activity is 40% lower (Cheryan and Escobar 1993).

2. Unhydrolyzable components of the feedstream invariably limit the period of operation. For example, vegetable proteins are more difficult to hydrolyze than animal proteins because of the presence of "hydrophobic cores" in their protein structure (Deeslie and Cheryan 1981b), which are resistant to hydrolysis, except by specific enzymes that are difficult and expensive to obtain. It could also contain fiber from the grain or plant source. In starch hydrolysis, it may be the "mud" components. This unhydrolyzable portion of the feedstream will build up in the system. Adequate pretreatment of the substrate, such as preheating (to improve reaction rates by denaturing protein substrates or to liquefy starch substrates) and prefiltration (to remove coarse insolubles that might foul or plug up the membrane module), are recommended to minimize these problems.

3. Control of concentration polarization is critical to the success of the bioreactor: almost all DESC reactors came to a dead stop or resulted in poor activities after a few hours. This is not only related to the first factor mentioned above, but also to the unsatisfactory design of DESC systems. On the other hand, there are numerous examples of successful application of the recycle bioreactor for enzyme hydrolysis (Cheryan 1986; Mehaia and Cheryan 1990a).

Protein hydrolysis has been studied extensively. Chemical or enzymatic modification of proteins has been practiced for centuries and is the basis for many traditional products such as cheese and soy sauce. Protein hydrolysates, which are proteins that have been broken down into their component peptides or amino acids, have many uses, such as flavor enhancers in formulated foods, improvement of certain functional properties of proteins, in defined formula or infant/medical diets, for those cases of clinical insufficiency that are unable to properly digest or absorb whole protein, and as a protein fortifier, e.g., for incorporation into soft drinks and acidic beverages and as a vehicle for improving nutritional status of specifically targeted populations.

The extent of reaction of protein hydrolysis must be carefully controlled, or else it gives rise to bitter flavors in the hydrolysate. In addition, functional properties of proteins are governed to a large extent by their molecular size. In principle, membrane bioreactors should help deal with these problems, since the molecular size of the product can be controlled to a limited extent by proper selection of membrane pore characteristics while the extent of reaction is controlled by the residence time in the reactor.

Figure 8.71 is a schematic flow diagram for continuous protein hydrolysis. Depending on the enzyme, its concentration, and the residence time, steady state can be achieved in under 1 h. Table 8.23 summarizes the reaction conditions and some results from the literature. The membrane recycle bioreactor is much more productive than the traditional batch process. If productivity is defined as in the table, then it is better by 20–30 times. If the time factor is included, it would be about 40–50 times better (Cheryan and Deeslie 1983, 1984). Yields were also much higher. Long-term stability of the bioreactor was found to be affected by temperature, leakage of active enzyme fractions through the membrane initially, and loss of activators. Functional properties of these hydrolysates were found

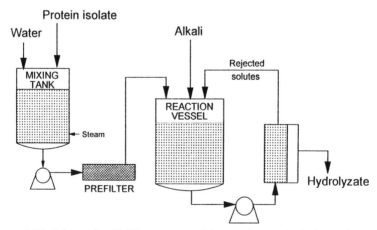

Figure 8.71. Schematic of MRB system used for continuous hydrolysis of proteins (adapted from Deeslie and Cheryan 1981b).

Table 8.23. Performance of membrane recycle bioreactors for protein hydrolysis.

	Soy Protein*	Soy Protein**	Casein[†]	Casein[‡]	Bovine Plasma Protein[§]
Enzyme	Pronase	Fungal acid protease	Alcalase	Alcalase	Alcalase
Bioreactor	MRB	MRB	MRB	DESC	MRB
pH	8.0	3.7	8.0	8.8	7.5
Temperature (°C)	50	60	50	40	45
Residence time (min)	17–305	60	71	375	240
Protein concentration (%)	0.5–3	1.5	5	5	2.1
Overall yield (%)	85–94	85–95	90	90	15
Productivity (grams of hydrolyzate per gram enzyme):					
Batch reactor	10–15	9.4	25	—	—
Membrane reactor	72–108	18–28	320	—	—

* *Source:* Data from Cheryan and Deeslie (1983, 1984); Deeslie and Cheryan (1981a, 1981b, 1992)
** *Source:* Data from Iacobucci et al. (1974)
† *Source:* Data from Mannheim and Cheryan (1990)
‡ *Source:* Data from Boudrant and Cheftel (1976)
§ *Source:* Data from Bressollier et al. (1988)

to be good and superior in some respects. Different pore size membranes resulted in peptide fractions with different average molecular weights (although with considerable overlap) and different functional properties of each fraction (Deeslie and Cheryan 1992).

Carbohydrate hydrolysis with membrane bioreactors has also evoked considerable interest, especially the hydrolysis of oligosaccharides (lactose, maltose, raffinose, cellobiose, sucrose), starch, and cellulose. Usually, higher productivity was observed with membrane reactors compared to batch reactors (Cheryan 1986; Mehaia and Cheryan 1990a). Several studies point out the importance of using the appropriate enzyme, operating conditions (which may be different than the batch reactor), and pretreatment.

Starch hydrolysates are used in the production of glucose, high-fructose sweeteners, brewing syrups, and as fermentation substrates. Figure 8.72 shows typical results in an MRB used for hydrolysis of corn starch by glucoamylase using a 5000 MWCO UF membrane. By charging the reactor initially with much higher levels of enzyme (10–15 times higher than normal), the reaction time could be reduced from the traditional 40–72 h to 3–5 h (Sims and Cheryan 1992a, 1992b). The reverse reaction (glucose → di- and trisaccharides) was minimized since the glucose was being removed from the system continuously. An added benefit was that the product stream was crystal clear dextrose solution, which means the filtration step used in traditional dextrose manufacture (Section 8.L.1.) could be eliminated.

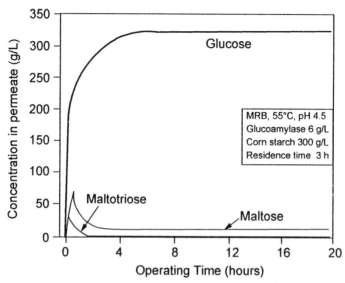

Figure 8.72. *Continuous hydrolysis of liquefied corn starch by glucoamylase in an MRB. The temperature was 55°C, pH = 4.5, residence time = 3 hours. Starch conversion to glucose was 96–98% during the experiment (adapted from Sims and Cheryan 1992b).*

Table 8.24 compares performance and economic data for starch hydrolysis. Operating at a lower temperature (45°C is probably the optimum between enzyme stability and activity) is important; otherwise, the loss of enzyme activity is too rapid to make the MRB economical. Extra enzyme is added to compensate for the lower activity at 45°C. Other starch-hydrolyzing enzymes (in addition

Table 8.24. Performance of the membrane recycle bioreactor for starch hydrolysis.

Process Parameter	Conventional Process	MRB Option 1	MRB Option 2
Temperature (°C)	60	55	45
Liquefied starch (%w/w)	30	30	30
Enzyme concentration (mL/kg starch)	0.83	14	20
Residence time (hours)	72	5	5
Substrate conversion (%)	98	98	98
Product composition			
Glucose (DP-1)	95	95	94
Maltose (DP-2)	3	3	3
Others (DP-3+)	2	2	3
Volumes processed per dose enzyme	1	>20	>40
Capacity (g · glucose/mL enzyme/h)	18	320	440

Source: Data from Cheryan and Escobar (1993)

to glucoamylase) can be added to ensure that the required product distribution is obtained. The major problems are membrane fouling due to the enzyme and nonstarch components (mud) in the feedstream. The added cost of the membrane system will be more than balanced by the reduction in reaction vessels and elimination of the filter aid filters. The operating cost of the MRB (enzyme, membrane replacement, cleaning, electric power for recirculation) will be balanced by the savings in enzyme cost with the conventional saccharification system.

An MRB was also used successfully for the production of maltose from soluble starch using a combination of β-amylase and pullulanase (Nakajima et al. 1990). A 30 K MWCO hollow fiber was used that completely retained the enzyme. Productivities were 5–25 times higher than immobilized enzyme reactors, and maltose conversions were 10–20% higher.

Almost all studies on cellulose hydrolysis were done with DESC membrane reactors. Ghose and Kostick (1970) operated their DESC system semicontinuously for 10 days, resulting in an overall 71% conversion of cellulose to glucose. Henley et al. (1980) improved upon this system by using a recycle system with a hollow fiber module as the separation device. Conversions of 87–92% were obtained in the membrane bioreactor compared to 62–67% in a batch CSTR. Adequate prefiltration of the cellulose slurry is necessary in order to prevent damage to the hollow fibers. A mixture of cellulase and cellobiase resulted in 95% cellulose hydrolysis in the MRB compared to 40% in batch hydrolysis over a 20-h reaction period (Ohlson et al. 1984). The amount of reducing sugars produced was 25.7 g/g enzyme compared to 4.7 g/g enzyme in a batch reactor.

Multiphase membrane reactors are useful for enzymatic reactions that involve sparingly soluble substrates or products or when the product exerts a feedback inhibition. The original concept involved a two-layer membrane sandwich (Matson and Quinn 1986): one was a hydrophobic membrane such as PTFE filled with an organic solvent and the other a hydrophilic membrane containing the cross-linked enzyme within its voids. The latest version is a single-membrane reactor (Lopez and Matson 1997). The enzyme is trapped within the asymmetric side (i.e., in the voids) of a solvent-resistant, hydrophilic, hollow fiber membrane. The interior skin of the hollow fiber has a pore size sufficient to retain the enzyme. The hollow fibers are placed in a conventional stainless steel housing. During operation, the voids are filled with the enzyme solution. An aqueous stream flows in the interior lumen of the hollow fibers, and an organic solvent stream flows in the shell side (Figure 8.73). This organic stream, at a slightly higher pressure than the aqueous phase, effectively "immobilizes" the enzyme within the voids. Substrate dissolved in the organic solvent is pumped through the shell side and partitions into the membrane where it is enzymatically converted to a water-soluble product, which subsequently diffuses out into the interior lumen stream. Enzyme solutions can be easily replaced by

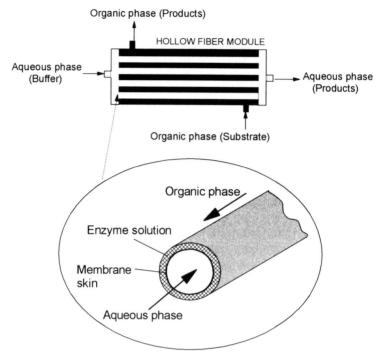

Organic phase (Products)

HOLLOW FIBER MODULE

Aqueous phase (Buffer)

Aqueous phase (Products)

Organic phase (Substrate)

Organic phase

Enzyme solution

Membrane skin

Aqueous phase

Figure 8.73. *Multiphase membrane bioreactor for simultaneous enzyme reaction and extraction. Top shows general flows through the reactor module. Bottom enlargement shows how different phases are localized in the hollow fiber (adapted from Lopez and Matson 1997).*

reversing the pressure differential and flushing out the old enzyme. This concept has been successfully demonstrated with the separation of racemic mixtures of chiral drugs and drug synthons. Over 75 tons per year of resolved glycidate ester used in the manufacture of Diltiazem, a calcium channel blocker, is being produced in a 1440-m^2 membrane reactor (Lopez and Matson 1997).

8.N.4.e.
FERMENTATIONS

Fermentations, using microbial cells as the biocatalyst, suffer from the same problems as enzymatic reactions, except that end product inhibition is more severe, resulting in fairly low product concentrations. For example, in the production of ethanol by yeast, alcohol productivity at 6% v/v alcohol is only half that at zero alcohol, while it is only 1/100 at 12% ethanol. This results in long times needed for the fermentation, sometimes measured in days, resulting in low fermenter productivities in conventional batch fermentations (productivity is measured as amount of product per unit volume per unit time, e.g., g/L/h). The following approaches can be taken for designing high-performance fermentation systems:

- Use a continuous process instead of a batch fermentation.
- Operate at high dilution rates (where dilution rate, D, is flow rate/volume).
- Maintain high cell densities at all times to allow the high dilution rates.
- Continuously remove inhibitory end products from the bioreactor.

The advantages of conducting fermentations in a continuous mode instead of a batch mode include reduction in capital cost, better process control, and improved productivity. However, the simple CSTR, operated "continuous-culture" fashion, has one major disadvantage. Dilution rate cannot exceed the maximum growth rate of the microorganism, or there will be cell "washout." Recycling the cells results in much higher productivity. Although modern centrifuges are fairly efficient for cell separation, they are expensive to buy, to operate, and to maintain (Section 8.N.1.). The other alternative, immobilization in the conventional sense, also suffers from several problems, as discussed earlier.

The use of semipermeable membranes with microorganisms dates back as early as 1896 when Metchnikoff et al. attempted to show the existence of diffusable cholera toxin in cultures of *Cholera vibrios* contained in collodion sacs (Schultz and Gerhardt 1969). Gerhardt and coworkers performed several pioneering experiments on the use of in vitro dialysis culture systems for a variety of applications (Abbott and Gerhardt 1970; Steiber and Gerhardt 1981). They showed that continuous removal of metabolic products would result in a superior fermentation process. However, the dialysis culture system may not be practical on a large scale, since the reaction rate will be limited by the rate at which substrate and product can diffuse through the membrane. Pressure-activated processes using MF or UF membranes should be more efficient. Michaels (1968) suggested this concept, and Budd and Okey (1969) may have been the first to apply cell recycle with membranes for sewage treatment. This is apparently the basis of the MARS anaerobic treatment process marketed by Dorr-Oliver (Figure 8.74). A similar concept was used to demonstrate the treatment of several types of wastewater as part of Japan's Aqua-Renaissance '90 project (Kimura 1991).

The most attention with membrane fermenters has been given to the production of ethanol and lactic acid. Both MRB and plug-flow reactors have been studied. A typical flow sheet for the conversion of cheese whey to ethanol and to lactic acid is shown in Figure 8.75. The cheese whey (containing 4.5% lactose, 0.4–0.6% true protein) first undergoes the necessary pretreatment, such as clarification by MF or centrifugation and/or pasteurization, prior to UF for the production of whey protein concentrates. The permeate, containing the lactose, nonprotein nitrogen, and dissolved salts, is concentrated by RO to the

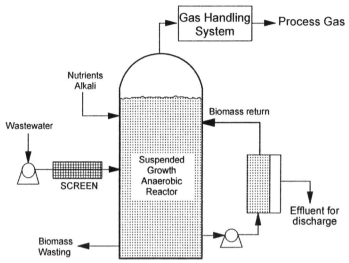

Figure 8.74. Schematic of the Dorr-Oliver MARS (Membrane Anaerobic Reactor System) process for waste treatment (adapted from Dorr-Oliver 1982).

appropriate sugar concentration for the fermentation ($2.3\times$ for lactic acid, 4–$5\times$ for ethanol). Nutrients or other additives can be incorporated at this stage and the feed clarified and sterilized by MF before being pumped into the reaction vessel of the membrane recycle fermenter, which contains a preconcentrated suspension of the microorganism. In the case of ethanol production by the yeast

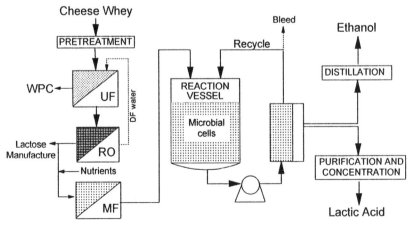

Figure 8.75. Membrane processing of cheese whey to produce whey protein concentrates (WPC), concentrated lactose, and fermented products using membrane recycle bioreactors.

Kluyveromyces fragilis, a cell concentration of 90 g/L is optimum. Lactic acid production with *Lactobacillus* sp. would require about 60 g/L. Excess cells, due to continued cell growth in the system, can be bled off as shown, through a microfilter if necessary.

Tables 8.25 and 8.26 are comparisons of the performance of membrane bioreactors with batch fermenters for the production of ethanol and lactic acid using various substrates and microorganisms. The apparent superiority of the membrane recycle bioreactor is obvious. The major reason is high cell concentrations [as much as 150 g/L of *S. cervisiae* used by Mehaia and Cheryan (1990b)] that allowed high dilution rates with complete substrate utilization. However, under optimal conditions, high productivity and high product concentration are mutually exclusive, as shown in Figure 8.76. Thus, the trade-off is between high productivity in the fermentation step and higher downstream purification costs caused by the lower product concentration. It is likely that membrane fermenters will be restricted to small improvements in productivity (e.g., two to five times higher) with the added benefit of a cell-free clarified product stream from the membrane unit. In the case of ethanol fermentation, it should also reduce stillage handling significantly.

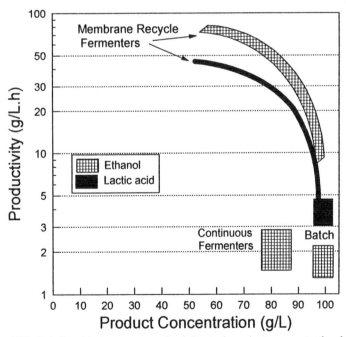

Figure 8.76. *Relationship between productivity and product concentration in high-rate bioreactors. Data shown are typical of that obtained in membrane recycle bioreactors for production of ethanol or lactic acid from glucose or lactose.*

Table 8.25. Comparison of bioreactors for producing ethanol. The substrate was glucose unless otherwise indicated.

Bioreactor	Organism	Initial Sugar (g/L)	Ethanol Produced (g/L)	Conversion (%)	Productivity (g/L · h)	Reference
Batch	S. cerevisiae	100	49	99	2–3	Cheryan and Mehaia (1984a)
		200	98	99	4–5	Cheryan and Mehaia (1984b)
	Z. mobilis	250	102	80	3–4	Rogers et al. (1982)
Hollow fiber	S. cerevisiae	89	12	27	26	Inloes et al. (1983)
		100	40	85	10	Mehaia and Cheryan (1984a)
Membrane recycle	S. cerevisiae	100	49	99	100	Cheryan and Mehaia (1984b)
		100	35	99	95	Lee et al. (1991)
		200	65	65	130	Cheryan and Mehaia (1984b)
		150	76	99	55	Groot et al. (1992)
		140	65	99	85	Lee and Chang (1987)
		195	90	94	32	Lee and Chang (1987)
		140*	65	99	31	Mehaia and Cheryan (1990b)
	S. bayanus	133	62	99	80	Mota et al. (1987)
	S. carlsbergen.	—	63	—	69	Asakura and Toda (1991)
	Z. mobilis	100	44	99	120	Rogers et al. (1982)
		120	49	90	44	Nipkow et al. (1986)
		100	45	99	44	Khorakiwala et al. (1987)
		150	72	99	29	Khorakiwala et al. (1987)

*Hydrolyzed whey permeate

466

Table 8.26. Comparison of bioreactors for producing lactic acid. The substrate was glucose unless otherwise indicated.

Bioreactor	Organism	Lactate (g/L)	Conversion (%)	Productivity (g/L·h)	Reference
Batch	L. bulgaricus*	49	99	2–3	Mehaia and Cheryan (1984b)
	L. amylovorus**	96	90	4.8	Zhang and Cheryan (1994)
Hollow fiber	L. bulgaricus*	12	26	2	Mehaia and Cheryan (1984b)
	L. delbrueckii	2	4	—	Vick Roy et al. (1982)
Membrane recycle	L. bulgaricus*	28	80	25	Hamilton and Howell (1983)
		43	99	85	Mehaia and Cheryan (1986)
		117	99	84	Mehaia and Cheryan (1987)
	L. delbrueckii	35	78	76	Vick Roy et al. (1983)
		59	99	151	Ohleyer et al. (1985)
	L. helveticus*	27	—	22	Boyaval et al. (1987)
		43	92	8	Aeschlimann and von Stockar (1991)
	L. amylovorus**	42	92	8.4	Zhang and Cheryan (1994)
	L. cremoris*	48	90	50	Bibal et al. (1991)

*Lactose, whey or whey permeate substrate
**Starch substrate

467

Hollow fiber fermenters (HFFs), where the cells are packed into the shell side of the cartridge (as shown in Figure 8.66) have not fared as well as membrane recycle fermenters (Tables 8.25 and 8.26). Either productivities have been much lower when compared on the same basis of substrate utilization, or the reactors have been quite short-lived. For example, substrate conversions of only 3–4% were obtained for lactic acid production in an HFF compared to essentially 100% substrate utilization in an MRB. The relatively poor performance of HFFs compared to MRB is a reflection of several practical problems that arise during operation. Because the microbial cells are separated from the substrate and product stream by a physical barrier (the membrane), the rate limiting step becomes the diffusion of substrate into and the diffusion of product out of the shell side of the hollow fiber cartridge. Visual observation of the shell side revealed that most of the cells were attached to the fibers or were localized to an annular volume just around the fibers. Thus, the cells far from the fibers were probably starved of nutrients and dying. These cells are probably continuously lysing, and so they must be removed from the system. Otherwise, the growing mass of cells may even rupture the relatively sensitive fibers, especially if a gas is also produced in the fermentation. In fact, gas accumulation in the shell side is one of the biggest problems in HFF; the high pressures, unless constantly relieved, could rupture the fibers or cause cell grow-through, thus contaminating the product stream. pH control is difficult in HFFs because of the lack of bulk mixing. By the same token, HFFs are ideal if one wishes to avoid contaminating the culture with other microorganisms or macromolecules. Hollow fiber bioreactors appear to be best suited for reactions involving nongrowing cells or tissue culture applications.

Other fermentations studied, mostly on laboratory-scale equipment, include

- Acetone-butanol-ethanol using *Clostridium acetobutylicum* (Minier et al. 1984; Pierrot et al. 1986)
- Conversion of D-sorbitol to L-sorbose by *Gluconobacter oxydans* and glucose to 2-ketogluconic acid by *Serratia marcescens* (Bull and Young 1981)
- Citric acid from glucose using the yeast *Candida* (Enzminger and Asenjo 1986; Rane and Sims 1995)
- Propionic acid (Colomban et al. 1993; Crespo et al. 1991)
- Acetic acid: In the anaerobic process from glucose using *Clostridium thermoaceticum* (Parekh and Cheryan 1994), the productivity was 10-fold higher than the batch fermenter. The aerobic process for vinegar production from ethanol uses *Acetobacter aceti*. In one case, the MRB gave a productivity of 8 g/L · h (Reed and Bogdam 1986). In another, it was 12.6 g/L · h which is 4.6 times the batch fermentation productivity (Park et al. 1989).

- 2,3-butanediol from glucose by *Klebsiella oxytoca*: Productivity was two to three times that of the batch reactor at comparable concentrations of product and glucose utilization (Qureshi and Cheryan 1989; Ramachandran and Goma 1988).

8.N.4.f.
COFACTOR IMMOBILIZATION IN MEMBRANE REACTORS

Cofactors, such as NAD, NADP, Co-A, ATP, etc., are required in many reactions involving covalent bond synthesis, energy transfer, group transfer, or redox reactions. Hence, if continuous processes that involve these reactions are to be developed, a system of cofactor regeneration, recycling, and reuse must also be developed. Membrane systems have been the object of many investigations, where either the NAD is bound to a water-soluble polymer and then entrapped in a membrane reactor with conjugated enzymes (Morikawa et al. 1978; Wichmann et al. 1981; Yamazaki et al. 1976), or native NAD is entrapped in a bioreactor with a membrane whose MWCO is lower than that of the cofactor (Chambers et al. 1981) or where the conjugated enzymes are immobilized but the cofactor is not (Fink and Rodwell 1975; Miyawaki et al. 1982).

Figure 8.77 shows a generalized flow scheme for the latter system. The conjugated enzymes were trapped in the tube side of the hollow fiber cartridge and the substrates and NAD were continuously fed through the shell side, where

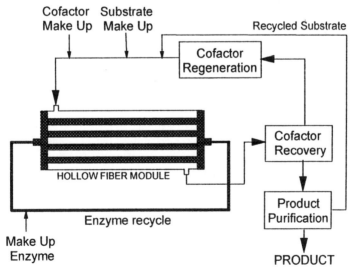

Figure 8.77. *Generalized scheme for a hollow fiber reactor with cofactor regeneration. The enzymes are trapped in the tubes, while substrate and cofactor are fed in from the shell side (adapted from Fink and Rodwell 1975).*

NAD was "dynamically recycled." It appears that NAD was concentrated locally in the hollow fiber tube as enzyme–coenzyme complexes. NAD recycle numbers of over 1000 with conversions of 50% are possible at residence times as low as 1 h, provided proper reactor design and operating conditions are employed. This method appears to be simpler than the other methods, especially those using conventional immobilization techniques, although a good system of recovery of the coenzyme is needed to reduce overall costs.

8.O.
FRUIT JUICES AND EXTRACTS

There are three primary areas where membranes can be applied in processing of fruit juices:

1. Clarification, e.g., in the production of sparkling clear beverages using MF or UF
2. Concentration, e.g., using RO to produce fruit juice concentrates of greater than 42° Brix
3. Deacidification, e.g., ED or NF to reduce the acidity in citrus juices

This section focuses on the first application (clarification): details on the other applications have been provided by Cheryan and Alvarez (1995).

Traditional methods of producing clear single-strength fruit juices involve several batch operations that are labor- and time-consuming (Figures 8.78 and 8.79). After preliminary sorting, washing, and peeling steps, depending on the fruit, the fruit is crushed and sent through presses and screens to remove large particulate matter. If a clear juice is to be produced, the press juice is pasteurized and then treated with an enzyme (pectinase) to hydrolyze the pectin and reduce cloudiness. The enzyme treatment also makes the subsequent filtration easier, presumably by lowering juice viscosity. The enzyme treatment is followed by a preliminary clarification step, where a fining agent such as gelatin is added and the juice held for 20–30 h. After decanting, the juice goes for precoat filtration, using diatomaceous earth DE as a filter aid to remove suspended solids, colloidal particles, proteins, and condensed polyphenols. The final clear juice may be pasteurized again before packaging.

The primary goal of membrane filtration in the fruit juice industry is to replace the holding, filtration, and decantation steps. Research in the late 1970s indicated that MF or UF could be used successfully to replace several of these operations. Figure 8.78 compares traditional and UF processes for producing clear juices, and Figure 8.79 shows a typical schematic of a membrane process. However, not all fruit extracts are candidates for UF processing. For example, of the four common types of apple juice produced—natural, crushed, clarified, and clear—only the clear juice is suitable for membrane processing.

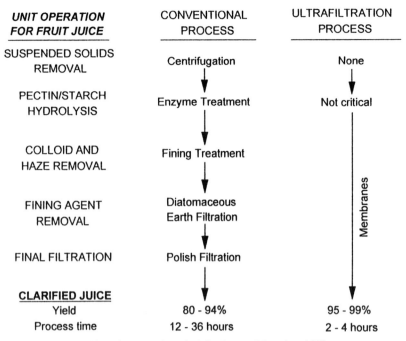

UNIT OPERATION FOR FRUIT JUICE	CONVENTIONAL PROCESS	ULTRAFILTRATION PROCESS
SUSPENDED SOLIDS REMOVAL	Centrifugation	None
PECTIN/STARCH HYDROLYSIS	Enzyme Treatment	Not critical
COLLOID AND HAZE REMOVAL	Fining Treatment	
FINING AGENT REMOVAL	Diatomaceous Earth Filtration	Membranes
FINAL FILTRATION	Polish Filtration	
CLARIFIED JUICE		
Yield	80 - 94%	95 - 99%
Process time	12 - 36 hours	2 - 4 hours

Figure 8.78. Manufacture of apple juice by traditional and UF processes.

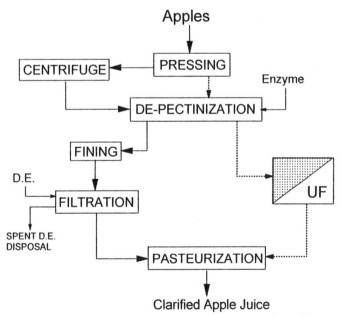

Figure 8.79. Typical schematic for producing clear fruit juices. Operations on the left side are used in the conventional method. The membrane method is depicted on the right side.

471

The MWCO of the membrane affects flux and the clarity of the juice, as well as browning compounds and total phenolics of the finished product (Figure 8.80). Nevertheless, a wide range of pore sizes are being used today in the industry, from 18,000 MWCO to 0.2 μ. Flux values in the literature (see Alvarez et al. 1996 for a listing) have ranged from 35 LMH with undepectinized juice with a 30,000 MWCO membrane to almost 300 LMH with depectinized apple juice with 0.2-μ ceramic membranes. The method used for extracting the juice affects the suspended solids in the raw juice, which affects the flux.

Historically, 1″ (25 mm) tubular modules were the first to be used in juice clarification. With these modules, the partially depectinized juice could go directly to the UF system, depending on the suspended solids in the raw juice. If thin-channel or hollow fiber equipment is used or if the press generates high suspended solids, some prefiltration or centrifugation is necessary. The suspended solids level in the raw juice is important since the practical limit of pumpability of the retentate is about 40% v/v. It is necessary to have flow diverters in the loops that are preset to divert and dilute the retentate at the first sign of loss of retenate flow, to minimize catastrophic blockage of membrane tubes.

In recent years, several fruit juice installations have incorporated ceramic membranes. The higher cost has been justified by the higher flux, much longer life, and its resistance to aggressive processing and cleaning conditions. The ability to backflush to unblock feed channels and backpulsing during operation are other advantages (Padilla and McLellan 1993).

Since fruit juices have a very low level of retained solids, the optimum mode of operation is the modified batch operation with partial recycle of retentate (Figure 7.10), i.e., where the bulk of the retentate is within the recycle loop and a small portion is used to top off the feed tank. This offers considerable savings over multistage recycle systems. Further savings in capital cost and reduction in holdup volume can be achieved by a judicious arrangement of modules in series and in parallel.

For the manufacture of clear single-strength juices, the permeate is the desired stream. Since UF membranes are being used, the final product is essentially sterile and, if handled carefully and in an aseptic manner, should not require a subsequent heat treatment prior to storage or bottling. The retentate can be subjected to a DF operation to recover more of the sugars and low molecular weight soluble compounds. Being dilute, however, the DF permeate is usually sent for concentration by evaporation or RO for incorporation into juice concentrates. The residue of the retentate contains potentially valuable pectin and enzymes. As a result of the high ratio of permeate to feed during the UF and the DF operation, yields of 98–99% are common.

Membrane filtration has several advantages over traditional methods:

- Clarification and fining done in one step: While traditional methods require fining agents (bentonite, gelatin, etc.), enzymes (pectinase,

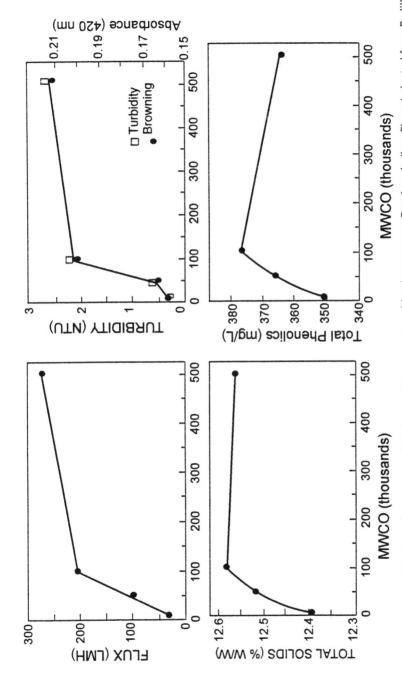

Figure 8.80. Effect of MWCO of membranes on apple juice quality parameters. Membranes were Romicon hollow fibers (adapted from Padilla and McLellan 1989).

473

amylase), centrifugation, and diatomaceous earth filtration that may take 12–36 h, membrane cross-flow filtration avoids these steps and can produce a clear and satisfactory product in 2–4 h.

- Increased juice yield: Juice recovery of 96–98% is typical without diafiltration or predilution. In contrast, conventional processes recover only 80–94% of the juice. This increased recovery comes from elimination of the diatomaceous earth and fining agent, which can absorb juice. A nominal 1% increase in recovery in a plant processing 350,000 L per day can increase revenue by $150,000 per season.

- Elimination of filter aid and filter presses: This alone can save about $10–25 per 1000 gal of juice, as well as eliminate the troublesome disposal of the diatomaceous earth and fining agents.

- Better product quality: The reliable and consistent removal of haze-forming components (suspended solids, colloidal particles, proteins, and polyphenols) results in superior juice quality, typically less than 0.1–0.3 NTU compared to the 2–5 NTU from conventional processing (Table 8.27). A clear sparkling juice is produced, free from cloudiness and sediment. This is not only due to the removal of pectins and other large carbohydrates, but also because of reduced tannin–protein complexes in the finished juice. However, if the membrane is $>25,000$ MWCO, tannins may pass into the permeate (the clarified juice), resulting in a brownish color and sharp flavor. High tannin passage could explain the eventual clouding of membrane-filtered samples on storage. Tighter membranes produce a light golden colored product. Polyphenol oxidase, responsible for the "browning" reaction, is also retained by the membrane and could be responsible for the post-bottling haze formation sometimes observed on storage of MF juice, especially with juice that

Table 8.27. Composition of raw and microfiltered apple juice. The MF was done with a Carbosep inorganic membrane of 0.2-μ pore size at 50°C.

	Raw Apple Juice	Microfiltered Juice
Total solids (% w/v)	12.46	11.62
Total sugars (% w/v)	11.2	11.0
°Brix	11.8	11.0
pH	3.45	3.50
Acidity (as % malic acid)	0.486	0.435
Density (g/mL)	1.048	1.044
Viscosity (cP)	1.88	1.42
Turbidity (650 nm)	1.2	0.025
Total phenols (g/L)	0.218	0.016
Starch	Positive	Negative

Source: Data from Alvarez et al. (1996)

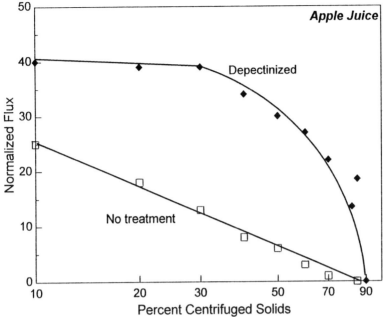

Figure 8.81. Effect of pectinase enzyme treatment on apple juice flux with Romicon hollow fiber membranes (adapted from Romicon, 1988).

has been processed at higher temperatures. Greater flavor losses have been observed with some polysulfone and polyamide UF membranes compared to conventional plate-and-frame filter presses for apple juice (Rao et al. 1987).

- Reduced enzyme usage: Although pectinase enzyme can be eliminated altogether, some pretreatment with enzymes help reduce membrane fouling and viscosity, resulting in higher flux and lower pumping energy (Figures 8.81 and 8.82). For example, untreated apple juice had a flux of 35 LMH at 28°C in a Romicon hollow fiber system, while "polished" feed had a flux 120 LMH, under otherwise identical operating conditions. The pectinase can be added to the feed tank of the UF unit during processing. It just recycles within the loop during operation (akin to a membrane recycle bioreactor), which adds to the savings in enzyme. In general, enzyme usage is reduced 50–75% depending on the juice.

All these advantages result in higher profits for the UF process (Figure 8.83). For example, a Cadbury-Schwepp's apple juice plant processing over 500,000 L per day reported savings of $350,000 per year caused by the elimination of more than 350,000 kg per year of diatomaceous earth, a major reduction in labor

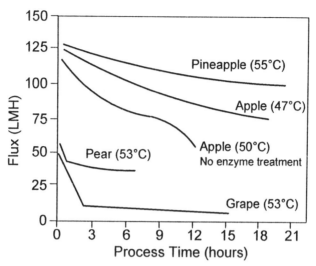

Figure 8.82. Typical flux with Koch SuperCor tubular UF membranes with various fruit juices (adapted from Blanck and Eykamp 1986).

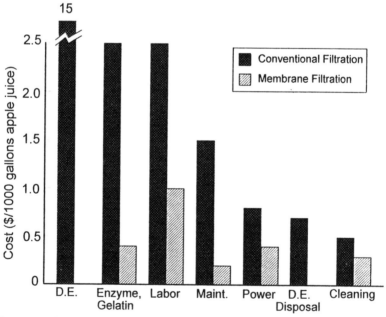

Figure 8.83. Cost advantages of cross-flow membrane filtration versus conventional filtration using diatomaceous earth (D.E.). Based on 1990 estimates by Koch Membrane Systems.

costs, and a 4% increase in yield (Koch 1988). Another apple juice processor installed a tubular UF system to handle 25,000 gal every 6 h, which resulted in a savings of $50,000/year in filter aid, reduced enzyme usage by 50%, increased juice yield by 5%, reduced process time by 75%, and enhanced product quality (Hackert and Swientek 1986).

Fruit and vegetable juice has been one of the most successful applications of membrane technology. In additional to apples, the following have also received considerable attention: apricot, carrot, cherry, cranberry (one of the earliest applications of ceramic membranes), (black) currant, grape (white and red), kiwi, lemon, lime, melon, orange, passion fruit, peach, pear, pineapple, plum, raspberry, strawberry, and tomato. With passion fruit juice, even a UF membrane (PCI BX6, 25K MWCO) retained over 80% of the flavor compounds (Yu et al. 1986). Sugars and organic acids had a retention of 10–40%. The high pressures (12 bar) and low temperature (20°C) could have created a dynamic fouling layer that increased retention during processing. Retentivities of these compounds increased with the concentration ratio. Membrane foulants were cellulose, hemicellulose, sugars, citric acid, and pectin.

Pear juice has been processed by UF since the mid-1980s. Membrane MWCOs between 10 K and 50 K have little influence on permeate juice color (Kirk et al. 1983). Optimum operating conditions were 157 kPa pressure, 0.15 m/sec, and 50°C. Permeate flux declined with the logarithm of the concentration.

Pineapple juice drainings from the skin and other discarded parts of the fruit are first clarified by UF prior to concentration for use as a natural sugar source in canned fruit products or as a drink base.

Kiwi fruit juice processing has been reported by Wilson and Burns (1983). A dramatic improvement in clarity was obtained, and it allowed the nonthermal sterilization of kiwi fruit juice. Storage for 6 months at 15°C resulted in only slight deterioration as browning and ascorbic acid loss. Ultrafiltration of strawberry juice apparently has a detrimental effect on color and appearance of the product, perhaps caused by a loss of anthocyanins by adsorption on the membrane (Rwabahizi and Wrolstad 1988).

Maple sap could be clarified with a 10 K membrane, using pressures of 190–200 kPa (Pouliot and Goulet 1987).

Enzyme-treated apricot puree could be microfiltered with a 0.45-μ ceramic membrane to obtain a sparkling, clear apricot juice at fluxes of 90–190 LMH (Hart et al. 1989). Concentrating the permeate subsequently by vacuum evaporation resulted in a concentrate with good retention of clarity, flavor, and aroma.

Citrus juices (orange, lemon, lime) are being ultrafiltered or microfiltered prior to debittering (Figure 8.84) or concentration (Figure 8.85). The "upgrading" of fruit juice combines UF and adsorbent resin technology to remove bitter compounds such as limonin (found in all citrus juices), naringin (grapefruit), hesperidin (orange, lemon, lime), polyphenols (found in all juices: its removal enhances cloud stability and reduces astringency and browning reactions), and

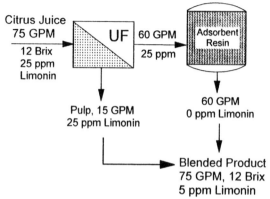

Figure 8.84. *Debittering citrus juices by ultrafiltration and adsorption (adapted from Romicon 1988; Koch 1991).*

many other off-flavor compounds. These compounds are in the aqueous phase of the juice. Before it is contacted with the adsorbent resin, the suspended solids and pulp are separated to minimize fouling of the resin (Koch 1991). Fresh or reconstituted citrus juice that has been deoiled and pasteurized is first ultrafiltered to 5–10× to separate the pulp. The clarified permeate, containing the sugars and bitter compounds, enters the adsorption column, which contains an adsorbent resin, e.g., polyaromatic macroreticular Amberlite XAD-16 with no functional groups from Rohm and Haas or divinylbenzene copolymer XU-43520 from Dow. These resins are specifically designed to remove these compounds. The debittered juice is then recombined with the pulp (the UF retentate) to give a product with less than 5 ppm limonin, which is its apparent taste threshold (400–500 ppm for naringin). The adsorbent resin has a 4-year life and can

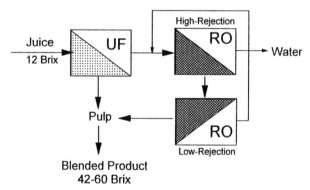

Figure 8.85. *Producing high-strength fruit juice concentrates by ultrafiltration and reverse osmosis (adapted from Cheryan and Alvarez 1995).*

Table 8.28. Comparison of operating costs ($ per year) for debittering orange juice. Estimates are for a 50 gpm line operating for 3600 hours per year.

	Conventional (Centrifuge + Hydrophilic Adsorbent)	Membrane (+ Adsorbent Resin)
Electric power ($0.05/kWh)	9,000	19,000
Steam ($4/1000 kg)	3,000	2,000
Water ($1/m^3)	9,000	2,000
Chemicals	45,000	4,000
Labor ($12/hour)	5,000	5,000
Solids loss (pulp in column)	40,000	0
Solids loss (regeneration)	40,000	9,000
Adsorbent ($40,000/m^3)	35,000	10,000
Membrane replacement	0	75,000
Total annual cost	$186,000	$126,000
Cost per ton of 68 °Brix concentrate	$34	$23

Source: Data from Norman (1994)

be steam cleaned. The membrane/adsorbent resin process is cheaper than the conventional centrifuge/hydrophilic adsorbent system (Table 8.28).

There are several UF plants clarifying lime juice at the point of production. Lime juice is very corrosive to plastic components, especially the seals and gaskets. UF also recovers lime oil, which is a valuable by-product. Yields are usually 80–95%. Lemon juice (8.4°Brix, 0.3% pulp) processed with Nitto-Denko's UF membranes (3250-C3R and 520-P18LP) resulted in a clear permeate (Table 8.29). Initial flux was 50 LMH at 15°C and 2 bar pressure and 20 LMH at VCR 6 (Yabushita 1989).

The second application of UF/MF in citrus juices is as a pretreatment for producing concentrates (Koseoglu et al. 1990). RO can produce highly concentrated (42–60°Brix) fruit juices using a combination of high- and low-retention

Table 8.29. Properties of lemon juice before and after ultrafiltration.

	Optical Density (@420 nm)			Total Nitrogen (mg/L)		
Feed Brix	Feed	Permeate (NTU-3250)	Permeate (NTU-3520)	Feed	Permeate (NTU-3250)	Permeate (NTU-3520)
8.4	2.808	0.087	0.089	550	380	410
15.4	5.832	0.193	0.177	960	740	750

NTU-3250: Capillary polysulfone membrane module of 20,000 MWCO
NTU-3520: Tubular polysulfone module with 20,000 MWCO
Source: Data from Yabushita (1989)

RO membranes (Cheryan and Alvarez 1995). As shown in Figure 8.85, UF is first used with the juice to separate out the pulp from the serum, which contains the sugars and flavor compounds. The pulp fraction contains some soluble solids, all the insoluble solids, pectins, enzymes, orange oils, and the microorganisms that would affect the stability of the concentrate. This UF retentate, about 1/10th to 1/20th the feed volume, is subjected to a pasteurization treatment that destroys spoilage microorganisms and improves stability of the finished product when blended back with the concentrated UF permeate.

The serum (UF permeate), which is 90–95% of the feed volume, is concentrated by RO with high-retention membranes in the early stages and low-retention membranes (i.e., leaky, "loose" membranes) in later stages. The concentrated serum can then be blended back with the pulp or bottom solids stream.

Honey has been successfully processed by UF and has been commercially available in Japan since 1988 (Kato 1996). The UF honey is used primarily as a sweetening agent in beverages. Raw honey is first diluted with water and its temperature adjusted before UF. The result is a microbe-free and clear product that does not cloud when mixed with other beverages.

8.P.
ALCOHOLIC BEVERAGES

8.P.1.
WINE

There are several potential benefits of MF and UF in winemaking (Figure 8.86). UF can be used either before the fermentation (i.e., for clarifying

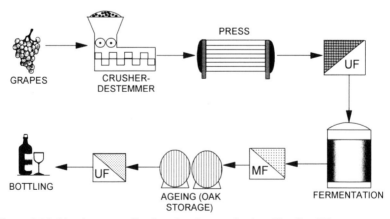

Figure 8.86. *Membrane applications in wine production. The first UF step removes microorganisms, colloids, and high molecular weight materials. The second membrane step (MF) removes yeast. The third membrane step is a final filter: it could also be a sterilizing microfilter.*

the "must") or after (for treating the finished wine). UF of the must results in removal of colloids, high molecular weight tannins, polysaccharides, haze proteins, suspended solids, gums, polyphenols, and all undesirable microorganisms. The latter may be particularly significant since wild microorganisms are traditionally controlled by adding bisulfite to the must. Not only does bisulfite cause undesirable off-flavors and increase costs, but it has recently been implicated in potentially life-threatening side reactions for asthmatics.

After fermentation and before storage, MF can be used to remove the yeast or UF to improve stability of the finished wine. Haze is a common problem in wines, especially white wines. Haze is caused by proteins that are unstable or insoluble at the acidic pH of the finished wine. This has traditionally been removed by adding fining agents such as bentonite. However, some of the flavor compounds may also be removed by the fining agent. UF can remove the haze-causing proteins while minimizing flavor loss (if UF was done to the must, this will be less of a problem).

Polyphenols (tannins) give wine its color and subtle, complex flavors. Excessive polyphenol levels lead to a chalky and astringent flavor. Normally, gelatin is added as a fining agent to remove polyphenols. However, gelatin is a colloidal protein, and the tannin–protein complex must then be removed by decanting or filtration. UF can reduce the tannin concentration but not completely eliminate it since some of the polyphenols may pass through the membrane. However, the enzymes (polyphenol oxidase) that cause undesirable brown compounds to form are removed by UF, thus improving the color/flavor stability. The final step before bottling could be sterile filtration with MF membranes.

Several studies have been carried out showing that membrane pore size plays a crucial part in the retention of color and aroma compounds in white and especially in red wines. Lower MW proteins (12,600–30,000 MW) and glycoproteins are the major fractions contributing to protein instability in wines, and 99% are retained by 10,000 MWCO UF membranes. This would reduce bentonite requirements by 80–95% (Hsu et al. 1987). Other low MW components such as ethanol, acidity, or reducing sugars were not modified by either conventional filtration or UF (Peri et al. 1988). Tight UF membranes resulted in a large reduction in colloids (Figure 8.87). However, colloid removal did not correlate with an improvement in the clarity of the wine, as shown by the turbidity data with the same membranes in Figure 8.87. On the contrary, the elimination of soluble colloids may affect the taste and impair the stability of the wine with regard to tartarate precipitation. Regulations in some countries permit the use of cross-flow filtration in wineries provided the residual colloid content remains above 100 mg/L.

With red wine, partial removal of polyphenols/tannins resulted in a better taste because of a lesser astringency. However, the 20 K UF membrane removed too much anthocyanins, resulting in an unacceptable loss of red wine color. The

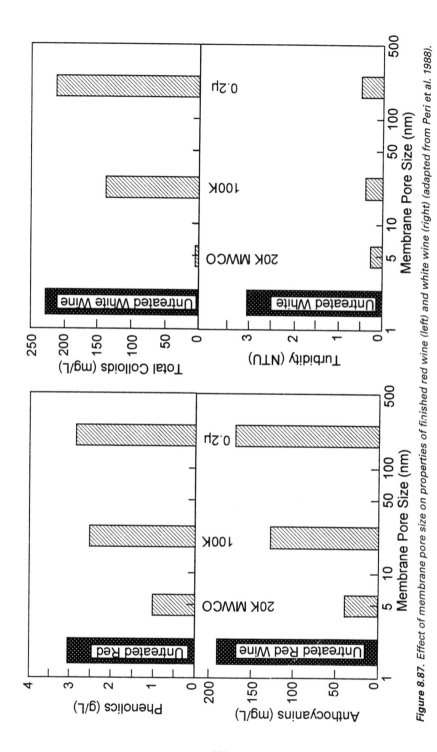

Figure 8.87. *Effect of membrane pore size on properties of finished red wine (left) and white wine (right) (adapted from Peri et al. 1988).*

best membrane to use that meets all these requirements may be an open UF membrane of 100 K–500 K MWCO.

Thus, UF and MF have the potential for considerable savings in capital and operating costs by eliminating filter presses, porous candle filters, cartridge filters, and centrifuges that are currently used in the wine industry. "Wine filters," built with Romicon hollow fibers of 1.8 mm diameter, are being used in Europe for final filtration. These relatively small units contain 2–32 MF cartridges capable of filtering 80–2600 gal per hour, depending on the wine. The smaller ones are priced at about $15,000–30,000 each, and there are over 500 of them in use today. Ceramic MF units are also being used for this purpose in Spain and Italy. Although capital costs of the ceramic system ($0.088/hL of wine) are double that of a diatomaceous earth filtration system ($0.042/hL), the savings in filters alone ($0.076 versus $0.263/hL) justifies the ceramic system (Short 1995b).

Vinegar is typically a product of fermentation of ethanol. The fermentation broth must be clarified to remove the microbial cells, vinegar eels, and residual nutrients. UF to a VCR of 30 (i.e., a 97% product recovery in the permeate) can easily be obtained at an average flux greater than 60 LMH. The membranes must be carefully selected to ensure the quality of the product.

8.P.2.
BEER

The application of membranes in the manufacture of beer takes an approach similar to wine and fruit juice. Membranes could be used for removal of yeast from the beer in the fermentation vessel and for removal of "haze" proteins. This will eliminate the need for diatomaceous earth (*Keiselguhr*) filtration and all associated problems and reduce or eliminate enzymes (e.g., papain) for chill-proofing and fining agents. Of special value is the ability to control microstability of the beer for extended periods. Spoilage organisms are ubiquitous and difficult to control in a brewery. If the beer is to be stored for several weeks unrefrigerated or kegged, then the presence of unwanted microorganisms is a concern, since they may cause undesirable flavors, aromas, and appearance in the form of precipitates, haze, or stringiness. Tunnel or flash pasteurization is the usual method of providing microstability, but in addition to the cost, there is the disadvantage of exposure of the beer to heat. MF membranes of 0.45 μ will retain almost all the troublesome bacteria (larger pores such as 0.65 μ may not stop *L. brevis* or *P. damnosus*). A good combination of prefilters, cross-flow, and final filters should preserve the beer's natural flavor with little or no exposure to heat and oxygen and at a cost comparable to or less than traditional pasteurization methods (Meier 1995; Membre et al. 1991; Ryder et al. 1988).

Tank bottoms recovery is another potential application of membranes. After fermentation, a large proportion of the yeasts flocculate and settle to the bottom of the vessels. These so-called fermentation vessels (FV) tank bottoms may account for 2–4% of the total volume in the tank. Tank bottoms contain beer and represent a significant loss to the brewers. Conventional equipment used for the recovery of that beer include vacuum filters, filter presses, and centrifuges. Membranes offer the advantage of obtaining a bright, clear beer without the use of filter aids. Up to 40–60% of the tank bottoms can be recovered as the permeate, which is microbe-free with a neutral flavor and aroma. The retentate can contain 20–26% total solids, which can be sold to yeast processors.

REFERENCES

ABBOTT, B. J. and GERHARDT, P. 1970. *Biotechnol. Bioeng.* 12: 577.

ABRAHAMSEN, R. K. and HOLMEN, T. B. 1980. *Milchwissenschaft.* 35: 399.

ADHAM, S. S., JACANGELO, J. G. and LAINE, J. M. 1996. *J. AWWA.* 88(5): 22.

AESCHLIMANN, A. and VON STOCKAR, U. 1991. *Enz. Microbial Technol.* 13: 811.

AIMAR, P., TADDEI, C., LAFAILLE, J. P. and SANCHEZ, V. 1988. *J. Membrane Sci.* 38: 203.

AKAY, G. and WAKEMAN, R. J. 1993. *Chem. Engr. Res. & Design.* 71: 411.

AKRED, A. R., FANE, A. G. and FRIEND, J. P. 1980. *Polymer Sci. Technol.* 13: 353.

ALMAS, K. A. 1985. *Desalination* 53: 167.

ALVAREZ, V., ANDRES, L. J., RIERA, F. A. and ALVAREZ, R. 1996. *Can. J. Chem. Engr.* 74: 156.

AMAR-REKIK, R. B., BEJAR, S. and ELLOUZ, R. 1994. *International Sugar J.* 96: 434.

ANG, H. G., KWIK, W. L., LEE, C. K. and THENG, C. Y. 1986. *Food Chemistry* 20: 183.

ASAKURA, T. and TODA, K. 1991. *Bioprocess Engr.* 7: 83.

AZHAR, A. and HAMDY, M. K. 1981. *Biotechnol. Bioeng.* 23: 1297.

BARBARIC, S., KOZULIC, B., RIES, B. and MILDNER, P. 1980. *Biochem. Biophys. Res. Communic.* 95: 404.

BARTLING, G. J. and BARKER, C. W. 1976. *Biotechnol. Bioeng.* 18: 1023.

BATSTONE, D. 1990. *Membrane News (UNSW, Australia).* 2(1): 2.

BAUDIN, I. 1993. *Aqua* 42 (5): 205.

BEATON, N. C. 1976. *I. Chem. E. Symp. Series* No. 51: 59.

BEER, V. 1979. *Fette Seifen Anstrichmittel* 81: 89.

BEMBERIS, I. and NEELY, K. 1986. *Chem. Engr. Progr.* 82(11): 29.

BERRY, S. E. and NGUYEN, M. H. 1988. *Desalination* 70: 169.

BERSILLON, J. L., ANSELME, C. and MALLEVAILLE, J. 1991. *Proceedings, AWWA Specialty Conference on Membrane Technology*, Orlando, FL.

BIBAL, B., VAYSSIER, Y., GOMA, G. and PAREILLEUX, A. 1991. *Biotechnol. Bioeng.* 37: 746.

BILSTAD, T., ESPEDAL, E. and MADLAND, M. 1994. *Water Sci. Technol.* 29: 251.

BLANCK, R. G. and EYKAMP, W. 1986. *AIChE Symp. Ser.* 82 (No. 250): 59.

BLATT, W. F., ROBINSON, S. M. and BIXLER, H. J. 1968. *Anal. Biochem.* 26: 151.

BOUDRANT, J. and CHEFTEL, C. 1976. *Biotechnol. Bioeng.* 18: 1735.

BOYAVAL, P., CORRE, C. and TERRE, S. 1987. *Biotechnol. Lett.* 9: 207.

BRESLAU, B. R. 1981. U.S. Patent 4,266,026.

BRESLAU, B. R. and BUCKLEY, R. G. 1995. Personal communication. Pall Filtron, Northborough, MA.

BRESLAU, B. R., TESTA, A. J., MILNES, B. A., MEDJANIS, G. 1980. *Polymer Sci. Technol.* 13: 109.

BRESSOLLIER, P., PETIT, J. M. and JULIEN, R. 1988. *Biotechnol. Bioeng.* 31: 650.

BUDD, W. E. and OKEY, R. W. 1969. U.S. Patent 3,472,765.

BUECHBJERG, E. and MADSEN, R. F. 1987. U.S. Patent 4,677,065.

BULL, D. N. and YOUNG, M. D. 1981. *Biotechnol. Bioeng.* 23: 373.

CHAKRABORTY P. 1985. *J. Food Sci. Technol. (India)* 22: 248.

CHAKRABORTY, B. and SINGH, D. P. 1990. *Desalination* 78: 279.

CHAMBERS, R. P., FORD, J. R., ALLENDER, J. W., BARICOS, W. H. and COHEN, W. 1981. *Enzyme Engineering* (Wiley, NY) 2: 195.

CHANG, I. S., CHOO, K. H., LEE, C. H., PEK, U. H., KOH, U. C., KIM, S. W. and KOH, J. W. 1994. *J. Membrane Sci.* 90: 131.

CHAO, A. C., TZOU, L. and GREEN, D. 1984. *Proc., 39th Purdue Industrial Waste Conference*, Ann Arbor Science Publishing, Ann Arbor, MI. p. 555.

CHARLEY, R. C., FEIN, J. E., LAVERS, B. H., LAWFORD, H. G. and LAWFORD, G. R. 1983. *Biotechnol. Lett.* 5: 169.

CHAUFER, B. and DERATINI, A. 1988. *Nuclear and Chem. Management* 8: 175.

CHERYAN, M. 1980. *CRC Critical Rev. Food Sci. Nutr.* 13(4): 297.

CHERYAN, M. 1986. *Ultrafiltration Handbook*, Technomic, Lancaster, PA.

CHERYAN, M. 1992. *Corn Utilization Conference IV*, National Corn Growers Association, St. Louis, MO.

CHERYAN, M. 1994. *Proceedings, The Membrane Conference on Technology/Planning*, BCC Inc., Newton, MA.

CHERYAN, M. and ALVAREZ, J. 1995. In *Membrane Separations. Technology, Principles and Applications*, R. D. Noble and S. A. Stern (eds.), Elsevier, Amsterdam. p. 415.

CHERYAN, M. and BOGUSH, G. H. 1995. In *Proceedings, 3rd International Conference on Inorganic Membranes*, Y. Y. Ma (ed.), Worcester Polytechnic Institute, Worcester, MA. p. 147.

CHERYAN, M. and DEESLIE, W. D. 1983. *JAOCS* 60: 1112.

CHERYAN, M. and DEESLIE, W. D. 1984. U.S. Patent 4,443,540.

CHERYAN, M. and ESCOBAR, J. 1993. *Proceedings, First Biomass Conference of the Americas, Volume II*, National Renewable Energy Laboratory, Golden, CO. p. 1068.

CHERYAN, M. and MEHAIA, M. A. 1984a. *Appl. Microbiol. Biotechnol.* 20: 100.

CHERYAN, M. and MEHAIA, M. A. 1984b. *Process Biochem.* 19(6): 204.

CHERYAN, M. and MEHAIA, M. A. 1985a. In *Food Engineering and Process Applications. Volume 2: Unit Operations*, M. Le Mageur and P. Jelen (eds.), Elsevier Applied Science Publishers, Barking, Essex, U.K. 1986. p. 335.

CHERYAN M. and MEHAIA M. A. 1985b. In *Membrane Separations in Biotechnology*, W. C. McGregor (ed.), Marcel Dekker, New York. p. 255.

CHERYAN, M. and PAREKH, S. R. 1995. *Process Biochem.* 30(1): 17.

CHERYAN, M. and SAEED, M. 1989. *J. Food Biochem.* 13: 289.

CHIANG, B.H., SU, C. K. and TSAI, G. J. 1993. *J. Food Sci.* 58: 303.

CHOE, T. B., MASSE, P. and VERDIER, A. 1986. *Biotechnol. Lett.* 8: 163.

CHRISTIAN, S. D., BHAT, S. N., TUCKER, E. E., SCAMEHORN, J. F. and El-Sayed, D. A. 1988. *AIChE J.* 34: 189.

COLOMBAN, A., ROGER, L. and BOYAVAL, P. 1993. *Biotechnol. Bioeng.* 42: 1091.

CRESPO, J. P. S. G., MOURA, M. J., ALMEIDA, J. S. and CARRONDO, M. J. T. 1991. *J. Membrane Sci.* 61: 303.

CULIOLI, J., MAUBOIS, J. L., CHOPIN, A. and PIOT, M. 1975. *Ind. Aliment. Agrie.* 92(9–10): 1029.

DAUFIN, G., MICHEL, F. and MERIN, U. 1992. *Australian J. Dairy Technol.* 47: 7.

DEESLIE, W. D. and CHERYAN, M. 1981a. *Biotechnol. Bioeng.* 23: 2257.

DEESLIE, W. D. and CHERYAN, M. 1981b. *J. Food Sci.* 46: 357.

DEESLIE, W. D. and CHERYAN, M. 1982. *Biotechnol. Bioeng.* 24 : 69.

DEESLIE, W. D. and CHERYAN, M. 1992. *J. Food Sci.* 57: 411.

DELANEY, R. 1975. *J. Sci. Food Agric.* 26: 303.

DELANEY, R. 1977. *J. Food Technol.* 12: 339.

DELANEY, R., DONNELLY, J. and BENDER, L. 1975. *Lebensm. Wiss. u. -Technol.* 8: 20.

DELAUNAY, J., TESSIER, J. and LOUVEAU, V. 1979. *Revue Generale du Froid* 11: 622.

DIOSADY, L. L., TZENG, Y. M. and RUBIN, L. J. 1984. *J. Food Sci.* 49: 768.

DIOSADY, L. L., RUBIN, L. J. and HUSSEIN, A. 1992. *INFORM (AOCS).* 3: 536.

DORNIER, M., PETERMANN, R. and DECLOUX, M. 1995. *J. Food Engr.* 24: 213.

Dorr-Oliver Company. 1982. Product catalog. Stamford, CT.

DRIOLI, E. and CORTESE, B. 1980. *Desalination* 34: 131.

DUNN, R. O., SCAMEHORN, J. F. and CHRISTIAN, S. D. 1985. *Separation Sci. Technol.* 20: 257.

DUNN, R. O., SCAMEHORN, J. F. and CHRISTIAN, S. D. 1989. *Colloids and Surfaces,* 35: 49.

DUTRE, B. and TRAGARDH, G. 1995. *J. Food Engineering* 25: 233.

EAKIN, D. E., SINGH, R. P., KOHLER, G. O. and KNUCKLES, B. 1978. *J. Food Sci.* 43: 544.

ELMALEH, S. and GHAFFOR, N. 1996. *J. Membrane Sci.* 118: 111.

ENZMINGER, J. D. and ASENJO, J. A. 1986. *Biotechnol. Lett.*, 8: 7.

EPRI. 1992. *Techapplication,* Electric Power Research Institute, Walnut Creek, CA.

ERICKSSON, G., ERRIKSSON, P., HALLSTROM, B. and WIMMERSTEDT, R. 1977. *Desalination* 27: 81.

ERICKSSON, G. and SIVIK, B. 1976. *Potato Res.* 19: 279.

ERICKSSON G. and VON BOCKELMANN, I. 1975. *Process Biochem.* 10(Sept.): 11.

FAUQUANT, J., VIECO, E., BRULE, G. and MAUBOIS, J. L. 1985. *Lait.* 65: 1.

FAUQUANT, J., MAUBOIS, J. L. and PIERRE, A. 1988. *Technique Laitiere & Marketing.* 1028: 21–23.

FERNANDO T. 1981. *Biotechnol. Bioeng.* 23: 19.

FiltratioNews. 1984. Filtration Division, Alfa-Laval, Sweden.

FINK, D. J. and RODWELL, V. W. 1975. *Biotechnol. Bioeng.* 16: 1029.

FINNIGAN, T. J. A. and LEWIS, M. J. 1989. *Lebensm. Wiss. Technol.* 22: 237.

FLASCHEL, E., RAETZ, E. and RENKEN, A. 1983. In *Enzyme Technology,* R. M. Lafferty and E. Maier (eds.), Springer-Verlag, Berlin. p. 285.

FREUND, P. and RIOS, G. M. 1992. *Can. J. Chem. Eng.* 70: 250.

FRONING, G. W., WEHLING, R. L., BALL, H. R. and Hill, R. M. 1987. *Poultry Sci.* 66: 1168.

GARRETSON, R. 1983. Presented at *IMTEC '83*, Sydney, Australia.

GECKELER, K. E. and VOLCHEK, K. 1996. *Environmental Sci. Technol.* 30: 725.

GESAN, G., DAUFIN, G. and MERIN, U. 1995. *J. Membrane Sci.* 104: 271.

GESAN, G., DAUFIN, G., MERIN, U., LABBE, J. P. and QUEMARAIS, A. 1993. *J. Dairy Res.* 62: 269.

GHOSE, T. K. and KOSTICK, J. T. 1970. *Biotechnol. Bioeng.* 12: 921.

GOLDBERG, M. and CHEVRIER, D. 1979. *Indust. Alim. et Agricoles* 9/10: 951.

GOLDSMITH, R. L. 1981. *Dairy Field.* 16(3): 88.

GOLDSMITH, R. L. and HORTON, B. S. 1972. *EPA Project No. 12060 DXF*, Office of Research and Monitoring, Environ. Protection Agency, Washington, DC.

GOULD, R. M. and NITSCH, A. R. 1996. U.S. Patent 5,494,566.

GOVIND, R. and ITOH, N. 1989. *Membrane Reactor Technology. AIChE Symp. Series Vol. 85*, Number 268. American Inst. Chem. Engrs., NY.

Graver Separations. 1996. *Lit. No. S-106*, Glasgow, DE.

GROOT, W. J., SIKKENK, C. M., WALDRAM, R. H., VAN DER LANS, R. G. J. M. and LUYBEN, K. C. A. M. 1992. *Bioprocess Engr.* 8: 39.

GUEGUEN, J., QUEMENER, B. and VALDEBOUZE, P. 1980. *Lebensm. Wiss. Technol.* 13: 72.

HACKERT, R. and SWIENTEK, R. J. 1986. *Food Processing* 47(1): 80.

HAMILTON, K. M. and HOWELL, J. A. 1983. In *Adv. Ferment. Proc. Conf.*, Wheatland, Rickmansworth, U.K. p. 171.

HANEMAAIJER, J. H. 1985. *Desalination* 53: 143.

HANSEN, C. 1983. *Food Technol.* 37(2): 77.

HANSSENS, T. T., VAN NISPEN, J. G. M., KOERTS, K. and DE NIE, L. H. 1984. *International Sugar Journal* 86: 227,240.

HART, M. R., HUXSOLL, C. C., TSAI, L. S., NG, K. C., KING, A. D., JONES, C. C., HALBROOK, W. U. 1990. *J. Food Process Engr.* 12: 191.

HART, M. R., NG, K. C. and HUXSOLL, C. C. 1989. *ACS Symp. Ser.* 405: 355–367.

HAYWARD, M. F. 1982. *Proceedings of World Filtration Congress III*, Uplands Press, U.K. p. 572.

HEINEN, W., and LAUWERS, A. M. 1975. *Arch. Microbiol.* 106: 201.

HENLEY, R. G., YANG, R. Y. K. and GREENFIELD, P. F. 1980. *Enz. Microbial Technol.* 2: 206.

HERVE, D., LANCRENON, X. and ROUSSET, F. 1995. *Sugar y Azucar* 5: 40.

HSU, J. C., HEATHERBELL, D. A., FLORES, J. H. and WATSON, B. T. 1987. *Amer. J. Enol. Vitic.* 38(1): 17.

HUANG, Y. C. and KOSEOGLU, S. S. 1993. *Waste Management* 13: 481.

HUMMEL, W., SCHUTTE, H. and KULA, M. R. 1981. *Eur. J. Appl. Microbiol. Biotechnol.* 12: 22.

Hydranautics. 1996. Hydrapaint bulletin. Oceanside, CA.

IACOBUCCI, G. A., MYERS, M. J., EMI, S. and MYERS, D. V. 1974. *Proc. IV Intern. Congress Food Science Technol.* 5: 83.

INLOES, D. S., SMITH, W. J., TAYLOR, D. P., COHEN, S. N., MICHAELS, A. S. and ROBERTSON, C. R. 1983. *Appl. Environ. Microbiol.* 46: 264.

IWAMA, A. 1989. In *Proceedings of World Conference on Edible Fats and Oils Processing*, American Oil Chemist's Society, Champaign, IL. pp. 244–250.

JACANGELO, J. G., LAINE, J. M., CARNS, K. E., CUMMINGS, E. W. and MALLEVIALLE, J. 1991. *J. AWWA* 83(9): 97.

JACANGELO, J. G., ADHAM, S. S. and LAINE, J. M. 1995. *J. AWWA* 87(9): 107.

JAMESON, G. W. 1987. *Food Technol. Australia.* 39: 560.

JONSSON, A. S., JONSSON, C., TEPPLER, M. and Wannstrom, S. 1996. *Filtr. & Separation (Elsevier),* 33(6): 453.

JONSSON, A. S. and JONSSON, B. 1991. *J. Membrane Sci.* 56: 49.

JUANG, R. S. and LIANG, J. F. 1993. *J. Membrane Sci.* 82: 175.

KARRS, S. R. and MCMONAGLE, M. 1993. *Plating and Surface Finishing* 80 (Sept.): 45.

KATO. 1996. Product literature, Kato Brothers Honey Company, Japan.

KATOAKA, H., SAIGUSA, T., MUKUTAKA, S. and TAKAHASHI, J. 1980. *J. Ferment. Technol.* 58: 431.

KAWAKAMI, K., HAMADA, T. and KUSUNOKI, K. 1980. *Enz. Microbial Technol.* 2: 295.

KAWASAKI, Y., KAWAKAMI, H., TANIMOTO, M., DASAKO, S., TOMIZAWA, A., KOTAKE, M. and MAKAJIMA, I. 1993. *Milchwiss.* 48: 91.

KEURENTJES, J. T. F. 1991. Ph.D. Thesis, Agricultural University of Wageningen, The Netherlands.

KEURENTJES, J. T. F., BOSKLOPPER, T. G. J., VAN DROP, L. J. and VAN'T RIET, K. 1990. *JAOCS* 67: 28.

KHORAKIWALA, K. H., CHERYAN, M. and MEHAIA, M. A. 1987. *Biotechnol. Bioeng. Symp. Ser.* 15: 249.

KIM, S. H., MORR, C. V., SEO, A. and SURAK, J. G. 1989. *J. Food Sci.* 54: 25.

KIMURA, S. 1991. *Water Sci. Technol.* 23: 1573.

KIRJASSOFF, W. R., PINTO, S. and HOFFMAN, C. 1980. *Chem. Engr. Progr.* 76(2): 58.

KIRK, D. E., MONTGOMERY, M. W. and KORTEKAAS, M. G. 1983. *J. Food Sci.* 48: 1663.

KISHIHARA, S., TAMAKI, H., FUJI, S. and KOMOTO, M. 1989. *J. Membrane Sci.* 41: 103.

KLEIN, E. 1991. *Affinity Membranes.* John Wiley, New York.

KLEPAC, J., SIMMONS, D. L., TAYLOR, R. W., SCAMEHORN, J. F. and CHRISTIAN, S. D. 1991. *Separation Sci. Technol.* 26: 165.

KOCH. 1984. Product Literature. Koch Membrane Systems, Wilmington, MA.

KOCH. 1988. *Case History 3.* Koch Membrane Systems, Wilmington, MA.

KOCH. 1989. Product Literature. Koch Membrane Systems, Wilmington, MA.

KOCH. 1991. *Case History 5: Sunkist's Bitterfree Bounty.* Koch Membrane Systems, Wilmington, MA.

KOCH. 1992. *Industrial Wastewater Treatment.* Koch Membrane Systems, Wilmington, MA.

KOCH. 1995. *Abcor tubular ultrafiltration. Effluent-free paper coating.* Koch Membrane Systems, Wilmington, MA.

KOCH. 1996. *Konsolidator 252, Bulletin KPN 0679191.* Koch Membrane Systems, Wilmington, MA.

KOCHERGIN, V. 1996. Personal communication, Amalgamated Research, Inc., Twin Falls, Idaho.

KOHLWEY, D. K. and CHERYAN, M. 1981. *Enz. Microbial Technol.* 3: 64.

KORUS, R. A. and OLSON, A. C. 1977a. *Biotechnol. Bioeng.* 19: 1.

KORUS, R. A. and OLSON, A. C. 1977b. *J. Food Sci.* 42: 258.

KOSEOGLU, S. S. 1991. *INFORM (AOCS)*, 2: 334.

KOSEOGLU, S. S. and ENGELGAU, D. E. 1990. *JAOCS* 67: 239.

KOSEOGLU, S. S., LAWHON, J. T. and LUSAS, E. W. 1990. *Food Technology (IFT, USA)*, 44 (12): 90.

KOSIKOWSKI, F. V. 1986. *Food Technol.* 40(6): 71–77, 156.

KROLL, J., KUJAWA, M. and SCHNAAK, W. 1991. *Fette Wissenschaft Technologie* 93(2): 61.

KRONER, K. H., SCHUTTE, H., HUSTEDT, H. and KULA, M. R. 1984. *Process Biochem.* 19: 67–74.

KUO, K. P. and CHERYAN, M. 1983. *J. Food Sci.* 48: 1113.

KUNIKANE, S., MAGARA, Y., ITOH, M. and TANAKA, O. 1995. *J. Membrane Sci.* 102: 149.

LAH, C. L. and CHERYAN, M. 1980a. *J. Agric. Fd. Chem.* 28: 911.

LAH, C. L. and CHERYAN, M. 1980b. *Lebensm. Wiss. u. -Technol.* 13: 259.

LAH, C. L., CHERYAN, M. and DEVOR, R. E. 1980. *J. Food Sci.* 45: 1720.

LAMONICA, D. A. 1994. U.S. Patent 5,310,487.

LANCRENON, X., THEOLEYRE, M. A. and KIENTZ, G. 1994. *Intern. Sugar Journal* 96: 365.

LAWHON, J. T. and LUSAS, E. W. 1987. U.S. Patent 4,645,677.

LAWHON, J. T., MULSOW, D., CATER, C. M. and MATTIL, K. F. 1977. *J. Food Sci.* 42: 389.

LE BERRE, O. and DAUFIN, G. 1996. *J. Membrane Sci.* 117: 261.

LEE, C. W. and CHANG, H. N. 1987. *Biotechnol. Bioeng.* 29: 1105.

LEE, J. H., SKOTNICKI, M. L. and ROGERS, P. J. 1982. *Biotechnol. Lett.* 4: 615.

LEE, S. H., SON, M. P., KWON, Y. J. and PYUN, Y. R. 1991. *Sanop Misaengmul Hakhoechi* 19: 419 (*Chemical Abstr.* 119: 26778).

LIU, F. K., NIE, Y. H. and SHEN, B. Y. 1989. *Proc. World Congress Vegetable Prot. Utiliz. Human Foods Anim. Feedstuffs*, p. 84.

LOPEZ, J. L. and MATSON, S. L. 1997. *J. Membrane Sci.* 125: 189.

LOWE, E., DURKEE, E. L., MERSON, R. L., IJICKI, K., and CIMINO, S. L. 1969. *Food Technol.* 23: 753.

LUONG, J. H. T., NGUYEN, A. L. and MALE, K. B. 1987. *Trends in Biotechnology*, 5: 281.

LUONG, J. H. T., MALE, K. B. and NGUYEN, A. L. 1988. *Biotechnol. Bioeng.* 31: 516.

MAARTENS, A., SWART, P. and JACOBS, E. P. 1996. *J. Membrane Sci.* 119: 9.

MADAENI, S. S., FANE, A. G. and GROHMANN, G. S. 1995. *J. Membrane Sci.* 102: 65.

MADSEN, R. F. 1973a. *Intern. Sugar J.* 75: 163.

MADSEN, R. F. 1973b. British Patent No. 1,330,037.

MAK, F. K. 1991. *Intern. Sugar J.* 93: 263.

MALMBERG, R. and HOLM, S. 1988. *North European Food Dairy J.* 1: 75.

MAMERI, N., ABDESSEMED, D., BELHOCINE, D., LOUNICI, H., GAVACH, C., SANDEAUX, J. and SANDEAUX, R. 1996. *J. Chem. Technol. Biotechnol.* 67: 169.

MANNHEIM, A. and CHERYAN, M. 1990. *J. Food Science* 55: 381.

MANNHEIM, A. and CHERYAN, M. 1993. *Cereal Chem.* 70: 115.

MATSON, S. L. and QUINN, J. A. 1986. *Annals N.Y. Acad. Sci.* 469: 152.

MATSUMOTO, K., KATSUYAMA, S. and OHYA, H. 1987. *J. Ferment. Technol. (Japan).* 65: 77.

MATTIASSON, B. and LING, T. G. I. 1986. In *Membrane Separations in Biotechnology*, W. C. McGregor (ed.), Marcel Dekker, New York. p. 99.

MATTIASSON, B. and RAMSTORP, M. 1984. *J. Chromatog.* 283: 322.

MAUBOIS, J. L. 1989. *North American Membane Society Annual Meeting*, Austin, TX.

MAUBOIS, J. L. 1991. *Australian J. Dairy Technol.* 46: 91.

MAUBOIS, J. L., MOCQUOT, G. and VASSAL, L. 1969. Brevet Français 2,052,121.

MAUBOIS, J. L., PIERRE, A., FAUQUANT, J. and PIOT, M. 1987. *International Dairy Federation Bulletin.* 212: 154–159.

MEHAIA, M. A. and CHERYAN, M. 1983. *Milchwissenschaft.* 38: 708.

MEHAIA, M. A. and CHERYAN, M. 1984a. *Appl. Microbiol. Biotechnol.* 20: 100.

MEHAIA, M. A. and CHERYAN, M. 1984b. *Enz. Microbial Technol.* 6: 117.

MEHAIA, M. A. and CHERYAN, M. 1986. *Enz. Microbial Technol.* 8: 289.

MEHAIA, M. A. and CHERYAN, M. 1987. *Process Biochem.* 22(6): 185.

MEHAIA M. A. and CHERYAN M. 1990a. In *Biotechnology and Food Processing*, H. G. Schwartzberg and M. A. Rao (eds.), Marcel Dekker, New York. p. 67.

MEHAIA, M. A. and CHERYAN, M. 1990b. *Bioprocess Engr.* 5: 57.

MEHAIA, M. A. and CHERYAN, M. 1991. *Enz. Microbial Technol.* 13: 257.

MEHAIA, M. A., ALVAREZ, J. and CHERYAN, M. 1993. *International Dairy J.* 3: 179.

MEIER, P. 1995. Personal communication. Cuno, Inc., Meriden, CT.

MELLING J. 1974. *Process Biochem.* 9 (Sept.): 7.

MEMBRE, J. M., PETIOT, P., RENE, F. and LALANDE, M. 1991. *Recents Progr. Genie Procedes.* 5: 91.

MEMBREX. 1989. *Taylor Applications. Technical Bulletins No. 1 and No. 2.* Membrex, Inc., Garfield, NJ.

MERIN, U., GORDIN, S. and TANNY, G. B. 1983. *J. Dairy Res.* 50: 503.

MERRY, A. 1995. *PCI Membrane Systems*, UK.

MICHAELS, A. S. 1968. In *Progress in Purification and Separation*, E. S. Perry (ed.), Interscience, New York. p. 297.

MICHAELS, S. L., MICHAELS, A. S., ANTONIOU, C., PEARL, S. R., GOEL, V., DE LOS REYES, G., KEATING, P., RUDOLPH, E., KURIYEL, R. and SIWAK, M. 1995. In *Separation Technology. Pharmaceutical and Biotechnology Applications*, W. P. Olson (ed.), Interpharm Press, Buffalo Grove, IL.

Millipore Corp. 1983. Catalog No. AB822, Bedford, MA.

MINIER, M., FERRAS, E., GOMA, G. and SOUCAILLE, P. 1984. Presented at the *VII International Biotechnology Symposium*, New Delhi, India.

MIYATA, Y. 1984. *Nippon Suisan Gakkaishi (Japan).* 50(4): 659.

MIYAWAKI, O., NAKAMURA, K. and YANO, T. 1982. *J. Chem. Engr. Japan*, 15: 224.

MORGAN, A. I., LOWE, E., MERSON, R. L. and DURKEE, E. L. 1965. *Food Technol.* 19(12): 1790.

MORIKAWA, Y., KARUBE, I. and SUZUKI, S. 1978. *Biochim. Biophys. Acta* 523: 263.

MOTA, N., LAFFORGUE, C., STREHAIANOI, P. and GOMA, G. 1987. *Bioprocess Engr.* 2: 65.

MSS. 1995. *Micro-Steel Caustic Recovery System; Mem-Brine System*, Membrane System Specialists, Wisconsin Rapids, WI.

MURALIDHARA, H. S., JIRJIS, B. F. and SEYMOUR, G. F. 1996. U.S. Patent 5,482,633.

MUTOH, Y., MATSUDA, K., OHSHIMA, M. and OHUCHI, H. 1985. U.S. Patent 4,545,940.

NABETANI, H., ABBOTT, T. P. and KLEIMAN, R. 1995. *Ind. Engr. Chem. Res.* 34: 1779.

NAKAJIMA, M., IWASAKI, K., NABETANI, H. and WATANABE, A. 1990. *Agric. Biol. Chem.* 54: 2793.

NGUYEN, Q. T., APTEL, P. and NEEL, J. 1980. *J. Membrane Sci.* 6: 71.

NICHOLS, D. J. and CHERYAN, M. 1981. *J. Food Sci.* 46: 357.

NINOMIYA, K., OOKAWA, T., TSUCHIYA, T. and MATSUMOTO, J. 1985. *Nippon Suisan Gakkaishi (Japan).* 51(7): 1133.

NIPKOW, A., SONNLEITNER, B. and FIECHTER, A. 1986. *J. Biotechnol.* 4: 49.

NIRO. 1994. *Technical Report No. 40: Niro chemical recovery (NCR) systems with Membralox ceramic membranes*, Niro Hudson, Inc., Hudson, WI.

NITTO-DENKO. 1983. Product Bulletin, Shiga, Japan.

NORMAN, S. I. 1994. *Membrane and Adsorbent Applications for Enhancement of Citrus Juices*. Dow Chemical Co., Midland, MI.

NUORTILA-JOKINEN, J. and Nystrom, M. 1996. *J. Membrane Sci.* 119: 99.

O'CONNOR, D. E. 1971. U.S. Patent 3,622,556.

OHLEYER, E., WILKE, C. R. and BLANCH, H. W. 1985. *Appl. Biochem. Biotechnol.* 11: 457.

OHLSON, I., TRAEGARDH, G. and HAHN-HAEGERDAL, B. 1984. *Biotechnol. Bioeng.* 26: 647.

OKUBO, K., WALDROP, A. B., IACOBUCCI, G. A. and MEYERS, D. V. 1975. *Cereal Chem.* 52: 263.

OLSEN, H. S. 1978. *Lebensm. Wiss. u. -Technol.* 11: 57.

OLESEN, N. and JENSEN, F. 1988. *Milchwissenschaft* 44: 476.

OMOSAIYE, O. and CHERYAN, M. 1979a. *J. Food. Sci.* 44: 1027.

OMOSAIYE, O. and CHERYAN, M. 1979b. *Cereal Chemistry* 56: 58.

OMOSAIYE, O., CHERYAN, M. and MATTHEWS, M. E. 1978. *J. Food Sci.* 43: 354.

OOSTEN, B. 1976. *Die Stärke.* 28: 135.

OSTROWSKI, H. T. 1979. *J. Food Proc. Preserv.* 3: 59.

PADILLA, O. I. and MCLELLAN, M. R. 1989. *J. Food Sci.* 54: 1250.

PADILLA, O. I. and MCLELLAN, M. R. 1993. *J. Food Sci.* 58: 369.

PAFYLIAS, I., CHERYAN, M., MEHAIA, M. A. and SAGLAM, N. 1996. *Food Research Intern.* 29: 141.

PAL, D. and CHERYAN, M. 1987. *Indian Dairyman.* 39: 373.

PALL. 1996. Bulletin PBB-DV50. Pall Ultrafine Filtration Company, East Hills, NY.

PAREKH, B. S. 1991. *Chem. Engr.* 98(1): 70.

PAREKH, S. R. and CHERYAN, M. 1994. *Enz. Microbial Technol.* 16: 104.

PARK, Y. S., OHTAKE, H., TODA, K., FUKAYA, M., OKUMURA, H. and KAWAMURA, Y. 1989. *Biotechnol. Bioeng.* 33: 918.

PARROTT, D. L. 1990. *Third International Congress on Membranes and Membrane Processes*, Chicago, IL.

Pasilac Company. 1982. Product Bulletin. Nakskov, Denmark.

PATEL, R. S., REUTER, H. and PROKOPEK, D. 1986. *J. Soc. Dairy Technol.* 39: 27.

PATEL, P. N., MEHAIA, M. A. and CHERYAN, M. 1987. *J. Biotechnol.* 5: 1.

PATOCKA, J. and JELEN, P. 1987. *J. Food Sci.* 52: 1241.

PAULSON, D. J., WILSON, R. L. and SPATZ, D. D. 1984. *Food Technology* 38 (12): 77.

PAYNE, R. E., HILL, C. G. and AMUNDSON, C. H. 1973. *J. Milk Food Technol.* 36: 359.

PEDERSEN, P. J. 1992. *International Dairy Federation Special Issue* 9201: 33.

PERI, C., RIVA, N. and DECIO, P. 1988. *Amer. J. Enol. Vitic.* 39: 162.

PERMIONICS. 1995. Personal communication, Vadodara, India.

PIERROTT, P., FICK, M. and ENGASSER, J. M. 1986. *Biotechnol. Lett.* 8: 253.

PIOT, P., VACHOT, J. V., VEAUX, M., MAUBOIS, J. L. and BRINKMAN, G. E. 1987. *Technique Laitiere & Marketing.* 1016: 42–46.

PLOTKA, A., SCHMIDT, J. and ZDZIENNICKI, A. 1977. *Prace Instytutow Lab. Bad. Przem. Spoz.t.* 27, zeszyt 1, s. 29.

POMPEI, C. and LUCISANO, M. 1978. *Lebensm.-Wiss. u. -Technol.* 9: 338.

PORTER, M. C. 1990. *Handbook of Industrial Membrane Technology.* Noyes, Park Ridge, NJ.

PORTER, M. C. and MICHAELS, A. S. 1971. *CHEMTECH.* 1: 440.

PORTER, J. J. and ZHUANG, S. 1996. *J. Membrane Sci.* 110: 119.

POULIOT, G. and GOULET, J. 1987. *J. Food Sci.* 52: 1394.

PUNGOR, E., AFEYAN, N. B., GORDON, N. F. and COONEY, C. L. 1987. *Bio/Technology,* 5: 604.

QURESHI, N. and CHERYAN, M. 1989. *Process Biochem.* 24(5): 172.

RAASKA, E. and KELLY, W. 1987. *Kem.-Kemi.* 14(3): 253–259.

RAJAGOPALAN, N. 1996. Personal communication. Hazardous Waste Research and Information Center, Champaign, IL.

RAJAGOPALAN, N. and CHERYAN, M. 1991. *J. Dairy Sci.* 74: 2435.

RAMACHANDRAN, K. B. and GOMA, G. 1988. *J. Biotechnol.* 9: 39.

RAMAN, L. P., CHERYAN, M. and RAJAGOPALAN, N. 1994a. *Chem. Engr. Progr.* 90(3): 68.

RAMAN, L. P., RAJAGOPALAN, N. and CHERYAN, M. 1994b. *Oils Fats Intern. (UK).* 6(10): 28.

RAMAN, L. P., CHERYAN, M. and RAJAGOPALAN, N. 1996a. *Fette Wiss. Technologie.* 98(1): 10.

RAMAN, L. P., CHERYAN, M. and RAJAGOPALAN, N. 1996b. *JAOCS* 73: 219.

RANE, K. D. and SIMS, K. A. 1995. *Biotechnol. Bioeng.* 20: 325.

RANE, K. D. and CHERYAN, M. 1996. Stillage processing with ceramic membranes (unpublished data). University of Illinois, Urbana.

RAO, M. A., ACREE, T. E., COOLEY, H. J. and ENNIS, R. W. 1987. *J. Food Sci.* 52: 375.

REED, W. M. and BOGDAM, M. E. 1986. *Biotechnol. Bioeng. Symp.* 15: 641.

RENNER, E. and ABD EL-SALAM, M. H. 1991. *Application of Ultrafiltration in the Dairy Industry.* Elsevier, New York.

RIESMEIER, B., KRONER, K. H. and KULA, M. R. 1990. *Desalination.* 77: 219.

RINN, J. C., MORR, C. V., SEO, A. and SURAK, J. G. 1990. *J. Food Sci.* 55: 510.

ROLCHIGO, P. M. 1995. Personal communication, Membrex Inc., Fairfield, NJ.

ROGERS, P. L., LEE, K. J. and SKOTNICKI, M. L. 1982. *Adv. Biochemical Engr.* 23: 37.

ROMICON. 1988. Product and Process Bulletins. Woburn, MA.

RWABAHIZI, S. and WROLSTAD, R. E. 1988. *J. Food Sci.* 53: 857.

RYDER, D. S., DAVIS, C. R., ANDERSON, D., GLANCY, F. M. and POWER, J. N. 1988. *Master Brew. Assoc. Am.—Tech. Quarterly* 25: 67.

SAGLAM, N. 1995. Ph.D. thesis, University of Illinois, Urbana.

SCHULTZ, J. S. and GERHARDT, P. 1969. *Bacteriol. Rev.* 33: 1.

SCHWERING, H., GOLLISC, P. and KEMP, A. 1993. *Plating and Surface Finishing* 80 (April): 56.

SEN GUPTA, A. K. 1978. U.S. Patent 4,093,540.

SEN GUPTA, A. K. 1985. U.S. Patent 4,533,501.

SEPROTECH. 1995. *Environmental Improvements in Hide Processing*, Ottawa, Canada.

SHAH, M. M. and CHERYAN, M. 1995. *Applied Biochem. Biotechnol.* 51/52: 413.

SHIH, J. and KOZNICK, M. 1980. *Poultry Sci.* 59: 247.

SHIMIZU, Y., MATSUSHITA, K., SHIMODERA, K. and WATANABE, A. 1992. In *Biochemical Engineering for 2001*, S. Furusaki (ed.), T. Endo and R. Matsuno, Springr-Verlag, Tokyo. p. 578.

SHORT, J. L. 1993. *Membrane Industry News*, Westford, MA. November issue.

SHORT, J. L. 1994. *Membrane Industry News*, Westford, MA. 2(4): 4.

SHORT, J. L. 1995a. *Membrane Industry News*, Westford, MA. 3(6): 5.

SHORT, J. L. 1995b. *Membrane Industry News*, Westford, MA. 3(1): 4.

SHORT, J. L. and SKELTON, R. 1991. In *Effective Industrial Membranes Processes: Benefits and Opportunities*, M. K. Turner (ed.), Elsevier, New York.

SIMS, K. A. and CHERYAN, M. 1986. *Biotechnol. Bioeng. Symp. Ser.* 17: 495.

SIMS, K. A. and CHERYAN, M. 1992a. *Biotechnol. Bioeng.* 39: 960.

SIMS, K. A. and CHERYAN, M. 1992b. *Starch/Stärke.* 44: 345.

SINGH, N. 1997. Ph.D. Thesis, University of Illinois, Urbana.

SINGH, N. and CHERYAN, M. 1997a. *Cereal Foods World* 42(7): 520.

SINGH, N. and CHERYAN, M. 1997b. *Cereal Foods World* 42(1): 21.

SNAPE, J. B. and NAKAJIMA, M. 1996. *J. Food Engr.* 30: 1.

STEIBER, R. W. and GERHARDT, P. 1981. *Biotechnol. Bioeng.* 23: 535.

STRATHMAN, H. 1980. *Sep. Sci. Technol.* 15: 1135.

SUZUKI, S., MAEBASHI, N., YAMANO, S., NOGAKI, H., TAMAKI, A. and NOGUCHI, A. 1992. U.S. Patent 5,166,376.

TABATABIA, A., SCAMEHORN, J. F. and CHRISTIAN, S. D. 1995. *J. Membrane Sci.* 100: 193.

TAKO, M. and NAKAMURA, S. 1986. *Agric. Biol. Chem.* 50: 833.

TANAHASHI, S. NAGANO, K., KASAI, M., TSUBONE, F., IWAMA, A., KAZUSE, Y., TASAKA, K. and ISOOKA, Y. 1988. U.S. Patent 4,787,981.

TANAGUCHI, M., KOTANI, N. and KOBAYASHI, T. 1987. *Appl. Microbiol. Biotechnol.* 25: 438, and *J. Ferment. Technol.* 65: 179.

TEJAYADI, S. and CHERYAN, M. 1988. *Appl. Biochem. Biotechnol.* 19: 1.

TONG, P. S., BARBANO, D. M. and JORDAN, W. K. 1988. *J. Dairy Sci.* 71: 2342.

TRAGARDH, C. 1974. *Lebensm.-Wiss. u.-Technol.* 1: 199.

TRAN, T. V. 1985. *Chem. Engr. Progr.* 81(3): 29.

TSAI, L. S., IJICHI, K. and HARRIS, M. W. 1977. *J. Food Protection.* 40: 449.

TURPIE, D. W. F., STEENKAMP, C. J. and TOWNSEND, R. B. 1992. *Water Sci. Technol.* 25: 127.

TZENG Y. M., DIOSADY, L. L. and RUBIN, L. J. 1988. *Canadian J. Food Sci. Technol.* 21: 419.

ULLOA, J. A., VALENCIA, M. E. and GARCIA, Z. J. 1988. *J. Food Sci.* 53: 1396.

USDA. 1978. *Handbook No. 8*, U.S. Department of Agriculture, Washington, DC.

US Filter. 1996. *Ceramic Membrane News*, Warrendale, PA. 2(6): 1.

VAN DER HORST, H. C. and HANEMAAIJER, J. H. 1990. *Desalination* 77: 235.

VAVRA, C. and KOSEOGLU, S. S. 1994. In *Developments in Food Engineering*, T. Yano, R. Matsuno and K. Nakamura (eds.). Blackie Academic, Glasgow, U.K. p. 683.

VERMA, S. K., SRIKANTH, R., DAS, S. K. and VENKIDACHALAM, G. 1996. *Indian J. Chem. Technology*, 3: 136.

VETIER, C., BENNASAR, M. and TARODO DE LA FUENTE, B. 1988. *J. Dairy Res.* 55: 381.

VICK ROY, T. B., BLANCH, H. W. and WILKE, C. R. 1982. *Biotechnol. Lett.* 4: 483.

VICK ROY, T. B., MANDEL, D. K., DEA, D. K., BLANCH, H. W. and WILKE, C. R. 1983. *Biotechnol. Lett.* 10: 665.

WAFILIN, B. V. 1983. Product Bulletins, Hardenberg, The Netherlands.

WAJNOWSKA, I., BEDNARSKI, W. and POZNANSKI, S. 1979. *Acta Aliment. Pol.* 5(3): 327.

WANDREY, C. and FLASCHEL, E. 1979. *Adv. Biochem. Engr.* 12: 147.

WATANABE, A., OHTANI, T., HORIKITA, H., OHYA, Y. and KIMURA, A. 1984. In *Engineering and Food. Volume 1*, B. M. McKenna (ed.), Elsevier Applied Science Publishers, London, U.K. p. 225.

WELSH, F. W. and ZALL, R. R. 1984. *Can. Inst. Food Sci. Technol.* 17: 92.

WICHMANN, R., WANDREY, C., BUCKMANN, A. F. and KULA, M. R. 1981. *Biotechnol. Bioeng.* 23: 2789.

WILSON, E. L. and BURNS, D. J. W. 1983. *J. Food Sci.* 48: 1101.

WU, Y. V., SEXSON, K. R. and LAGODA, A. A. 1985. *Cereal Chem.* 62: 470.

YABUSHITA, T. 1989. *Nitto-Denko Technical Reports*, 4: 47.

YAMAZAKI, Y., MAEDA, H. and SUZUKI, H. 1976. *Biotechnol. Bioeng.* 18: 1761.

YU, Z. R., CHIANG, B. H. and HWANG, L. 1986. *J. Food Sci.* 51: 841.

ZAHKA, J. and MIR, L. 1977. *Chem. Eng. Progr.* 73(12): 53.

ZANTO, L. T., CHRISTIFFER, L. M. and BIRSCHEL, S. E. 1970. *J. Am. Soc. Sugar Beet Technologists.* 16(1): 26.

ZHANG, S. Q., KUTOWY, O., KUMAR, A. and MALCOLM, I. 1997. *Canadian Agricultural Engineering.*

ZHANG, D. X. and CHERYAN, M. 1994. *Process Biochem.* 29: 145.

APPENDIX A

Manufacturers and Suppliers of Membrane Systems

Name	Address	Comments
Abcor, Inc.	See Koch Membrane Systems	
Acumem	See USFilter	
Advanced Membrane Technology (AMT)	10350 Barnes Canyon Road San Diego, CA 92121 (619) 549-4488	RO, NF, UF, MF. Polymeric tubular and spiral modules.
A/G Technology Corporation	34 Wexford Street Needham, MA 02194 (617) 449-5774	Hollow fiber UF and MF
Akzo Nobel (Membrana)	See Microdyn	
Alcoa	See USFilter	
Alfa-Laval Separation	Hans Stahles vag S-14780 Tumba Sweden	
Amicon	24 Cherry Hill Drive Danvers, MA 01923 (617) 777-3692	Laboratory and process scale UF and MF hollow fibers and spirals (Division of Millipore)
Anotec Separations Ltd.	Wildmere Road, Daventry Road Industry Estate Banbury, Oxon OX16 7JU (UK)	Inorganic flat disc MF membranes: lab use (available from Whatman)
Applied Membranes, Inc.	110 Bosstick Boulevard SanMarcos, CA 92069 (619) 727-3711	RO, NF, and UF systems manufacturer
APV	395 Fillmore Avenue P.O. Box 366 Tonawanda, NY 14151 (716) 692-3000	OEM for several manufacturers
AquaTech Systems	See Graver	Bipolar membranes

(continued)

495

Name	Address	Comments
Aqualytics	See Graver	
Asahi Kasei (Asahi Chemical Industry Co., Ltd)	1-1, 1-Chome, Uchisawai-Cho, Chiyoda-ku, Tokyo 100 Japan (813) 3507-2255	MF and UF hollow fibers. Marketed as Microza UF by Pall
atech innovations gmbh	Teilungsweg 28 D-45329 Essen Germany (49) 201 3410-2425	Ceramic membranes for MF and UF
Bend Research Inc.	64550 Research Road Bend, Oregon 97701	Contract research; liquid transport membranes, hollow fibers
Berghof	Harretstrasse 1 D-72800 Eningen Germany (49) 711 456-7753	Hollow fiber polyamide modules, ceramic membranes, ED, NF, RO, MF, UF
Carbone Lorraine	Tour Manhattan F-92095 la Defense Cedex France (33)-1-4762-8800	Carbon composite for MF, GFT pervaporation
CARRE	See Graver	
CeraMem Separations	20 Clematis Avenue Waltham, MA 02154 (617) 899-0467	Ceramic membranes
Ceraver	See USFilter	
Cer-Wat Corporation	10425-B Codgill Road Knoxville, TN 37932 (423) 675-1574	Belt press MF membrane systems
Cuno, Inc.	400 Research Parkway Meriden, CT 06450 (203) 237-5541	UF and MF membranes and equipment
DDS	See Dow	
Daicel	3-8-1 Toranomon Building Kasumigasiki Chiyoda-Ku Tokyo, Japan	Tubular and spiral cellulosic and noncellulosic membranes and modules (MOLSEP tubular products also marketed by Hoechst)

Name	Address	Comments
DESAL	Desalination Systems, Inc. 1238A Simpson Way Escondido, CA 92025 (619) 746-8141	Spiral membranes for RO, NF, UF, MF. Now part of Osmonics
Domnick Hunter Ltd.	Durham Road, Birtley, Co. Durham DH3 2SF (UK) (44) 191-410 5121	MF and UF membranes
Dow Chemical Company	Liquid Separations Systems Midland, MI 48674 (800) 447-4369	FilmTec membranes for RO, NF, and UF
	FilmTec Corporation 7200 Ohms Lane Edina, MN 55439 (612) 897-4386	
Downstream Technologies Inc.	11 Cambridge Road Tenafly, NJ 07670 (201) 569-6866	Plate UF and MF
E.I. du Pont de Nemours	1007 Market Street Wilmington, Delaware 19898	RO hollow fine fibers and spirals for sea water and brackish water
Elga Group	High Street, Lane End, High Wycombe, Buckinghamshire HP14 3JH England 44 (1494) 881-393	RO water systems
	Elga Inc. 430 Old Boston Road Topsfield, MA 01983 (508) 887-6300	
Enka	See Microdyn	
EnviroPure Solutions	100 Bridge Street Wheaton, IL 60187 (630) 871-1001	OEM. Wastewater treatment
Epoc	3065 N. Sunnyside Fresno, CA 93727 (209) 291-8144	Exxflow MF systems
Fairey Industrial Ceramics Ltd	Filleybrooks, Stone Staffordshire ST15 0PU (UK) (44) 1785 813241	Ceramic membranes (star-shaped channels)
FilmTec	See Dow	
Filtron	See Pall Filtron	

(continued)

Name	Address	Comments
Fluid Systems Corporation	10054 Old Grove Road San Diego, CA 92131 (619) 695-3840	Spiral RO, NF, and UF membranes
Gelman Sciences	600 South Wagner Road Ann Arbor, MI 48103 (313) 665-0651	MF and UF membranes and cartridges (acquired by Pall)
GFT (USA)	460 Main Avenue Wallington, NJ 07057 (201)773-2900	Carbon–carbon composite membranes for MF. Pervaporation (Division of Le Carbone-Lorraine)
Gore	W. L. Gore and Associates, Inc. 551 Paper Mill Road Box 9206 Newark, DE 19714 (302) 738-4880	PTFE MF membranes
Graver Separations, Inc.	200 Lake Drive Newark, DE 19702 (302) 731 3539 7 Powder Horn Drive Warren, NJ 07059	Stainless steel MF membranes, (Scepter) FIP UF, NF, and RO
Hoechst AG	Rheingaustrasse 190/196 D-65174 Wiesbaden Germany (49) 611-962 6237	UF and NF equipment
Hydranautics, Inc.	401 Jones Road Oceanside, CA 92054 (619) 901-2500	Division of Nitto Denko
Illinois Water Treatment Co. (IWT)	4669 Shepherd Trail Rockford, IL 61105 (815) 877-3041	Division of USFilter
Ionics, Inc.	65 Grove Street Watertown, MA 02172 (617) 926-2500	Primarily ED, but also RO, NF, and UF
Kiryat Weizmann Ltd.	See Membrane Products	
Koch Membrane Systems	850 Main Street Wilmington, MA 01887 (508) 657-4250	UF and MF spirals, tubular and Romicon hollow fibers. Markets Kiryat Weizmann NF
Kuraray Company Ltd.	12-39, 1-chome Umeda, Kita-ku Osaka 530, Japan	Polyvinyl alcohol UF and MF systems; tubular nonasymmetric membranes
Kurita Water Industries Ltd.	4-7 Nishi-Shinjuku 3-chome Shinjuku-ku Tokyo 160, Japan	Primarily spiral-wound RO systems.

Name	Address	Comments
Luxx UltraTech	148 Ledge Road Medina, OH 44256 (216) 659 4001	Restores used membranes
Membrane Products Kiryat Weizmann Ltd.	P.O.Box 138 Rehovot 76101 Israel (972) 8-407557	UF, NF, RO membranes in tubular and spiral. NF marketed by Koch
Membrane Systems Specialists (MSS)	3998 Wood Ridge Trace Wisconsin Rapids, WI 54494 (715) 421-2333	OEM. Systems supplier (MF, UF, NF, RO) for the dairy industry
Membrane Technology and Research, Inc.	1360 Willow Road Menlo Park, CA 94025 (415) 328-2228	Pervaporation, gas separations
Membrex, Inc.	155 Route 46 West Fairfield, NJ 07004 (201) 575-8388	Rotary vortex flow modules, spiral modules, modified PAN membranes
Memtec Ltd.	1 Memtec Parkway Windsor NSW 2756 Australia (61) 45-776 800	Hollow fiber MF, primarily for water and wastewater
	1750 Memtec Drive DeLand, FL 32724 (904) 822-8000	
Memtek/Wheelabrator	See USFilter	
Microdyn	Microdyn Modulbau GmbH Ohder Strasse 28 D-42289 Wuppertal Germany (49) 202-602092	Accurel polypropylene hollow fibers
	Microdyn Technologies Inc. 1204 Briar Patch Lane Raleigh, NC 27624 (919) 872-9375	
Microgon	23152 Verdugo Drive Laguna Hills, CA 92653 (714) 581-3880	Dyna-Fibre (cellulose ester), Kros-Flo capillary modules, lab. & process scale
Millipore Corporation	Ashby Road Bedford, MA 01730 (800) 225-1380	Broad line of membrane filters, laboratory and process equipment.
Minntech	14605 28th Avenue North Minneapolis, MN 55447 (612) 553-3300	Hollow fibers (FiberFlo)

(continued)

Name	Address	Comments
New Logic International	1155 Park Avenue Emeryville, CA 94608 (510) 655-7305	V-SEP vibratory module
NGK Filtech Ltd.	Shin Maru Building 1-5-1 Marunouchi Chiyoda-ku Tokyo 100, Japan	Cefilt ceramic membranes
Niro Hudson, Inc.	1600 O'Keefe Road Hudson, WI 54016 (800) 367-6476	OEM. RO, NF, UF and MF systems engineering and design
Nitto-Denko Corporation	3rd floor, Mori Building 31, 5-7-2 Kojimachi Chiyoda-ku Tokyo 102, Japan	Tubular, hollow fiber, spiral MF, UF, NF, RO membranes and systems See Hydranautics
North Carolina SRT, Inc.	1018 Morrisville Parkway Morrisville, NC 27560 (919) 469-5848	Systems manufacturer. Manufactures plate modules
Nuclepore Corporation	See Poretics for track-etch MF membranes	
NWW Acumem	See Acumem	
Osmonics, Inc.	5951 Clearwater Drive Minnetonka, MN 55343 (800) 848-1750	RO, NF, UF and MF spiral-wound membranes, systems, prefilters.
Pall Ultrafine Filtration Company	2200 Northern Boulevard East Hills, NY 11548 (516) 484-5400	Flat sheet membranes, cartridge filters, rotary and vibratory membrane modules
Pall Filtron Corporation	50 Bearfoot Road Northborough, MA 01532 (800) 345-8766	Flat sheet membranes for MF and UF; analytical, laboratory and process scale systems
Parker Hannifin Corporation	P.O.Box 1300 Lebanon, IN 46052 (317) 482-8437	MF and UF membranes
PCI Membrane Systems Inc.	Laverstoke Mill, Whitchurch, Hampshire RG28 7NR (UK) (44) 1256-896-966	Tubular RO, NF, UF and MF membranes and systems
	623 Grace Avenue Fond du Lac, WI 54935 (414) 923-5869	

Name	Address	Comments
Permionics	Gorwa Road 5/11 Industrial Estate Vadodara 390 016 India	Manufacturer of spiral UF, NF and RO membranes and systems. OEM for several
Poretics	151 Lindberg Avenue, Suite 1 Livermore, CA 94550 (415) 373-0500	Silver membranes, track-etch membranes (Division of Osmonics)
Rochem Separation Systems	3904 Del Amo Boulevard Suite 801 Torrance, CA 90503 (310) 370-3160	Plate systems for MF, NF, UF, and RO
Romicon	See Koch	
Sartorius AG	Weender Landstrasse 94-108 D-37075 Gottingen Germany	Lab scale devices, depth filters, process cross-flow plate systems
	Sartorius Corporation 131 Heartland Boulevard Edgewood, NY 11717-9957 (800) 368-7178	
Schleicher & Schull GmbH	Hahnestrage 3 D-3354 Dassel Germany	Laboratory MF and UF membranes and apparatus
Schumacher	Zur Fluglau 70 D-74564 Crailsheim Germany (49)-7951-3020	Ceramic membranes
SCT (Societe des Ceramiques Techniques)	BP 1 F-65460 Bazet France	Ceramic membranes. Division of USFilter
Sepracor Inc.	33 Locke Drive Marlborough, MA 01752 (617) 460-0412	Membrane bioreactor for optically active compounds
Seprotech Systems, Inc.	2378 Holly Lane Ottawa, Ontario K1V 7P1 Canada (613) 523-1641	OEM. Systems design and fabrication
SFEC	See Tech-Sep	
Spectrum-Microgon	See Microgon	
Spin-Tek Systems	16421 Gothard Street, Unit A Huntington Beach, CA 92647 (714) 848-3060	Centrifugal/rotary devices and spiral membranes

(continued)

Name	Address	Comments
Tech-Sep	5 chemin du Pilon F-01703 Mirabel, Cedex France (33) 7201-2727	Division of Rhone-Poluenc; Carbosep and Kerasep inorganic membranes, Pleiade plate polymeric systems
	Applexion-Tech Sep 15700 Lathrop Avenue Harvey, IL 60426 (708) 210-5047	
Ultrafilter Gmbh	Bussingstrasse 1 D-42781 Haan Germany (49) 2129-5690	MF and UF
	5555 Oakbrook Parkway Norcross, GA 30093 (800) 543-3634	
U.S. Filter Corporation	181 Thornhill Road Warrendale, PA 15086 (412) 772-1337	Ceramic membranes (Membralox) and systems
Wafilin BV	Bruchterweg 88 Postbox 5 7700 AA Hardenberg The Netherlands	Tubular RO and UF systems. Marketed by Stork Friesland
Wheelabrator	See USFilter	
X-flow B.V.	Bedrijvenpark Twente 289 7602 KK Almelo The Netherlands (31)-546-575202	Hydrophilic hollow fiber MF and UF
Zenon Environmental, Inc.	845 Harrington Court Burlington, Ontario L7N 3P3 Canada (905) 639-6320	UF and RO systems, primarily for wastewater treatment

APPENDIX B

Conversion Factors

Area

$1\,\text{ft}^2 = 0.0929\,\text{m}^2 = 929\,\text{cm}^2$

$1\,\text{m}^2 = 10^4\,\text{cm}^2 = 10.764\,\text{ft}^2$

$1\,\text{cm}^2 = 0.0011\,\text{ft}^2 = 0.1550\,\text{in}^2$

Density

$1\,\text{lb(mass)/ft}^3 = 16.0185\,\text{g/L}$

Diffusivity

$1\,\text{ft}^2/\text{h} = 2.581 \times 10^{-5}\,\text{m}^2/\text{s}$

$1\,\text{m}^2/\text{s} = 3.875 \times 10^4\,\text{ft}^2/\text{h}$

Flux

$1\,\text{gallon/ft}^2/\text{day}$ (GSFD or GFD) $= 1.7\,\text{liters/m}^2/\text{hour}$(LMH)

Force and Pressure

$1\,\text{dyne} = 1\,\text{g} \cdot \text{cm/s}^2$

$1\,\text{kg} \cdot \text{m/s}^2 = 1\,\text{Newton(N)}$

$1\,\text{bar} = 10^5\,\text{Pascal (Pa)} = 10^5\,\text{N/m}^2 = 14.5\,\text{psi (pounds/square inch)}$

$1\,\text{psi (lb}_\text{f}/\text{in}^2) = 6.8947\,\text{kPa} = 6.8947 \times 10^3\,\text{Newton/m}^2$

$1\,\text{psi} = 2.036''\,\text{Hg} = 51.715\,\text{mm Hg}$

$1\,\text{psi} = 6.8947 \times 10^4\,\text{g/cm/sec}^2 = 6.8947 \times 10^4\,\text{dyne/cm}^2$

$1\,\text{atmosphere (atm)} = 14.696\,\text{psi} = 1.01325\,\text{bar} = 101.3251\,\text{kPa} = 760\,\text{mm Hg}$
$= 33.9\,\text{ft water}$

$1\,\text{dyne/cm}^2 = 2.0886 \times 10^{-3}\,\text{lb}_\text{f}/\text{ft}^2$

1 kPa = 0.1450383 psi

1 kg_f/cm^2 = 14.2234 psi

Length

1 Angstrom (Å) = 10^{-10} m = 10^{-4} microns = 0.1 nm

1 inch (″) = 2.54 centimeters (cm)

1 meter = 39.37 inches

1 micron (μ or μm) = 10^{-6} m = 10^{-4} cm

1 mil = 0.001 inch = 25.4 μm

1 nanometer (nm) = 10 Å = 10^{-9} m

Mass

1 pound (lb) = 16 ounces (oz.) = 453.59 grams (g)

1 kilogram (kg) = 2.2046 lb

1 ton, short (U.S.) = 2000 lb

1 ton, long = 2240 lb

1 ton, metric = 1000 kg

Mass transfer coefficient

1 cm/sec = 0.01 m/sec

1 ft/h = 8.4668×10^{-5} m/sec

Power, work, and energy

1 horsepower (hp) = 0.74570 kilowatts (kW) = 550 ft · lb/sec = 0.7068 Btu/sec

1 watt(W) = 14.3 cal/min

1 joule/sec (J/sec) = 1 Watt

1 Btu = 1055.06 J

1 kilocalorie (kcal) = 4.1840 kJ

1 hp-hour = 0.7457 kWh = 2544 Btu

1 ft · lb_f = 1.35582 J

Temperature

T (°F) = [T(°C) × 1.8] + 32

T (°C) = [T(°F) − 32]/1.8

Velocity

1 ft/sec = 0.3048 m/sec

1 m/sec = 3.281 ft/sec

Viscosity

1 centipoise (cP) $= 0.01$ poise $= 10^{-3}$ Pa $\cdot$ s $= 10^{-3}$ kg/m/sec $= 10^{-3}$ N $\cdot$ sec/m^2

1 lb/ft/sec $= 1488.16$ cP

1 cP $= 2.4191$ lb/ft/h $= 6.72 \times 10^{-4}$ lb/ft/sec

Volume

1 gallon (U.S.) $= 4$ quarts (qt) $= 3.78541$ liters (L)

1 British gallon $= 1.20094$ U.S. gal

1 liter $= 1000$ cm^3

1 m^3 $= 1000$ L

1 in^3 $= 16.387$ cm^3

1 ft^3 $= 28.317$ L

1 m^3 $= 264.17$ U.S. gal

APPENDIX C

Books and General References

AMJAD, Z. 1992. *Reverse Osmosis. Membrane Technology, Water Chemistry and Industrial Applications*. Van Nostrand Reinhold, New York.

BELFORT, G. (Ed.). 1984. *Synthetic Membrane Processes*. Academic Press, New York.

BHAVE, R. R. (Ed.). 1991. *Inorganic Membranes. Synthesis, Characteristics and Applications*. Van Nostrand Reinhold, New York.

BIER, M. (Ed.). 1971. *Membrane Processes for Industry and Biomedicine*. Plenum Press, New York.

BITTER, J. G. A. 1991. *Transport Mechanisms in Membrane Separation Processes*. Plenum, New York.

BOEN, D. F. and JOHANNSEN, G. L. 1974. *Reverse Osmosis of Treated and Untreated Secondary Sewage Effluent*. National Environmental Research Center.

BROCK, S. D. 1983. *Membrane Filtration: A User's Guide and Reference Manual*. Science-Tech, Inc., Madison, WI.

BUNGAY, P. M., LONSDALE, H. K. and DEPINTO, M. N. (Eds.). 1986. *Synthetic Membranes: Science, Engineering and Applications*, D. Reidel Pub. Co.

BURGGRAAF, A. J. and COT, L. (Eds.). 1996. *Fundamentals of Inorganic Membrane Science and Technology*. Elsevier Science, Amsterdam.

CECILLE, L. and TOUSSAINT, J.-C. (Eds.). 1989. *Future Industrial Prospects of Membrane Processes*. Elsevier, New York.

CHENOWETH, M. B. 1986. *Synthetic Membranes*. Harwood Academic Publishers, New York.

CHERYAN, M. 1986. *Ultrafiltration Handbook*. Technomic Publishing Company, Lancaster, PA. (*Handbuch Ultrafiltration*, in German. Behr's Verlag, GmbH, Hamburg. 1990)

COOPER A. R. (Ed.). 1980. *Ultrafiltration Membranes and Applications*. Plenum Press, New York.

COSTA, C. A. and CABRAL, J. S. (Eds.). 1991. *Chromatographic and Membrane Processes in Biotechnology*. Kluwer Academic Publishers, The Netherlands.

DAUBNER, P. 1974. *Membranfilter in der Mikrobiologie des Wassers*. Wleter de Gruyter & Co., Berlin, Germany.

DRIOLI, E. and NAKAGAKI, M. (Eds.). 1986. *Membranes and Membrane Processes*.

Plenum Press, NY, *Proceedings of the Europe–Japan Congress*, Stressa, Italy, June 1984.

DUTKA, B. J. (Ed.). 1981. *Membrane Filtration. Applications, Techniques and Problems*. Marcel Dekker, New York.

EISENBERG, T. N. and MIDDLEBROOKS, E. J. 1986. *Reverse Osmosis of Drinking Water*. Butterworth Publishers, Stoneham, MA.

FLINN, J. E. (Ed.). 1970. *Membrane Science and Technology: Industrial, Biological and Waste Treatment Processes*. Plenum Press, New York.

GLOVER, F. A. 1985. *Ultrafiltration and Reverse Osmosis for the Dairy Industry*. NIRD Technical Bulletin No. 5, National Institute for Research in Dairying, Reading, U.K.

GOVIND, R. and ITOH, N. (Eds.). 1989. Membrane Reactor Technology. *AIChE Symposium Series* Vol. 85, #268. Amer. Inst. Chem. Engrs., New York.

GUTMAN, R. G. 1987. *Membrane Filtration. The Technology of Pressure-Driven Crossflow Processes*. Adam Hilger, IOP Publishing Ltd., Bristol, U.K.

HALLSTROM, B., LUND, D. B. and TRAGARDH, H. (Eds.). 1981. *Fundamentals and Applications of Surface Phenomenon Associated with Fouling and Cleaning in Food Processing*. Lund University Press, Sweden.

HO, W. S. W. and SIRKAR, K. K. (Eds.). 1992. *Membrane Handbook*. Chapman and Hall, New York.

HOORNAERT, P. 1984. *Reverse Osmosis*. Pergamon Press, McLean, Virginia [Survey of European and Japanese patent literature].

HOWELL, J. A. (Ed.). 1990. *The Membrane Alternative: Energy Implications for Industry*. Elsevier, London, U.K.

HOWELL, J. A., SANCHEZ, V. and FIELD, R. W. (Eds.). 1993. *Membranes in Bioprocessing*. Chapman & Hall, New York.

HSIEH, H. P. 1996. *Inorganic Membranes for Separation and Reaction*. Elsevier, Amsterdam.

HWANG, S. and KAMMERMEYER, K. 1975. *Membranes in Separations*. Wiley-Interscience, New York.

JOHNSTON, J. R. 1992. *Fluid Sterilization by Filtration*. Interpharm Press, Buffalo Grove, IL.

KELLER, P. R. (Ed.). 1976. *Membrane Technology and Industrial Separation Techniques*. Noyes Data Corporation, Park Ridge, NJ [Survey of patent literature].

KESTING, R. E. 1971. *Synthetic Polymeric Membranes*. McGraw-Hill, New York.

KESTING, R. E. 1985. *Synthetic Polymeric Membranes. A Structural Perspective*. 2nd edition. Wiley-Interscience, New York.

KESTING, R. E. and FRITZSCHE, A. K. 1993. *Polymeric Gas Separation Membranes*. Wiley, New York.

KLEIN, E. 1991. *Affinity Membranes*. John Wiley, New York.

LACEY, R. E. and LOEB, S. (Eds.). 1972. *Industrial Processing with Membranes*. Wiley-Interscience Publishers, New York.

LAKSHMINARAYANAIAH, N. 1969. *Transport Phenomena in Membranes*. Academic Press, New York.

LAKSHMINARAYANAIAH, N. 1985. *Equations of Membrane Biophysics*. Academic Press, New York.

LLOYD, D. R. (Ed.). 1985. *Materials Science of Synthetic Membranes*. American Chemical Society, Washington, DC.

LONSDALE, H. K. and PODALL, H. E. (Eds.). 1972. *Reverse Osmosis Membrane Research.* Plenum, New York.

MADSEN, R. F. 1977. *Hyperfiltration and Ultrafiltration in Plate-and-Frame Systems.* Elsevier, Amsterdam.

MADSEN, R. F., MASTERS, K., WEIGAND, B. ET AL. (Eds.). 1986. *Evaporation, Membrane Filtration and Spray Drying for Milk Powder and Cheese Production*, Technical Dairy Publishing House, Denmark.

MATSUURA, T. 1994. *Synthetic Membranes and Membrane Separation Processes.* CRC Press, Boca Raton, FL.

MATSUURA, T. and SOURIRAJAN, S. (Eds.). 1989. *Advances in Reverse Osmosis and Ultrafiltration.* National Research Council Canada, Ottawa.

MEARES, P. (Ed.). 1976. *Membrane Separation Processes*, Elsevier, Amsterdam.

McDERMOTT, J. (Ed.). 1972. *Industrial Membranes. Design and Applications*, Noyes Data Corporation, Park Ridge, NJ [Description of patent literature].

McGREGOR, W. C. (Ed.). 1985. *Membrane Separations in Biotechnology.* Marcel Dekker, New York.

MELTZER, T. H. (Ed.). 1987. *Filtration in the Pharmaceutical Industry*, Marcel Dekker, New York.

MERTEN, U. (Ed.). 1966. *Desalination by Reverse Osmosis*, MIT Press, Cambridge, MA.

MORR, C. M., LEEPER, S. A., ENGELGAU, D. E. and CHARBONEAU, B.L. 1988. *Membrane Applications and Research in Food Processing: An Assessment*, Noyes Publications, Park Ridge, NJ.

MULDER, M. 1991. *Basic Principles of Membrane Technology.* Kluwer Academic Publishers, Norwell, MA.

MURKES, J. and CARLSSON, C.G. 1988. *Cross-flow Filtration.* John Wiley, New York.

ORR, C. (Ed.). 1977. *Filtration. Principles and Practices, Part I.* Marcel-Dekker, New York [Chapter 6, "Ultrafiltration" by R. P. deFilippi].

ORR, C. (Ed.). 1979. *Filtration. Principles and Practices, Part II.* Marcel-Dekker, New York [Chapter 3, "Filtration in the food, beverage and pharmaceutical industries" by J. L. Dwyer].

OSADA,Y. and NAKAGAWA, T. (Eds.). 1992. *Membrane Science and Technology.* Marcel Dekker, New York.

OUELLETTE, R. P., KING, J. A., CHEREMISINOFF, P. N. (Eds.). 1978. *Electrotechnology. Vol. 1.Wastewater Treatment and Separation Methods.* Ann Arbor Science Publishers, Ann Arbor, MI [Chapter 2: "Electrodialysis," Ch.3: "Reverse osmosis," Ch.4: "Ultrafiltration"].

PAREKH, B. P. (Ed.). 1988. *Reverse Osmosis Technology*, Marcel Dekker, New York.

PASSINO, R. (Ed.). 1976. *Biological and Artificial Membranes and Desalination of Water*, Elsevier, Amsterdam.

PORTER, M. C. (Ed.). 1990. *Handbook of Industrial Membrane Technology*, Noyes Publications, Park Ridge, NJ.

RAUTENBACH, R. and ALBRECHT, R. 1989. *Membrane Processes.* John Wiley, New York.

RENNER, E. and EL-SALAM, M. H. 1991. *Application of Ultrafiltration in the Dairy Industry.* Elsevier, New York.

ROGERS, C. E. (Ed.). 1971. *Permselective Membranes.* Marcel Dekker, New York.

ROUSSEAU, R. W. 1987. *Handbook of Separation Process Technology.* John Wiley, New York [Chapters 18–22 on membrane separations].

SCHWEITZER, P. A. (Ed.). 1979. *Handbook of Separation Techniques for Chemical Engineers.* McGraw-Hill, New York [Chapter 2, "Membrane filtration," by M. C. Porter].

SCHWEITZER, P. A. (Ed.). 1988. *Handbook of Separation Techniques for Chemical Engineers.* McGraw-Hill, New York [Chapter on "Membrane Filtration" by M. C. Porter].

SCOTT, J. (Ed.). 1980. *Membrane and Ultrafiltration Technology: Recent Advances.* Noyes Data Corporation, Park Ridge, New York [Survey of patent literature].

SCOTT, K. 1995. *Handbook of Industrial Membranes.* Elsevier, U.K.

SCOTT, K. and HUGHES, R. (Eds.). 1995. *Industrial Membrane Separation Technology.* Blackie Academic & Professional, Chapman and Hall, London, U.K.

SHARPE, A. N. and PETERKIN, P. I. 1988. *Membrane Filter Food Microbiology*, Research Studies Press Ltd., Herts, U.K. Wiley, NY.

SIRKAR, K. K. and LLOYD, D. R. (Eds.). 1988. New Membrane Materials and Processes for Separation. *AIChE Symposium Series Number 261*, Vol. 84. Amer. Inst. Chem. Engrs., New York.

SOURIRAJAN, S. (Ed.). 1970. *Reverse Osmosis.* Academic Press, New York.

SOURIRAJAN, S. (Ed.). 1977. *Reverse Osmosis and Synthetic Membranes. Theory—Technology—Engineering.* National Research Council Canada, Ottawa, Canada.

SOURIRAJAN, S. and MATSUURA, T. (Eds.). 1985. *Reverse Osmosis and Ultrafiltration.* American Chemical Society, Washington, DC.

SOURIRAJAN, S. and MATSUURA, T. 1985. *Reverse Osmosis/Ultrafiltration Principles.* National Research Council Canada, Ottawa, Canada.

SPIEGLER, K. S. and LAIRD, A. D. K. 1980. *Principles of Desalination, Part B.* Academic Press, New York [Chapter on "Hyperfiltration" by L. Dresner and J. S. Johnson].

TIMASHEV, S. F. 1991. *Physical Chemistry of Membrane Processes.* Ellis Horwood, London, U.K.

TITUS, J. B. (Ed.). 1973. *Reverse Osmosis Bibliography.* Plastics Technical Evaluation Center, Dover, NJ.

TORREY, S. (Ed.). 1984. *Membrane and Ultrafiltration Technology: Developments Since 1981.* Noyes Publications, Park Ridge, NJ.

TURBAK, A. F. (Ed.). 1970. *Membranes from Cellulose and Cellulose Derivatives.* Wiley-Interscience, New York.

TURBAK, A. F. (Ed.). 1981. *Synthetic Membranes. Vol. I. Desalination.* American Chemical Society, Washington, DC.

TURBAK, A. F. (Ed.). 1981. *Synthetic Membranes: Vol. II. Hyperfiltration and Ultrafiltration Uses.* American Chemical Society, Washington, DC.

TURNER, M. K. (Ed.). 1991. *Effective Industrial Membrane Processes: Benefits and Opportunities*, Elsevier, London, U.K.

TWINNER, S. B. 1962. *Diffusion and Membrane Technology.* Reinhold Publishing, New York.

VIETH, W. R. 1988. *Membrane Systems: Analysis and Design.* Hanser Publishers, Munich, Germany.

WHITE, R. E. and PINTAURO, P. N. (Eds.). 1986. Industrial Membrane Processes. *AIChE Symp. Series, No. 248*, Volume 82. American Inst. Chem. Engrs., New York.

ZEMAN, L. J. and ZYDNEY, A. L. 1996. *Microfiltration and Ultrafiltration. Principles and Applications.* Marcel Dekker, New York.

GLOSSARY OF TERMS

Anisotropic. Defines a particular type of ultrastructure of microporous membranes. The pore diameter increases in a direction perpendicular to the membrane surface, with the pore opening near the separation surface being smaller than the pore opening on the bottom of the membrane.

Area. Surface area of the membrane that will be exposed to the feed. Expressed as square feet (ft^2) or square meters (m^2). $1\ m^2 = 10.764\ ft^2$.

Asymmetric. Defines a particular type of ultrastructure of the membrane in the plane perpendicular to the membrane surface. The surface of the membrane where separation occurs is more dense than the rest of the membrane body. This "skin" layer is typically present in polymeric membranes made by the phase-inversion process. Some asymmetry can also be observed with many inorganic membranes.

ATD. Antitelescoping Device. A circular ring with spokes fitted to the ends of a spiral module to prevent the membrane from pushing itself outwards under high pressure drops.

Backflush, Backwash, Backpulse. Brief reversal of flow from permeate to feed side of the membrane.

Bioreactor. Vessel for conducting biological reactions.

BOD. Biochemical Oxygen Demand. A measure of the amount of oxygen required to support growth of bacteria during the breakdown of organic compounds in a sample. Usually expressed in parts per million.

Boundary Layer. The region near a solid surface where fluid motion is affected by the surface. The boundary layer is a major resistance to transport (e.g., by heat, mass, or momentum). To improve transport, the thickness of this boundary layer must be reduced.

cfu. Colony forming units, a measure of the number of microorganisms in a sample that are capable of forming colonies on a plate.

Channel. The space in a membrane module where feed flows, e.g., the inside of hollow fibers or tubes, the space between parallel plates.

Chemical Potential (μ). Defined in terms of Gibbs free energy, it is a measure of how the Gibbs function for a system changes when a specified amount of component is added to the system.

511

C.O.D. Chemical Oxygen Demand. A chemical measure of the amount of oxygen required to breakdown organic matter in a sample. Expressed in parts per million.

Colloid. Fine particles $(0.001–0.1\mu)$ suspended in a gas or liquid.

Compaction. Compression of a membrane across its thickness under high pressures normally associated with RO.

Concentration Polarization (CP). The buildup of solutes close to or on the membrane surface. Solute is brought to the membrane surface by convective transport; solutes larger than the nominal MWCO of the membrane are retained by the membrane, while solutes smaller than the pores will freely or partially permeate through the membrane. Solutes not passing through the membrane will accumulate on the membrane surface, causing either an increased resistance to solvent transport or an increase in local osmotic pressure (either of which may decrease flux) and possibly a change in the sieving characteristics of the membrane.

Concentrate. That portion of the feed solution that is retained (on the high-pressure side) of the membrane. As a result, the retained components are usually more concentrated than they were in the original feed solution. The terms concentrate and retentate are used interchangeably.

Concentration Profile. The relationship between solute concentration (C) and distance (y) from the membrane surface.

Concentration Factor or Concentration Ratio. The ratio of the initial feed volume (or weight or flow rate) to the retentate volume (or weight or flow rate). Also referred to as the X value.

Continuous Diafiltration (CD). A mode of processing where water is added continuously to the feed tank or the retentate loop at the same rate as permeate flux. The total volume of the system and the concentration of the retained solute remain constant during CD. The concentration of the permeable solutes decrease in proportion to the volumes diluted and their individual rejections.

Cross-flow. The flow of fluid across a membrane surface, parallel to its surface. These modules have one port of entry (for the feed) and two outlets (for retenate and for permeate). This is in contrast to dead-end modules.

DE. This could have two meanings: diatomaceous earth, which is the fossilized skeletal remains of microscopic prehistoric plants; used as a filter aid in conventional filtration. Also could refer to dextrose equivalence, which is a measure of the degree of conversion of starch to glucose molecules; values range from 0–100.

Dead-end. Mode of operation where there is only one feedstream and one outlet stream (the filtrate or permeate). Refers to modules that do not have a means for cross-flow, e.g., stirred cells. Cross-flow modules could be operated in the dead-end mode by shutting off the outlet of the module.

Delta-P (ΔP). Pressure drop, defined as inlet pressure minus outlet pressure. Not to be confused with TMP, or transmembrane pressure.

Diafiltration (DF). The convective elimination of permeable solutes by the addition of fresh solvent (water) to the retentate. Two modes of operation may be used in diafiltration: continuous or discontinuous.

Diffusivity/Diffusion Coefficient (D). A measure of a solute's diffusive mobility in a solvent.

Discontinuous Diafiltration (DD). Permeable solutes are first eliminated by conventional UF or MF. Water is then added to the concentrated retentate to dilute it back to a certain volume and reprocessed by MF or UF. This process of repetitive MF/UF and dilution to eliminate permeable solutes is called discontinuous diafiltration.

Element. Usually refers to a finished spiral-wound membrane (without the housing).

Feed-and-bleed. A continuous mode of operation where the feed is pumped into the recirculation loop of the membrane system at the same rate as the sum of the retentate flow and permeate flux out of that loop. The feed concentration within the recirculation loop will be determined by the concentration factor or X value.

Filter aid. A material added to a liquid to be filtered or on to the filter medium to improve the separation of solids from the liquid.

Flow Rate (Q). The volumetric rate of flow of fluid parallel to the membrane surface. This is expressed in terms of volume/time (e.g., liters/min or gal/min). Flow rate is velocity (V) × cross-sectional area of the feed channel. Also sometimes termed *recycling rate* or *recirculation rate*. This is the major determinant of the state of turbulence in a module.

Flux (J). Amount of fluid passing through the membrane, i.e., the volumetric rate of flow of the permeate through the membrane. It is usually given in terms of volume per unit membrane area per unit time, e.g., liters/m^2/hour (LMH) or gallons/ft^2/day (GFD).

Fouling. Phenomenon in which the membrane adsorbs or interacts in some manner with solutes in the feedstream, resulting in a decrease in membrane performance (lowering of the flux and/or increase in rejection of solutes). This is usually irreversible and time-dependent, which distinguishes it from concentration polarization. Fouling effects can usually be offset by cleaning the membrane.

Gel Layer. Extreme case of concentration polarization, describing a physical state of solids accumulating on the membrane surface, e.g., proteins.

GFD or GSFD. *See* Flux.

Housing. The vessel in which a membrane element is placed with ports and fittings to direct the feed, retentate and permeate streams through the membrane element. Same as pressure vessel.

Hydrophilic. A material that is compatible with or preferentially interacts with water (water loving).

Hydrophobic. A material that is incompatible with and cannot interact with water (water hating). These materials preferentially interact with oils and other hydrophobic components.

Isotropic. Refers to the ultrastructure of a membrane, where the pores are essentially uniform in size from the top to the bottom of the membrane body, and the body is more or less of a uniform density.

LMH. *See* Flux.

Mass Transfer Coefficient (k). A measure of the solute's mobility due to forced or natural convection in the system. Analogous to a heat transfer coefficient, it is measured as the ratio of the mass flux to the driving force. In ultrafiltration, the driving force is the difference in solute concentration at the membrane surface and at some arbitrarily defined point in the bulk fluid. When using the film theory to model mass transfer, k is also defined as D/δ, where D is solute diffusivity and δ is the thickness of the concentration boundary layer.

Membrane (bio)Reactor. System for simultaneous reaction and separation with a membrane. Could be conducted in the same physical enclosure or in separate enclosures if the reaction vessel and membrane unit are coupled in a semi–closed-loop configuration.

Mesh. Number of strands in a linear inch of filter material. Higher numbers mean finer particles are retained.

Micron (μ). Indication of size of microscopic particles. Also used to classify pore sizes of MF membranes. Used interchangeably with micrometers (μm).

Mil. One-thousandth of an inch. 1 mil $= 0.0254$ mm. Common way of expressing the thickness of spacers in flat-sheet membrane elements.

Module. Usually refers to the membrane and its housing. Sometimes the membrane element alone is called the module (e.g., with spiral-wound membranes).

MW. Molecular weight.

MWCO. Molecular weight cut-off, a term used to describe the potential separating capabilities of a UF membrane. Molecular weight of a theoretical solute with a 90% rejection of that membrane.

NMWCO. Nominal molecular weight cut-off.

Oleophilic. Same as hydrophobic.

ppm. Parts per million.

Permeate. That portion of the feed solution that passes through the membrane.

Pore Size. Provides an idea of the average or smallest particle that will be retained by MF membranes. Units are usually microns (μ).

Pressure Vessel. *See* Housing.

Pyrogen. Any substance capable of producing a fever when injected into mammals. Produced by bacteria and not necessarily inactivated by conventional heat treatment methods.

Rejection (R). A measure of how well a membrane retains or allows passage of a solute. When based on concentrations in the bulk of the permeate and retentate streams, it is the "apparent" rejection. The "intrinsic" rejection is based on concentrations at the membrane surface. Sometimes used interchangeably with retention factor.

Retentate. That portion of the feed solution that is retained on the high-pressure side of the membrane. The terms *retentate* and *concentrate* are used interchangeably.

Retention Factor. Defined in the same manner as rejection, except retention refers to the feed, permeate, and retentate streams entering and leaving a module, while rejection should ideally refer to local concentrations close to the membrane surface.

Reynolds Number (Re). A measure of the state of turbulence in a fluid system. It is calculated as the ratio of inertia effects to viscous effects. Reynolds number is dimensionless. Systems with Re values less than 1800 are considered to be laminar flow, while Re > 4000 is considered to be turbulent.

Schmidt Number (Sc). A measure of the ratio of momentum transfer to mass transfer.

Sherwood Number (Sh). A dimensionless measure of the ratio of convective mass transfer to molecular mass transfer. If the mass transfer coefficient k is defined in terms of the film theory, then Sh is a measure of the ratio of hydraulic diameter to the thickness of the boundary layer.

Spacers. A mesh-like material used in flat-sheet modules (e.g., plate, spirals, pleated sheet) to separate successive layers of membranes. Spacers control the feed channel dimensions in these modules.

TMP. *See* Ttransmembrane pressure.

Transmembrane Pressure (P_T). The driving force for flux. In cross-flow systems, it is measured as the average of the inlet and outlet pressures, minus permeate back-pressure.

Turbulence Promoter. Devices that are inserted into the feed/retentate channel of a module to improve mixing characteristics, increase turbulence, and thus improve flux. Typical turbulence promoters are spacers, spiral wires, static kenics-type mixers, spheres, or balls.

Velocity (V). The linear rate of flow of fluid parallel to the membrane, expressed in units of length/time (e.g., m/sec). This is calculated as flow rate/cross-sectional area of feed channel.

Velocity Profile. The relationship between fluid velocity parallel to the membrane surface (V) and distance from the membrane surface (y).

Voids. The openings or spaces in the body of a membrane, underneath its surface.

Volume Concentration Ratio (VCR). The ratio of the initial feed volume to the volume of retentate. Same as concentration factor or X value, but expressed in terms of volumes.

Volumes Diluted (V_D). The ratio of the volume of liquid permeated to the initial feed volume. This term is used in continuous diafiltration operations.

Weight Concentration Ratio (WCR). The ratio of the initial weight of feed solution to the weight of retentate. Same as concentration factor or X value but expressed in terms of weights.

X or X Value. Ratio of feed volume to retentate volume. Same as concentration factor, VCR, and WCR.

For additional reference, see

Koros, W. J., Mah, Y. H. and Shimidzu, T. 1996. Terminology for Membranes and Membrane Processes, *Pure and Applied Chemistry* 68: 1479–1489 (a publication of the International Union of Pure and Applied Chemistry's Working Party on Membrane Nomenclature).

INDEX

517

ABOUT THE AUTHOR

Dr. Munir Cheryan is Professor of Food and Biochemical Engineering at the University of Illinois, Urbana-Champaign, USA. He has a bachelor's degree in Chemical Engineering from the Indian Institute of Technology at Kharagpur an M.S. in Chemical Engineering, and a Ph.D. in Food Science from the University of Wisconsin (1974). He has been working on membrane separations since the mid-1970s, with emphasis on biological, food, agricultural, and waste treatment applications. He has published over 160 papers and has over 100 conference abstracts. He has received numerous awards and honors, including

- Archer-Daniels-Midland award from the American Oil Chemists Society in 1984
- Paul Funk award from the Funk Foundation in 1987 for research
- Gardner award in 1988 from the Association of Food Technologists (India)
- 1991 Soybean Utilization Research Team Award from the American Soybean Association/ICI Americas, Inc.
- 1993 Research and Commercialization award from the National Corn Growers Association
- Commendation from the State of Illinois Legislative Assembly in 1987
- "Outstanding Professor" award from university students for his teaching

Dr. Cheryan has received international fellowships and lectureships in several countries and provided consulting services to companies, government, and international agencies. He is on the editorial board of three research journals and a trade magazine and is a member of several professional societies, including Institute of Food Technologists, American Institute of Chemical Engineers, American Chemical Society, and North American Membrane Society. He can be contacted at *mcheryan@uiuc.edu*.